Practical Flatfish Culture and Stock Enhancement

Practical Flatfish Culture and Stock Enhancement

Editors

H.V. Daniels and W.O. Watanabe

WILEY-BLACKWELL
A John Wiley & Sons, Ltd., Publication

U.S. Aquaculture Society
A Chapter of the World Aquaculture Society

Edition first published 2010

Blackwell Publishing was acquired by John Wiley & Sons in February 2007. Blackwell's publishing program has been merged with Wiley's global Scientific, Technical, and Medical business to form Wiley-Blackwell.

Editorial Office
2121 State Avenue, Ames, Iowa 50014-8300, USA

For details of our global editorial offices, for customer services, and for information about how to apply for permission to reuse the copyright material in this book, please see our website at www.wiley.com/wiley-blackwell.

Library of Congress Cataloging-in-Publication Data

Practical flatfish culture and stock enhancement / editors, H.V. Daniels, W.O. Watanabe.
p. cm.
Includes bibliographical references and index.
ISBN 978-0-8138-0942-7 (hardback : alk. paper)
1. Flatfishes I. Daniels, H. V. (Harry V.) II. Watanabe, Wade O.
SH167.F55P73 2010
639.3′769–dc22

2009050270

A catalog record for this book is available from the U.S. Library of Congress.

Set in 10/12.5 pt Sabon by Aptara® Inc., New Delhi, India
Printed in Singapore

1 2010

Contents

Section 3: Asia and Australia Culture

Section 7: Flatfish Worldwide

Contributors

Masato Aritaki
National Center for Stock Enhancement
Fisheries Research Agency
Sakiyama Miyako, Iwate
Japan

Sungchul C. Bai
Department of Aquaculture/Feeds and
Foods Nutrition Research Center (FFNRC)
Pukyong Nat'l University
Busan, Republic of Korea

David Bengtson
Department of Fisheries, Animal and Veterinary Science
University of Rhode Island
Kingston, RI

Russell J. Borski
Department of Biology
North Carolina State University
Raleigh, NC

Nick Brown
Center for Cooperative Aquaculture Research
University of Maine
Franklin, ME

John Selden Burke
Center for Coastal Fisheries and Habitat Research
National Oceanic and Atmospheric Administration
Beafort, NC

Douglas Conklin
Department of Animal Science
UC Davis
Davis, CA

Harry Daniels
Department of Biology
North Carolina State University
Raleigh, NC

Christopher Dumas
University of North Carolina Wilmington
Wilmington, NC

Elizabeth A. Fairchild
Department of Zoology
University of New Hampshire
Durham, NH

Yuichiro Fujinami
Miyako Station
National Center for Stock Enhancement
Fisheries Research Agency
Sakiyama, Miyako, Iwate
Japan

John Godwin
Department of Biology
North Carolina State University
Raleigh, NC

Piers R. Hart
Lewes, East Sussex, BN

Kotaro Kikuchi
Biological Environment Sector
Environmental Science Research Laboratory
CRIEPI
Tokyo, Japan

Seunghyung Lee
Department of Fisheries Biology
Pukyong National University
Daeyeon dong, Namgu
Busan, Republic of Korea

Ji-Lin Lei
Yellow Sea Fisheries Research Institute
Chinese Academy of Fishery Sciences
Qingdao, Shandong
People's Republic of China

Xin-Fu Liu
Yellow Sea Fisheries Research Institute
Chinese Academy of Fishery Sciences
Qingdao, Shandong
People's Republic of China

Thomas Losordo
Department of Biological and Agricultural Engineering
North Carolina State University
Raleigh, NC

John Adam Luckenbach
School of Aquatic and Fishery Sciences
University of Washington
Seattle, Washington, DC

Reji Masuda
Maizuru Fisheries Research Station
Kyoto University
Nagahama, Maizuru
Kyoto, Japan

John M. Miller
Department of Biology
North Carolina State University
Raleigh, NC

Ryan Murashige
Castle International
Honolulu, HI

George Nardi
GreatBay Aquaculture
Portsmouth, NH

Edward J. Noga
Department of Clinical Sciences
College of Veterinary Medicine
North Carolina State University
Raleigh, NC

Oddvar H. Ottesen
Bodø University College
Department of Fisheries and Natural Sciences
Bodø, Norway

Raul Piedrahita
Department of Agricultural Engineering
UC Davis
Davis, CA

Jeanine Person-Le Ruyet
Unité Mixte Nutrition, Aquaculture, Génomique
Laboratoire Adaptation Reproduction Nutrition des Poissons,
IFREMER Centre de Brest
Plouzané, France

Tadahisa Seikai
Fukui Prefectural University
Obama City
Obama, Fukui, Japan

Alfonso Silva
Departamento de Acuacultura
Universidad Catolica del Norte
Casilla, Coquimbo, Chile

Stephen A. Smith
Department of Biomedical Sciences and Pathobiology
Virginia-Maryland Regional College of Veterinary Medicine
Virginia Tech University
Blacksburg, VA

C. R. Sparrevohn
Section for Coastal Ecology
National Institute of Aquatic Resources
Technical University of Denmark,
Charlottenlund Castle
Charlottenlund
Denmark

Josianne G. Støttrup
Technical University of Denmark
National Institute of Aquatic Research (DTU Aqua)
Charlottenlund Castle
Charlottenlund, Denmark

Robert Vega
Texas Parks and Wildlife Marine Development Center
Corpus Christi, TX

Wade O. Watanabe
Center for Marine Science
University of North Carolina Wilmington
Wilmington, NC

Yoh Yamashita
Maizuru Fisheries Research Station
Kyoto University
Nagahama, Maizuru
Kyoto, Japan

Preface

The United States Aquaculture Society

The United States Aquaculture Society (USAS) is a chapter of the World Aquaculture Society (WAS), a worldwide professional organization dedicated to the exchange of information and the networking among the diverse aquaculture constituencies interested in the advancement of the aquaculture industry, through the provision of services and professional development opportunities. The mission of the USAS is to provide a national forum for the exchange of timely information among aquaculture researchers, students, and industry members in the United States. To accomplish this mission, the USAS will sponsor and convene workshops and meetings, foster educational opportunities, and publish aquaculture-related materials important to U.S. aquaculture development.

The USAS membership is diverse, representing researchers, students, commercial producers, academics, consultants, commercial support personnel, extension specialists, and other undesignated members. Member benefits are substantial and include issue awareness, a unified voice for addressing issues of importance to the United States Aquaculture Community, networking opportunities, business contacts, employment services, discounts on publications, and a semiannual newsletter reported by regional editors and USAS members. Membership also provides opportunities for leadership and professional development through service as an elected officer or board member, chair of a working committee, or organizer of a special session or workshop, special project, program, or publication as well as recognition through three categories of career achievement (early career, distinguished service, and lifetime achievement). Student members are eligible for student awards and special accommodations at national meetings of the USAS, and have opportunities for leadership through committees, participation in Board activities, sponsorship of social mixers, networking at annual meetings and organization of special projects.

At its annual business meeting in New Orleans in January 2005, the USAS under the leadership of President LaDon Swann, voted to increase both the diversity and quality of publications for its members through a formal solicitation process for sponsored publications, including books, conference proceedings, fact sheets, pictorials, hatchery or production manuals, data compilations, and other materials that are important to United States Aquaculture development and that will be of benefit to USAS members. As aquaculture becomes increasingly global in scope, it is important for USAS members to gain an international perspective on the reasons for successful aquaculture developments at home and abroad. Flatfish (also known as flounder) are a group of marine or brackishwater finfish that support important recreational and commercial fisheries throughout the world and they are among the few finfish species that are the subject of significant marine stock enhancement efforts in Europe, Asia, and North America. In this book, *Practical Flatfish Culture and Stock Enhancement*, international experts provide comprehensive (i.e., from egg to market) reviews of the different species that are being researched or already being produced for commercial cultivation and for hatchery-based fisheries enhancement.

Through collaboration with Wiley-Blackwell on books projects such as these, the USAS Board aims to serve its membership by providing timely information through publications of the highest quality at a reasonable cost. The USAS thanks the editors Harry Daniels and Wade Watanabe for sharing royalties which will help provide the benefits and services to members and to the aquaculture community and Justin Jeffryes and Shelby Allen (Wiley-Blackwell) for their cooperation. The USAS Publications Committee members include Drs. Wade O. Watanabe (Chair), Jeff Hinshaw, Jimmy Avery, and Christopher Kohler, with Rebecca Lochmann and Douglas Drennan as immediate past and current Presidents, respectively.

Wade O. Watanabe, Ph.D.
Director and Publications Chair, United States Aquaculture Society
Research Professor and Aquaculture Program Coordinator
Mariculture Program Leader, Marine Biotechnology in North Carolina
University of North Carolina Wilmington, Center for Marine Science
Wilmington, North Carolina, USA

Preface

Vision for the project

This book is aimed to provide a valuable reference for members of the aquaculture and fisheries communities. The goal is to provide a practical perspective on culture methods for the aquaculturist while simultaneously providing key biological information, from the culturist's perspective, that is necessary for the fisheries manager. With the development of technologies for mass propagation of juveniles flounder, both stock enhancement and production aquaculture may allow a sustainable supply of flatfish for the foreseeable future. It is for this reason that we are including several chapters on flatfish stock enhancement. We feel that this approach will provide the first comprehensive treatment of these two issues as they relate to each other and will be useful to biologists for making proactive management decisions.

The biology of flatfish was comprehensively covered in a previous book by Robin N. Gibson, entitled *Flatfishes: Biology and Exploitation* (2005), which primarily covered flatfish biology, ecology, behavior, and fisheries, and included a chapter on flounder Aquaculture and Stock Enhancement (B. R. Howell and Y. Yamashita). However, there have been no comprehensive reviews of the practical aspects of flatfish culture and stock enhancement, with a detailed review of the different species that are being researched or already being produced for commercial cultivation or for stock enhancement. Furthermore, there are no book publications on the subject of flatfish that address the culture and stock enhancement issues simultaneously. We anticipate that this type of discussion will be particularly valuable to the practicing aquaculturist and should also provide a unique perspective to the student interested in fisheries management as well as aquaculture.

The primary audience for this book is intended to be researchers and state and federal fisheries biologists who use flatfish as their research model or are struggling with a lack of information on flatfish biology and culture practices and how they may affect enhancement decisions. This latter group is seen as a tremendously underserved group. Recent international, federal, and state-mandated quotas on flatfish harvest have increased the interest in stock enhancement of hatchery-cultured flatfish to supplement declining stocks. We see this book as a timely contribution to the debate about these issues. The secondary audience for this volume would be students who are interested in aquaculture and/or fisheries management. In this area would be the advanced undergraduate and graduate students. This book may serve as a particularly valuable reference for this latter group.

Scope and contents of the book

In this book, a summary of the "state-of-the-art" for the culture of each species is provided, including life history and biology, broodstock husbandry, larval culture, nursery culture and growout, harvesting and processing, marketing, and hatchery and production economics, and stock enhancement. For chapters on species, the book was structured to facilitate interspecific comparisons and contrasts, with the objective of summarizing available technology while accelerating technology development for the culture of all these species. Since each species has reached a different level of research and commercial development, the available information for each species is not necessarily uniform in coverage.

In addition to species coverage, there is detailed coverage of the diseases that have afflicted different species of flatfish and those that are likely to emerge as industrial flatfish culture develops. This includes general principles of fish disease diagnosis from the standpoint of what the culturist must do to enhance the ability of the fish health specialist or veterinarian to be able to provide a diagnosis of a disease problem. Special emphasis is placed on the treatment of fish prior to release into the natural environment and the types of screening processes or protocols that are needed to certify disease-free status. The latest information on flatfish stock enhancement, including release technologies for efficient stocking, particularly as it pertains to flatfish species, is provided and future perspectives for the management of the flounder stocks are discussed.

Because of the availability of wild broodstock and their ease of larval culture, relevance to research on marine ecotoxicology and their asymmetric metamorphic development, flatfish are increasingly being used as models for basic research on mechanisms of sex determination, cold tolerance, growth, and osmoregulation. A chapter on flatfish as research animals provides a brief overview of the biology of metamorphosis and sex determination and its regulation in flatfishes, including both environmental and genetic sex determining mechanisms, the primary hormones involved in regulating metamorphosis, and how these flatfish provide valuable research models to better understand how these developmental stages are controlled in vertebrates.

The final chapter crosscuts across species to uncover the similarities and differences in knowledge and technologies for flatfish at each phase of the culture process and to emphasize those technologies that are gaining commercial importance and the important areas for future research. We hope that flatfish culturists will be able to use the information in this book to accelerate progress in technology development for both culture and stock enhancement of this economically valuable and important group of fish species.

Harry V. Daniels and Wade O. Watanabe

Acknowledgments

The editors sincerely appreciate the efforts and dedication of the chapter authors for providing the basis for this work. We also wish to acknowledge the assistance we received from Claire and Will Daniels, who have helped at various stages in the production of this book.

Harry V. Daniels and Wade O. Watanabe

Section 1

North and South America Culture

Chapter 1

Halibut aquaculture in North America

Nick Brown

1.1 Life history and biology

The Atlantic halibut is a large pleuronectid flatfish distinguishable from other right-eyed flatfishes by its large mouth, which opens as far back as the anterior half of its lower eye, its concave caudal fin, and the distinctive arched lateral line. Dorsally, the adult fish is more or less uniformly chocolate brown or olive and the blind side is usually white, though in some cases, it may be partially brown (Collette and Klein-Macphee 2002). This species is among the commercially important groundfish of the Gulf of Maine where it has been harvested since the early part of the nineteenth century. The fishery was quickly depleted and has not been of economic importance since the 1940s. Annual catches after 1953 have been less than 100 metric tons on an average. The Atlantic halibut is one of the largest fish in the region. The largest individual caught on record was 280 kg (head on gutted) and was estimated to weigh 318 kg (live weight).

In the western North Atlantic, older juvenile and adult halibut undergo extensive migrations between feeding grounds and spawning areas (McCracken 1958; Cargnelli *et al.* 1999; Kanwit 2007). Coastal shelf areas of Browns Bank and the southwestern Scotian Shelf are thought to be important nursery grounds (Stobo *et al.* 1988; Neilson *et al.* 1993). Atlantic halibut are known to spawn at great depths where temperatures are generally stable and are between 5 and 7°C (Haug 1990; Neilson *et al.* 1993). The Atlantic halibut is a batch spawner, producing several batches of eggs during the spawning season in relatively regular intervals of 3–4 days (Smith 1987; Haug 1990; Holmefjord and Lein 1990; Norberg *et al.* 1991). The clear eggs are quite large for a marine fish (3 mm in diameter) and are bathypelagic during development, floating close to the ocean floor, and are neutrally buoyant at relatively high salinity of around 36 ppt.

After hatching, the larva hangs in a head down position exhibiting very little swimming activity (Pittman *et al.* 1990a). Halibut larvae hatch in a very primitive developmental state and organogenesis proceeds at a slow pace (Lonning *et al.* 1982; Blaxter *et al.* 1983; Pittman *et al.* 1990a). At around 150°C days, the

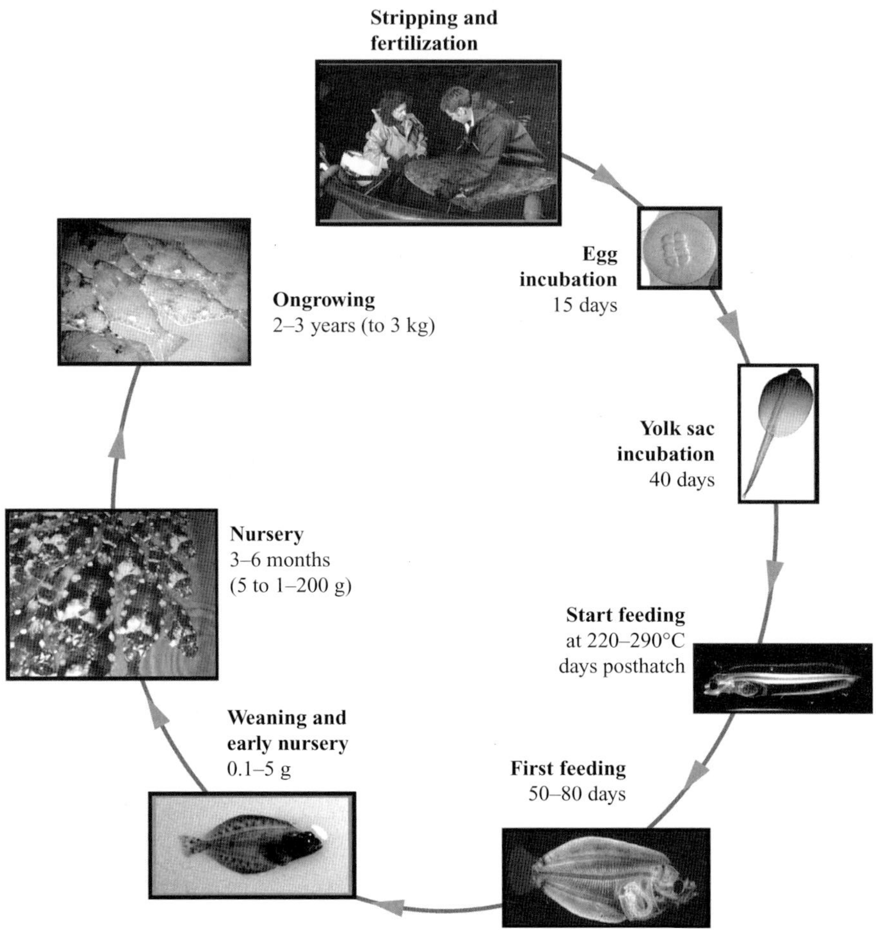

Figure 1.1 Production cycle of the Atlantic halibut.

eyes, mouth, and intestine become functional and the eye takes on pigmentation (Blaxter *et al.* 1983; Pittman *et al.* 1990b; Kvenseth *et al.* 1996).

Exogenous feeding can begin from around 240°C days and metamorphosis occurs around 80 days posthatch. At this point, the stomach is formed, the left eye migrates to the right side of the head, and the fish becomes fully pigmented. For aquaculture purposes, this represents the end of the hatchery phase and coincides with the establishment onto formulated feeds that will continue until harvest.

Capture of early life stages in the wild is very rare, little is known about their distribution and for researchers attempting to close the life cycle (Figure 1.1) in the hatchery, there has been a lot of trial and error.

Apart from the earliest trials (e.g., Rollefsen 1934), research into the techniques for the culture of halibut began in the 1980s and a few juveniles were reared past metamorphosis in the first attempts (Blaxter *et al.* 1983).

The Atlantic halibut has a number of attributes that make it an excellent candidate for aquaculture. These characteristics include firm, white, mild tasting flesh with a good shelf life, a high fillet yield, efficient feed conversion rates, and resistance to many common marine diseases. However, challenges with juvenile production and diversion of research resources and investment capital to other marine fish species, such as cod, have resulted in slow growth of this industry.

1.2 Broodstock

1.2.1 Acquisition of broodstock

Captive broodstock populations were first set up in Scotland and Norway in the early 1980s (Blaxter *et al.* 1983; Rabben *et al.* 1986; Smith 1987). Mature wild fish are caught using longlines or "tub trawls." A size 14/0 or larger circle hook is recommended to reduce injuries to the fish (Kanwit 2007). Fish for the University of Maine program, based at the Center for Cooperative Aquaculture Research (CCAR), were caught between 2000 and 2002. These 112 fish ranging in size from 9 to 40 kg were brought into the fishing ports of Jonesport, Stonington, and Steuben by fishermen participating in an experimental tagging program run by the Maine Department of Marine Resources (DMR) (Kanwit 2007). The fish were transferred from holding tanks on the boats to live transport tanks supplied with oxygen and driven by truck overland to the facility. Additional fish from research hatcheries in Canada were recruited to this founding population to result in a total population of 120 mature fish. An additional 150 fish reared at the CCAR hatchery were selected from the 2006 production run for broodstock. Additional wild fish from a DMR tagging study were also added in 2007. All mature hatchery reared (F1) fish have been genotyped using microsatellite markers developed in Canada (Jackson *et al.* 2003) to establish pedigree for future breeding programs.

Halibut may take up to 3 years to acclimate sufficiently to spawn in captivity following capture. Weaning onto a nonliving food item can be improved by using live fish such as mackerel as an intermediate step in the tanks. The use of large tanks, low light levels, good water quality, and temperature regimes that follow the natural environment of the halibut will all help to ensure successful acclimation.

1.3 Biosecurity

Fish recruited to a broodstock population are very valuable animals once weaned onto feed and acclimated to spawn in captivity. They are hard to replace and can give viable gametes for many years. It is therefore essential to use good biosecurity practices to help prevent the introduction of pathogens into a facility holding these fish. Quarantine of new fish from the wild should be done in a separate facility, for up to 6 months, preferably with a higher level of biosecurity in place. For example, there should be thorough disinfection of effluent water from such a facility through appropriate levels of ozonation, ultraviolet sterilization (or both), water pasteurization, or chlorination. Movement of personnel, equipment, water quality test samples, and handheld meters should be restricted. Mortalities occurring during quarantine should be quickly tested for pathogens and should be handled separately from other stocks or facilities. A quarantine system at the CCAR was used recently to receive wild halibut that were the subject of a tagging study run by the Department of Marine Resources. The system comprises four tanks that are 4 m in diameter and 2 m in depth. The 5% daily makeup water that leaves this facility is disinfected with a high level of ozone and then ultraviolet sterilization.

Established broodstock fish should be kept in a separate, designated facility. Water supplies should be filtered and treated with ozone or a UV sterilizer. Feed given to broodstock fish ideally should be in a dry form; although for halibut, the lack of knowledge of the nutritional requirements and suitable replacements for raw or frozen ingredients is an ongoing problem. Effective hygiene barriers should be in place at all entrances to broodstock facilities to ensure staff and visitors clean and sterilize footwear and hands.

Although broodstock facilities, which contain wild fish, should be near the incubation and the larval rearing facilities so that gametes can be conveniently carried over, it is necessary to ensure that effective hygiene barriers exist between broodstock and incubation systems. It is particularly important to disinfect the eggs before incubation.

1.3.1 System design and requirements

Broodstock Atlantic halibut are generally large fish that need to be housed in large tanks between 5 and 15 m in diameter. The broodstock at the CCAR are held in a designated facility, which comprises two recirculation systems, each with three tanks of 6.5 m in diameter and 1.5 m in depth (see Figure 1.2).

The recirculation system includes a moving bed biofilter, an UV sterilizer, a submersible circulating pump, and a drum filter (90 μm screen). The two systems are temperature controlled via titanium heat exchangers connected to oil-fired heating and electrical chillers. The room temperature and humidity are controlled via a dedicated HVAC unit. The optimum water temperature for

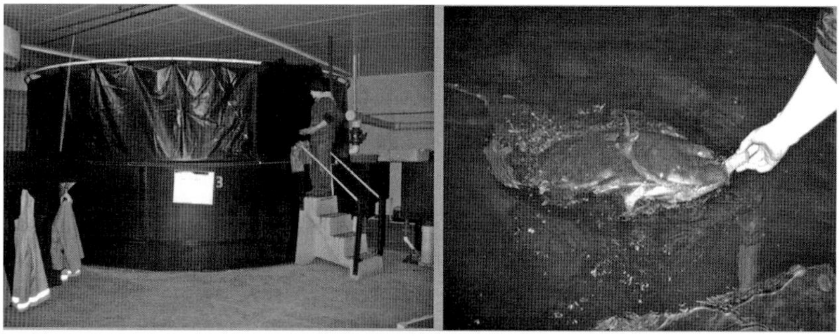

(a) (b)

Figure 1.2 One of the six 6.5-m diameter halibut broodstock tanks at the CCAR (a) and hand feeding with sausage diet (b).

broodstock halibut ranges from around 6°C in the winter to around 10°C in the summer. Water exchange is relatively slow at around 0.5 exchanges per hour. To enable the monitoring of egg releases during the spawning season, egg collectors are installed in the side box outlet where side and bottom drains meet before running to the treatment system.

The recommended stocking density for halibut is around 15 kg/m^2. Tank bottoms should be textured to prevent the formation of papillomas that are common in halibut kept in smooth-bottomed tanks at low densities (Ottesen and Strand 1996; Ottesen *et al.* 2007). An essential piece of equipment for the halibut broodstock facility is a table on which fish can be handled for manual stripping. All facilities have this and there are as many designs as there are broodstock managers. Some tables are power assisted (hydraulic or pulley block) to help lift what can be very large fish out of the water. Most are covered with some sort of soft pad such as neoprene rubber to help prevent injury to the valuable fish. The eyes of broodstock halibut are vulnerable and cataracts, gas bubbles, or other types of eye traumas are seen in some facilities. The cause of these problems is not clear and may be related to handling, in tank injury, gas supersaturation, or nutritional deficiencies.

1.4 Photothermal conditioning

The spawning season occurs between November and April under natural photoperiod (Kjorsvik *et al.* 1987; Haug 1990; Neilson *et al.* 1993). However, year-round egg production is possible using altered photoperiod (Smith *et al.* 1991; Holmefjord *et al.* 1993; Naess *et al.* 1996).

Manipulation of photoperiod is routinely used to influence natural spawning cycles enabling the production of the out-of-season eggs and, when multiple broodstocks are used, year-round production (Smith *et al.* 1991; Holmefjord *et al.* 1993; Naess *et al.* 1996). Delays of up to 6 months can be achieved in a single year. Advancing spawning time is more difficult and more than 3 months per year is not recommended since the fish need to build up reserves over the summer months for the subsequent spawning season. Halibut are sensitive to changes in light levels and good light proofing around holding tanks is necessary to ensure clear photoperiod signals. With photoperiod shifted stocks, attention must be paid to water temperature in out-of-season spawning groups to ensure good egg quality (Brown *et al.* 2006).

In the broodstock facility at the CCAR, the light to each tank is controlled via PLC and can simulate dawn/dusk via programmable dimming. The light source is from a dimmable compact fluorescent lamp suspended above the water in the center of the tank.

1.5 Monitoring gonad development

Captive halibut are generally stripped by hand although natural spawning can occur (Holmefjord and Lein 1990). The natural spawning period in the North

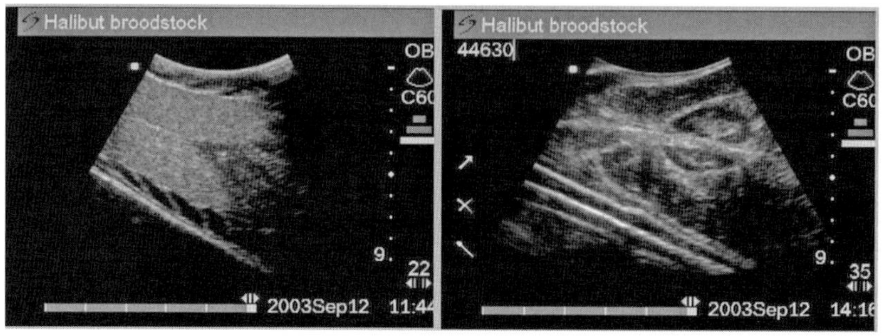

(a) (b)

Figure 1.3 Ultrasound scans of broodstock halibut showing an example of female (a) and male (b).

Atlantic occurs between late December and late March (Kjorsvik *et al.* 1987; Jákupsstovu and Haug 1988; Haug 1990).

Halibut are determinate batch spawners ovulating at intervals of 70–90 hours over the spawning season (Holmefjord 1991; Norberg *et al.* 1991). During the maturation process, batches of oocytes are sequentially hydrated. Adult female halibut have large gonads and are highly fecund. Adult female fish, weighing between 20 and 60 kg, are capable of producing between 6 and 16 batches, each of 10 to 200 $\times$ 10^3 eggs in a spawning season (Haug and Gulliksen 1988; Brown *et al.* 2006).

Egg collectors installed on each tank to intercept egg releases are checked regularly during the spawning season, often many times per day. Fish are usually allowed to spawn in the tank for the first two ovulations to give an indication of spawning interval. A marked reduction in viability can occur if fertilization is delayed longer than 4–6 hours after ovulation (Bromage *et al.* 1994). It has been shown that close observation of individual female ovulatory cycles can help to pinpoint the timing of stripping and improve viability and fertilization rates for halibut (Norberg *et al.* 1991; Holmefjord 1996) though this can be very time-consuming and potentially stressful for the fish. Egg quality can be highly variable in halibut and predicting the correct timing for manual stripping is one of the most difficult challenges remaining for halibut culture.

Ultrasound can be used to sex the fish (see Figure 1.3) and estimate the stage of development of the gonad (Shields *et al.* 1993; Martin-Robichaud and Rommens 2001). Individual fish are marked by PIT tags, FLOY tags, and/or sheep tags. The latter are easiest to use and are rarely lost.

1.5.1 Diet and nutrition

The natural diet of Atlantic halibut caught in various North Atlantic fishing grounds was described by MacIntyre (1953). Prey composition includes a wide variety of fish, mollusks, and crustaceans. The current lack of knowledge of broodstock halibut nutritional requirements means that the practice of feeding raw fish and shellfish is still quite common. This carries serious health risks for the broodstock and resulting eggs, larvae, and juveniles. Diseases found in the wild components can be transmitted to the captive broodstocks. The feeding of raw fish has been implicated in the transmission of such viral diseases as nodavirus (VNN) and viral hemorrhagic septicemia (VHS) (Dannevig *et al.* 2000).

Atlantic halibut broodstock nutrition studies are very challenging for a number of reasons. Egg quality is highly variable due to many confounding factors such as timing of stripping and replicated studies are hard to set up with such large, valuable fish. It has been shown that broodstock Atlantic halibut can be conservative in the levels of nutrients, in particular essential fatty acids, that they sequester to the eggs (Bruce *et al.* 1993) and despite varying levels in the diet, it may take months or years for deficiencies to emerge.

In two recent studies in Scotland (Mazorra *et al.* 2003; Alorend 2004), it took 3 years for dietary changes in fatty acid composition to have any effect. These studies did indicate that formulated feeds have the potential to replace raw fish components, though survival rates were not particularly high for resulting eggs and larvae. These investigators tested different dietary levels of the fatty acid arachidonic acid (ARA), an essential fatty acid thought to be important in broodstock nutrition due to its role as a precursor for prostaglandins which are involved in egg development and maturation (Bell and Sargent 2003). Mazorra *et al.* (2003) showed an improvement in egg quality when ARA levels were boosted to 1.8% and the authors suggest that the ratio of docosahexanoic acid (DHA) to eicosapentanoic acid (EPA) to ARA should be 8:4:1. The work of Alorend (2004) suggested that dietary levels of >4 mg/g ARA over the long term have a negative impact on egg quality and she suggested an optimum level of 3 mg/g of ARA.

It is important to ensure that broodstock feeds are formulated with the highest quality ingredients and often include components such as squid meal, squid hydrolysate, and krill meal. Broodstock nutrition studies have been ongoing at the CCAR for over 5 years in what is probably the longest running experiment of its kind with this species. Three different diets are under evaluation; two of these are formulated feeds that are compared to the traditional raw fish and squid diet. The formulated diets are mixed as a semi-moist paste and extruded into a 30-mm sausage skin. Given the variable quality of eggs from captive broodstock halibut, varying forms of reproductive dysfunction, and difficulties associated with accurate timing of manual egg collection, it is still unclear whether formulated feeds can match wet fish ingredients.

1.5.2 Controlled spawning

The reproductive endocrinology of this species has been studied in relatively little detail. Methven *et al.* (1992) studied the seasonal changes in vitellogenin and sex steroid levels in captive male and female halibut. They observed the typical pattern of increasing levels of estradiol 17β and testosterone during gonadal recrudescence followed by a drop coinciding with the first release of eggs. Subsequent fluctuating levels of estradiol 17β, testosterone, and vitellogenin were thought to correspond to sequential maturation and release of egg batches. More recently, Kobayashi *et al.* (2008) using advanced molecular techniques has shed more light on follicular expression of gonadotropic receptors FSH-R and LH-R.

Very few attempts have been made to control spawning using steroid hormones in halibut. Spermiation in male halibut generally starts before the females are ready to spawn and in captive males, spermiation may stop before all female broodstock have completed spawning. Though milt can be cryopreserved (Rana *et al.* 1995) or extended, the application of gonadotropin-releasing hormone agonist (GNRHa) implants has proved useful in synchronizing spermiation (Vermeirssen *et al.* 1999; Martin-Robichaud *et al.* 2000; Vermeirssen *et al.* 2004). The application of GNRHa implants also reduces spermatocrit and the resulting milt is easier to collect and use during artificial fertilization. Induction of spawning in female Atlantic halibut has not been documented and it is likely that this technique may be worth exploring in the future.

1.5.3 Egg collection and incubation

Eggs and milt are collected manually by hand stripping the fish out of the water raised on stripping tables. Fertilization is generally achieved using the wet method whereby milt is mixed into seawater then poured over and mixed gently with the eggs. This should be done quickly as the milt remains motile for only a couple of minutes. The motility of sperm is checked under a low power objective on a microscope prior to fertilization to confirm viability. A typical ratio in this mixture would be 1 mL to 1,000 mL to 1,500 mL (milt:eggs:water). The eggs are left to "water harden" for 20 minutes then rinsed of excess milt and ovarian fluid. After a sample is taken for fertilization checks, which are best done at the 8-cell stage after about 16 hours at 6°C, the eggs are stocked to upwelling incubators. A typical stocking density is up to 300 eggs per liter.

Blastomere morphology is easily examined in this species owing to the peripheral displacement of the large cells during early cell divisions and the lack of opacity of the egg. A strong link between the gross morphology of these blastomeres and egg viability has been demonstrated (Shields *et al.* 1997) which enables the hatchery manager to make decisions about which egg batches are worthwhile.

In general, the eggs of the Atlantic halibut have a relatively high specific gravity owing to their high inorganic content (Riis-Vestergaard 1982) and they will sink at ambient salinities found in most coastal marine hatcheries. To counteract this, the eggs are incubated in upwelling tanks. These are usually cylindroconical tanks of volume between 100 and 1,000 liters. A gentle flow enters through a bottom inlet and leaves via a surface outlet which is often a "banjo filter" with a 1-mm screen. This screen must have a large surface area to reduce velocity at the outlet to prevent collection of eggs at the outlet. Bunching of eggs here will cause high mortality. Room temperature is maintained at 6°C with an air chiller and the room is light proof, all procedures being carried out using low intensity light.

Bacterial contamination of halibut eggs may lead to a reduction in viability and it is common practice to use surface disinfectants, for example, glutaraldehyde (400 ppm, 10 minutes) (Harboe *et al.* 1994a). Increased survival rates during first feeding have been attributed to such treatments; however, this practice is not universally adopted. An alternative and less toxic egg disinfectant, peroxyacetic acid (200 ppm, 1 minute) initially tested in the United Kingdom with promising results (Kristjansson 1995) has been adopted by many hatcheries. Outbreaks of nodavirus in Norwegian hatcheries led to the development of ozone disinfection techniques. An exposure to a concentration of 2 mg/L with a contact time of 2 minutes is effective against this pathogen (Grotmol and Totland 2000; Grotmol *et al.* 2003).

Once per day, dead and nonviable eggs are removed from the tanks using the "salt plug technique" developed in Norway (Jelmert and Rabben 1987). The flow is turned off and about 10–20 liters of high salinity (40 ppt) seawater is injected into the bottom of the tank. Live eggs generally float on the resulting halocline and nonviable eggs drop to the bottom where they can be tapped off with the salt plug. The flow is then restored and the volume of dead eggs is recorded. Hatching takes place in the incubators after approximately 75–80°C days postfertilization. Hatched larvae will usually float in the surface layer and can be removed using plastic jugs. Larvae are transferred in jugs to yolk sac incubators in lightproof, insulated containers. Light can delay hatching (Helvik and Walther 1993) and this fact is used in some hatcheries to synchronize hatching of a batch. Eggs can be moved to the yolk sac incubation system just prior to hatching or immediately after hatching, in which case empty egg cases and hatching debris are left behind.

1.6 Larval culture

1.6.1 System design and requirements

The long yolk sac absorption phase in halibut (220–290°C days) necessitates a separate yolk sac incubation system. Usually housed in a light proof, temperature-controlled room set at the temperature between 5 and 6°C, the tanks are similar to egg incubation tanks but much larger (see Figure 1.4).

These cylindroconical tanks range in volume from 700 liters to large silos of 3–13 m^3 favored by Norwegian operators (Harboe *et al.* 1994b; Berg 1997). Incubators at the CCAR have a volume between 700 and 1,000 liters. The Canadian hatchery uses large, Norwegian/Icelandic style silos. Flows are upwelling and the outlet is set close to the top of the tank. A filter with a large surface area prevents entrapment of the larvae. Incubators in use at the CCAR have one inlet for salt water and do not use oxygen or aeration.

Prior to first feeding, larvae are moved to larger volume rearing tanks which are typically 2–10 m^3. These are circular fiberglass or plastic tanks, generally dark in color, with bottom drains, and often with additional side drains. Overhead lighting is provided either by fluorescent or incandescent lighting and the light intensity can be relatively high. Tanks are provided with aeration to create turbulence and prevent crowding of larvae under the light source, particularly at the start of feeding. Many facilities now incorporate self-cleaning equipment in the larval rearing tanks to reduce labor associated with siphoning out settled organic matter (Van der Meeren *et al.* 1998).

Figure 1.4 Yolk sac larvae incubator.

1.6.2 Hatchery protocols

The period from hatching to first feeding, when the endogenous reserves stored in the yolk sac are absorbed, can last up to 50 days depending on the temperature. During this period, larvae are held in upwelling cylindroconical incubators. Reported stocking densities in the larger silos are in the region of 1–20 larvae/liter (Olsen *et al.* 1999). Densities of around 45 larvae/liter are typical in the yolk sac incubation tanks used at the CCAR that compensates somewhat for the smaller volume. In practical terms, this means that larvae from an average single batch of hatched larvae can usually be accommodated in one incubator. Typical survival rates in these incubators range from 50 to 80%, similar to those reported in Norwegian installations (Mangor-Jensen *et al.* 1998).

Strict temperature control is necessary during this phase since suboptimal temperatures can cause developmental abnormalities or high mortality (Bolla and Holmefjord 1988; Lein *et al.* 1997a). Salinity must also be within a narrow range (Lein *et al.* 1997b; Bolla and Ottesen 1998) and maintenance of good water quality is required. The larvae are generally kept in near or complete darkness because they are strongly attracted to a light source at the later stages of this phase. The transition to exogenous feeding can occur between 200 and 290°C days and the duration of the live feed stage is typically 50–70 days (Harboe *et al.* 1990; Lein and Holmefjord 1992). Current practice at CCAR is that at about 240°C days posthatching, the larvae are moved out to covered larval rearing tanks. The larvae are strongly positively phototactic toward the end of the yolk sac period (Naas and Mangor-Jensen 1990) and this fact is used to attract the larvae to the surface for collection. Generally, the larvae are transported at high density to larval rearing tanks as quickly as possible and numbers are estimated from sample counts. The larvae are maintained in clear water in the larval rearing tank in complete darkness while the temperature is raised gradually to 10°C. First feeding begins at 290°C days posthatching when viable larvae should initiate feeding within a few hours of the first addition of feed.

Larvae are fed live prey, which in intensive hatcheries means *Artemia*, although rotifers, cultured copepods, wild zooplankton, or a mixture of these have been used (Holmefjord *et al.* 1993; Naess *et al.* 1995). "Green water" is generally used in intensive systems since it has been found to be beneficial for first feeding success (Naas *et al.* 1992; Holmefjord *et al.* 1993; Guldbransen *et al.* 1996).

Mass production of halibut was initially achieved in Norway using semi-intensive techniques and these have been reviewed by Mangor-Jensen *et al.* (1998). Larvae reared in indoor incubators are moved to outdoor bag enclosures prior to first feeding and fed harvested wild zooplankton and *Artemia*. Though this technique can potentially generate large numbers of fry and was the mainstay of production up until the mid-1990s, output from these systems fell drastically in 1995 and it is now accepted that the method has drawbacks. Seasonal variations in wild zooplankton harvests can result in shortages of live prey. There is also a greater risk of exposure to pathogens, for example, nodavirus (VNN) or infectious pancreatic necrosis (IPN), which can cause serious mortalities in halibut (Grotmol *et al.* 1997). Large size variations are also a characteristic of fry reared in these systems and this can cause problems at weaning (Berg 1997). The development of methods for hatchery production in Iceland and the United Kingdom focused on intensive techniques using *Artemia* as the primary live food source. Larvae are reared exclusively in tanks through the entire rearing process (see Figure 1.5).

U.S. and Canadian techniques for halibut culture evolved from technology transfers from commercial hatcheries in Norway and Iceland, and from research institutions in the United Kingdom (Seafish Industry Authority and the Institute of Aquaculture in Stirling). Semi-intensive production methods using wild harvested zooplankton were in use in some Canadian hatcheries up until the late 1990s but this culture methodology is no longer practiced in Canada.

1.6.3 Water quality

Methods currently used in Maine at the CCAR make extensive use of marine recirculation technology. This has resulted in a greater degree of control of water quality and important physical parameters of temperature, gas saturation levels, and salinity. It also has resulted in, as yet, unexplained benefits of consistency in larval survival thought to be associated with biofiltration. The possible probiotic effects of stable bacterial populations in the biofilters, pipes, and tanks could actually limit the impact of opportunistic pathogenic bacteria so commonly implicated in crashes of populations of larvae in the first feeding stage (Verner-Jeffreys *et al.* 2003). Makeup water supplies for egg incubation, yolk sac larval

Figure 1.5 Late pelagic phase halibut larvae feeding on Artemia.

rearing, first feeding, and for live food production are filtered to 1 μm and UV treated to 200 μWsec/cm^2.

Flow rates to the larval rearing tanks should start at a low rate of exchange and increase gradually as feed inputs and biomass increase. At the CCAR, tanks of 2–8.5 m^3 are used, depending on batch size. Water exchange rates start at once per 24 hours and by day 50 post first feeding, reach up to a 6-hour turnover.

Microalgae are commonly used for halibut rearing (Naas *et al.* 1992). The use of algae, or the "green water technique" as it is commonly known, has been in use since early times in the development of marine fish hatchery techniques and the practice is still almost universal. However, the use of algae pastes and algae substitutes is becoming more widespread. Experiences over the last few years at the CCAR with the use of powdered clay suggest that this is a very cost-effective alternative and in terms of juvenile quality, there have been no detrimental effects (Brown, unpublished data).

1.6.4 Food and feeding

Halibut larvae have a relatively large mouth size and can start to feed on *Artemia* nauplii from the outset. It is common practice to begin with freshly hatched nauplii and feed at a density of 1 per mL. Feeding should occur in the majority of the population within 4 hours with a vigorous batch of larvae. Nauplii are given for 2–3 days before switching to a 24-hour enriched *Artemia*. As the larvae grow, larger, ongrown *Artemia* should be given. While some hatcheries will grow *Artemia* for up to 4 days (Olsen *et al.* 1999), this requires a great deal of tank space. Experience at CCAR has demonstrated that a 48-hour enriched *Artemia* will supply the energetic and nutritional needs of halibut larvae from around 450°C days posthatch. Good results in terms of survival (average 35%), pigmentation (>95% normal), and eye migration (>95% normal) can be achieved with an *Artemia*-only diet using commercially available enrichment products.

The early trials using *Artemia* as the sole source of food prior to weaning demonstrated that nutritional deficiencies in this prey organism compared to wild copepods resulted in poor rates of normal metamorphosis (Naess *et al.* 1995). U.K. trials, however, indicated that by manipulating the nutritional profile of *Artemia* through enrichment strategies, fry quality could be improved (Gara *et al.* 1998) though rates of normal development were still relatively low. A compromise strategy to achieve good rates of growth, pigmentation, and eye migration was devised whereby *Artemia* were used as the main prey organism but copepods were fed during a critical period which became known as the "copepod window" (Naess *et al.* 1995). Breakthroughs by commercial hatcheries, in particular Fiskey in Iceland, demonstrated that with the correct enrichment regime, well pigmented fry with good eye migration morphology could be produced with *Artemia*-only feeding strategies. Work to develop diets which can mimic the biochemical profile of copepods, based partly on the detailed work of Van der Meeren *et al.* (2008) has resulted in proprietary enrichment products that produce juveniles with acceptable rates of normal morphology.

However, problems with eye migration and pigmentation still remain in some hatcheries (Hamre *et al.* 2007; Hamre and Harboe 2008) and the causes are still the subject of considerable debate. The requirement for essential fatty acids (EFAs) is often the focus of studies to find the cause of these abnormalities and advice on levels EFAs, in particular DHA, EPA, and ARA, is abundant (McEvoy *et al.* 1998; Sargent *et al.* 1999; Hamre and Harboe 2008). Other possible factors include overall energy intake (Gara *et al.* 1998), iodine and thyroid levels (Solbakken *et al.* 2002), and even photoperiod (Solbakken and Pittman 2004).

Multiple feedings help to ensure that *Artemia* presented to the larvae are freshly enriched and that valuable nutrients are not lost or catabolized. It is common practice to feed 3–4 times daily. This also enables the hatchery manager to keep a close track of how much a population of larvae is eating. Automated feeding systems help to reduce the need for staff to feed at night but in commercial hatcheries, night checks are common practice anyhow. Lights are left on for 24 hours and feed should also be available round the clock.

1.6.5 Formulated feeds

Despite many trials testing early weaning of halibut prior to metamorphosis, including work conducted by Brown (1998), lower survivals and poor growth are generally the result of most formulated feeds when this is attempted too early. Once larvae are through metamorphosis, a good batch of fish will wean very quickly, usually within 2–3 days. Protracted cofeeding strategies are not necessary and weaker fish unable to wean rapidly should be removed from the population at this point. The accumulation of uneaten feed at this stage presents a challenge for hatchery staff and self-cleaning tanks are desirable. Fast circular, self-cleaning flows are possible once fish are settled onto the bottom thus helping particle movement, which in turn helps to attract fish to feed.

1.6.6 Microbial environment

One of the principal reasons for inconsistent output from marine hatcheries is early die offs during larval rearing, often associated with changes in microbial flora. Water for larval rearing is often highly filtered and sterilized. Larvae emerge from incubation systems with guts which are largely uncolonized with bacteria. Added to the larval rearing tanks is a cocktail of bacteria coming from the live feed cultures; microalgae, *Artemia*, or rotifers and these bacteria have often shown up as dominant microbial flora in the larval gut in studies that monitor changes in bacterial flora through the rearing cycle for a number of species, including halibut (Verner-Jeffreys *et al.* 2003). Added to this environment is the build up of organic material in the form of dead larvae, fecal wastes, and dead prey items, which all act as substrate for colonization.

Microbial conditions do tend to be more stable in recirculating hatchery systems and this was demonstrated for Atlantic halibut by Verner-Jeffreys *et al.* (2004). Recirculation systems are used for all stages of hatchery production at the CCAR halibut hatchery.

It is important to control the build up of waste in the larval rearing systems to deprive the microbial food web of substrate. One of the most labor-intensive tasks in a commercial hatchery is the removal of organic wastes, which gets collected in the slow-moving tanks, used for larval rearing. This is often simply done by manually siphoning or with the use of a squeegee. If this is not done carefully, this material can easily be resuspended. The automation of tank cleaning is commonly quoted by hatchery managers as a priority and there are some systems available commercially. A group working on marine fish culture in Austevoll, Norway, at the Institute of Marine Research designed a system that incorporates a rotating squeegee arm to collect debris which is sucked up through outlet holes in the arm. The design was described in Van der Meeren *et al.* (1998) and equipment based on a variation of this design has since been commercialized and is in use in Atlantic halibut hatcheries in Norway.

1.6.7 Harvest

Halibut larvae are robust by the time they reach 150 mg wet weight, close to metamorphosis and can be harvested by net or siphon in the pelagic phase or after settling to demersal habit. At this point, they are moved to weaning tanks with a treatment system of sufficient scale that dry diet can be fed to excess without major disruption of water quality caused by the build up of uneaten feed. Water exchange at this stage should be at least once per hour. Smaller individuals will be targeted by larger, dominant fish and cannibalism is common. The smaller fish at this point are best removed from the population as they will tend to be the slower growers throughout the growout stage and will not help the farmer's bottom line.

1.6.8 Hatchery economics

The market price of Atlantic halibut juveniles remains quite high ($5 or more per 5 g fish) and is a barrier to entry for many would-be halibut growers. With relatively few major players and what are still relatively small production runs, there are few hatcheries that benefit from economy of scale. The biology of the halibut results in a long hatchery cycle. This fact cannot be changed and will always mean that a halibut juvenile will cost more than a salmon, cod, or turbot juvenile. However, the market price at harvest is also higher and for some farmers, halibut is already an attractive option for growout.

1.7 Potential for stock enhancement

There are a number of features that make the Atlantic halibut a good candidate for stock enhancement. Most hatcheries are still rearing a significant proportion of their fish from eggs spawned from wild origin fish. This could be seen as a benefit for stock enhancement in terms of genetics. Atlantic halibut juveniles are generally robust and transport at relatively high densities with little mortality. In the hatchery, they spend a long time on live feeds and could be stocked out at the end of this phase without affecting their instinct for predation. One major disadvantage for this species is the cost of rearing which would make any restocking effort expensive. Also, their relatively slow growth, and thus time to legal landing size, would make assessment of their recapture rate and recruitment to the fishery very complicated. Other factors that would need to be considered would be juvenile quality, especially eye migration and pigmentation. Presumably, albino fish would be more susceptible to predation. No restocking efforts for this species have been attempted so far.

1.8 Growout

1.8.1 System design

There are two basic approaches to ongrowing; using land-based tanks or raceways (Adoff *et al.* 1993; Blanquet and Lygren 1997; Brown 2002) or at sea in cages (Martinez-Cordero *et al.* 1994; Brown 2002).

Atlantic halibut juveniles will spend at least part of the growout cycle in a land-based system whether they stay in tanks all the way through to harvest, or move out to net pens for the latter part of the growout cycle. Nursery systems differ from growout system only in scale of tanks. Halibut are grown in a variety of tank types; shallow raceways, large circular tanks, semisquare tanks, and in tanks made of a variety of materials; fiberglass, concrete, polyethylene, glass

Figure 1.6 Halibut juveniles using shelving.

fused to steel, PVC-formed concrete, etc. This author conducted trials in Canada to test the use of shallow raceways compared to deeper tanks and found that growth during the juvenile stages (25–250 g) was significantly better in deeper tanks. Access to slow-sinking pellets in shallow water is impeded and aggression and collisions during feeding were more frequently observed. Added to this is the problem of deteriorating water quality from one end of the tank to the other and accumulation of uneaten feed. Deep circular tanks (4–10 ft deep) are very effective for halibut and while most flatfish species do not voluntarily occupy shelving if provided, Atlantic halibut will (see Figure 1.6).

The extent to which they do use it will depend on shelf spacing, hydrodynamics, lighting, and overall stocking density. Multiple layers of fish utilizing shelving in deep tanks can make very efficient use of space and tank volume. At the time of writing, there were only three growout operations using land-based facilities in North America (two in Nova Scotia and one in Maine). These facilities all employ recirculation technology of various descriptions and all report good growth and survival in these systems.

Cage designs for flatfish have evolved from round fish-type cage designs and are often simply modified from existing units. An important consideration for flatfish cages is the provision of a rigid base to ensure that when stocked with fish, the net pen will not distort or sag. This can cause aggregation of fish, dead spots with poor water exchange, poor feed distribution, and uneven loads on the net panels and frame. Cage bases are generally net panels tensioned to a rigid frame constructed of steel or plastic. Surface cages are most commonly used and a variety of designs, constructed of steel, plastic, or even wood have been used in Scotland, Norway, and Canada. Cages are generally 3–7 m in depth and net bases are fabricated from 6 to 15 mm netting and sides can be manufactured from larger mesh (e.g., 22 mm). Use of smaller mesh bases can reduce feed waste since halibut tend to take some feed off the bottom. Predator netting is very important since the fish in the cage spend a large proportion of the time adjacent to the netting, within easy reach of predators that are not excluded by additional barriers. Occasional losses of halibut to predators including seals and otters have been reported in Canada. Surface cages have been in use for small-scale ongrowing halibut for many years in New Brunswick, Canada. Submersible cage designs have been tested successfully for halibut at the University of New Hampshire offshore aquaculture site near the Isles of Shoals, New Hampshire (Howell and Chambers 2005). Early trials in Maine explored the use of a submersible cage for halibut (Duym 1996).

The use of lobster pounds was investigated in New Brunswick between 1999 and 2003. These types of facilities were found suitable for halibut initially but high mortalities were observed during extremely cold winter conditions in these shallow enclosures which are only flushed twice daily by tidal exchange. Following heavy losses due to extremely low temperatures, the use of these pounds was abandoned.

The use of cameras, which are already widely used in surface cages to allow visual observation, particularly during feeding, will be an essential component of submersible cage systems for flatfish as will the development of suitable feed delivery systems.

1.8.2 Environmental conditions

The optimum water temperature for halibut decreases with increasing fish size (see Table 1.1). The upper lethal temperature limit is around 18–20°C depending upon feeding and dissolved oxygen levels and the lower limit is near—1.3°C. Halibut can tolerate a wide range in salinity and in fact growth can be higher at salinities lower than full strength sea water (Imsland *et al.* 2008) and this opens up the possibility to utilize ground water sources or geothermally heated water sources. Recommendations for other water quality parameters are similar to other marine species. Ammonia nitrogen (unionized NH_3–N) should be maintained below 0.05 mg/L, pH range should be 7.2–8.0, dissolved O_2 kept

Table 1.1 Recommended stocking density and water temperature for Atlantic halibut.

Size range (g)	Recommended stocking density (kg/m^2)	Optimum temperature range (°C)
0–10	5	11–14
11–20	10	11–14
21–50	15	11–13
50–150	20	11–13
150–400	30	10–12
400–1,000	40	9–11
1,000+	50	7–11

above 6 mg/L and dissolved CO_2 kept below 20 mg/L. Total gas supersaturation should be avoided.

For halibut in open water pens, high current and wave motion cause increased swimming activity and sheltered sites are required for halibut. Exposure to high levels of UV light in cages (and outdoor tanks) under strong sunlight can cause health problems. Halibut are particularly susceptible and may develop fat cell necrosis, which may eventually lead to high levels of mortality following secondary infection (Bricknell *et al.* 1996). This problem is avoided with the use of shade netting (>80% is recommended) over the cages, particularly for juvenile fish or where shallow nets (<4 m) are used.

1.8.3 Diet and nutrition

Most major commercial feed companies are making diets tailored specifically for halibut farmers. Reported optimal levels of protein range from 45 to 63% (Hjertnes and Opstvedt 1990; Aksnes *et al.* 1996; Hamre *et al.* 2003). Berge *et al.* (1999) demonstrated that a significant fraction of the protein (44% of nitrogen) could come from soy protein concentrate. Increasing dietary lipid can have a protein sparing effect but may also result in elevated carcass fat deposition (Aksnes *et al.* 1996). Lipid utilization was recently examined by Martins *et al.* (2007) and their study revealed that halibut can tolerate up to 25% lipid in the diet but no beneficial effect is gained from levels higher than 14%. Halibut have a limited capacity for digestion and utilization of carbohydrates (Hatlen *et al.* 2005), indicative of their carnivorous feeding habits in the wild. While halibut are chiefly visual feeders, there is good evidence that higher levels of attractants in the diet will stimulate an increase in appetite (Yacoob and Browman 2007). Slow-sinking pellets are preferred, as fast-sinking pellets may remain uneaten in a crowded tank under layers of fish. One cost-saving feature in common with most non-salmonid diets is the fact that the expensive artificial pigments are not needed.

In general, feeding frequency decreases with increase in fish size and decrease in temperature. Hand feeding is still commonly used in halibut farms and automatic feeders are utilized, however, as a method to ensure satiation and during the fry/nursery stage to reduce aggression.

1.8.4 Stocking density and grading

As with all flatfish, surface area rather than volume is the critical factor in determining capacity of a tank or cage, assuming that water exchange is adequate for removal of metabolites and supply of oxygenated water. As the fish grow in size, the thickness of the fish increases and with it, the maximum recommended stocking density would increase. Table 1.1 presents guidelines on stocking density for various size ranges of fish.

Regular grading of halibut is required to reduce size variation, which if left unchecked can result in cannibalism and increased aggression. Aggression has been shown to lead to a high incidence of eye damage in halibut (Greaves and

Tuene 2001). Grading can be done with hand nets or mesh sorters during the nursery stage and grading tables during growout. Effective mechanical graders have been developed for halibut and are in use in most large facilities. If normal good husbandry practices are employed, halibut generally survive handling extremely well.

1.8.5 Harvesting and processing

Harvest of halibut or any flatfish from a net pen is slightly more challenging than for round fish as they are not easily captured in a seine net. Fish pumps can be used for smaller halibut. Starvation of fish for 24 hours is advisable prior to harvest though the removal of guts immediately postslaughter will negate the effect of a full stomach during processing. Rapid slaughter is desirable both from a welfare standpoint and to avoid rapid onset of rigor, drop in pH, reduced shelf life, and the development of "chalkiness" in the flesh (Kramer and Paust 1985). Halibut can be slaughtered in the same way as other farmed fish by automated percussive, electrical stunning, or anesthesia with CO_2 or ice/water slurry. Manual percussive stunning can be used in the area near the eyes, though this will cause damage and may affect the marketability of smaller fish sold "head on." Bleeding immediately postmortem by incision of a major artery during gutting or removal of gill arches is recommended. The presence of blood veins and spots in the fillet detracts from both appearance and taste. Bleeding is less effective following anesthesia than percussive stunning (Akse and Midling 2001). The fillet yield of halibut can be as high as 60% but is typically around 55%. Fat content in the fillet is around 3–4% and slightly lower near the tail (Nortvedt and Tuene 1998). Fat storage along the fin margins can result in fat contents of close to 45% in this region. A shelf life of up to 3 weeks on ice has been reported in chilled products without a loss in quality (Akse and Midling 2001).

1.8.6 Marketing

The traditional market for halibut has been based on the wild harvest of relatively large fish (>10 kg). Most of what is sold at retail is fresh and in the form of steaks. With the availability of farmed fish to the market, new product forms are emerging. Farmed halibut are usually marketed above 3 kg and fish over 5 kg fetch the best prices. However a small niche market for "plate sized" halibut does exist and these smaller whole halibut can also obtain very good prices. Small volumes (<100 MT/year) are sold into U.S. markets from Norway and Canada currently. The 4-month off-season for the Pacific halibut fishery is generally targeted for much of the farmed halibut sales in North America in order to fetch the highest prices.

1.9 Production economics

Very few studies have attempted to look at the economics of farming marine species in the United States. Efforts at predicting overall production costs for halibut have been made by a number of industry analysts around the world for the purposes of assessing the potential economic feasibility of halibut farming. Most halibut farms are still in Norway and only the larger scale farms there, which produce 200–300 MT/year are actually profitable with production costs approaching 5–6 €/kg ($3.3–4.0/lb) (Engelsen *et al.* 2004). A report prepared for the State of Alaska (Forster 1999) predicted U.S. production costs of between $3.19/kg ($1.45/lb) and $4.09/kg ($1.89/lb) based on cage growout of large halibut. An analysis of Canadian halibut farming (Penney 1999) found that sea cage farming of halibut would be profitable with an internal rate of return (IRR) of 9% but predicted that a land-based operation, using raceways and flow through would not be profitable. In contrast, McCallum (2000) predicted that at certain economies of scale (≥100 MT per annum), land-based recirculating growout of halibut in Canada was expected to give a good return on investment (IRR of 15%) with a production cost between $7.75/kg ($3.53/lb) at 100 MT/year and $7.19/kg ($3.27/lb) at 300 MT/year. In a study commissioned by the Maine Department of Marine Resources, Gardner Pinfold (2003) modeled IRRs of 9% for land-based recirculating production and 15% for sea-based growout on the basis of an ex-farm price of $4.50/lb.

1.10 Summary: industry constraints and future expectations

Long growout time is one of the main drawbacks for this species. As with many species, the halibut exhibits sexual dimorphism whereby males tend to grow more slowly than females and mature at a smaller size and younger age (Haug 1990). Triploidy has already been successfully induced in halibut using cold shocks (Holmefjord and Refstie 1997) or pressure shock (Brown 1998) and this may be one solution to this problem. Canada has taken the lead with production of all female fish reared from fertilizing eggs with milt from sex reversed males (Hendry *et al.* 2003). This technology is under development in Norway, Scotland, and the United States as well and could help to improve the economics of growout significantly. Breeding programs are under way in Scotland, Canada, Norway, and the United States though most eggs still come from wild caught broodstocks and there is little published information on performance of F1 stocks. Slower growing fish may be marketed at a smaller size and markets for fish under 1 kg are being developed by farms in North America and prices for these fish sold direct to the restaurants are high (>$10/lb).

Pilot halibut farming efforts have been underway elsewhere in North and South America. A facility in Hawaii with access to deep cold water has a small number of fish to growout for local markets (Jim Parsons, personal communication) and hatchery facilities in Punta Arenas, Chile, have been under development for some time (Alvial and Manriquez 1999). This project has been supplied with eggs, broodstock, and juvenile halibut from private and government entities in Canada.

Atlantic halibut has been on the lists of new, promising species for aquaculture for many years. As a result of a long-term research effort in many countries, many of the technical hurdles have been overcome. However, levels of production remain modest. Farmed halibut is generally well received in the marketplace and larger farming operations, particularly those which are vertically integrated, are likely to remain profitable and grow over the next few years.

Literature cited

Adoff, G.R., Andersson, T., Engelsen, R., and Kvalsund, R. 1993. Land-based farm for ongrowing of halibut. In: Reinertsen, H., Dahle, L.A., Jorgensen, L., and Tvinnereim, K. (eds) *Fish Farming Technology*. Balkema, Rotterdam, pp. 329–331.

Akse, L., and Midling, K.Ø. 2001. Slaughtering of Atlantic halibut (*Hippoglossus hippoglossus*) effects on quality and storage capacity. In: Kestin, S.C., and Warriss, P.D. (eds) *Farmed Fish Quality*. Blackwell Science, Oxford, UK, pp. 381.

Aksnes, A., Hjertnes, T., and Opstvedt, J. 1996. Effect of dietary protein level on growth and carcass composition in Atlantic halibut (*Hippoglossus hippoglossus* L.). *Aquaculture* 145:225–233.

Alorend, E. 2004. The effect of dietary arachidonic acid concentration on Atlantic halibut (*Hippoglossus hippoglossus*) broodstock performance. Assessment of egg, milt and larval quality. PhD thesis, University of Stirling, Stirling, Scotland, UK.

Alvial, A., and Manriquez, J. 1999. Diversification of flatfish culture in Chile. *Aquaculture* 176:65–73.

Bell, J., and Sargent, J. 2003. Arachidonic acid in aquaculture feeds: current status and future opportunities. *Aquaculture* 218:491–499.

Berg, L. 1997. Commercial feasibility of semi-intensive larviculture of Atlantic halibut (*Hippoglossus hippoglossus* L.). *Aquaculture* 155:333–340.

Berge, G., Grisdale-Helland, B., and Helland, S. 1999. Soy protein concentrate in diets for Atlantic halibut (*Hippoglossus hippoglossus*). *Aquaculture* 178:139–148.

Blanquet, D., and Lygren, E. 1997. Cultivation of marine fish in closed raceway system. In: Jorgensen, L. (ed.) *Aquaculture Trondheim 97: Cultivation of Cold Water Species; Production, Technology and Diversification*. European Aquaculture Society, Oostende, Belgium, pp. 8–9.

Blaxter, J.H.S., Danielssen, D., Moksness, E., and Oiestad, V. 1983. Description of the early development of halibut (*Hippoglossus hippoglossus*) and attempts to rear the larvae past first feeding. *Marine Biology* 73:99–107.

Bolla, S., and Holmefjord, I. 1988. Effect of temperature and light on development of Atlantic halibut larvae. *Aquaculture* 74:355–358.

Bolla, S., and Ottesen, O. 1998. The influence of salinity on the morphological development of yolk sac larvae of Atlantic halibut, *Hippoglossus hippoglossus* (L.). *Aquaculture Research* 29:203–209.

Bricknell, I.R., Bruno, D.W., Bowden, T.J., and Smith, P. 1996. Fat cell necrosis syndrome in Atlantic halibut, *Hippoglossus hippoglossus* L. *Aquaculture* 144:65–69.

Bromage, N., Bruce, M., Basavaraja, N., Rana, K., Shields, R., Young, C., Dye, J., Smith, P., Gillespie, M., and Gamble, J. 1994. Egg quality determinants in finfish: the role of overripening with special reference to the timing of stripping in the Atlantic halibut (*Hippoglossus hippoglossus*). *Journal of the World Aquaculture Society* 25:13–21.

Brown, N.P. 1998. Egg quality, triploidy induction and weaning of the Atlantic halibut *Hippoglossus hippoglossus*. PhD thesis, Institute of Aquaculture, University of Stirling, Scotland, UK, 325 pp.

Brown, N. 2002. Flatfish farming systems in the Atlantic region. *Reviews in Fisheries Science* 10:403–419.

Brown, N., Shields, R.J., and Bromage, N.R. 2006. The influence of water temperature on spawning patterns and egg quality in the Atlantic halibut (*Hippoglossus hippoglossus* L.). *Aquaculture* 261(3): 993–1002.

Bruce, M.P., Shields, R.J., Bell, M.V., Bromage, N.R., 1993. Lipid class and fatty acid composition of eggs of Atlantic halibut, *Hippoglossus hippoglossus* (L.), in relation to egg quality in captive broodstock. *Aquaculture and Fisheries Management* 24:417–422.

Cargnelli, L.M., Griesbach, S.J., Morse, W.W. 1999. Atlantic halibut, *Hippoglossus hippoglossus*, life history and habitat characteristics. NOAA Technical Memorandum NMFA-NE-125. U.S. Department of Commerce, Washington, DC, 25 pp.

Collette, B.B., and Klein-MacPhee, G. 2002. *Fishes of the Gulf of Maine*. Smithsonian Institution Press, Washington, DC, 748 pp.

Dannevig, B.H., Nilsen, R., Modahl, I., Jankowska, M., Taksdal, T., and Press, C.McL. 2000. Isolation in cell culture of nodavirus from farmed Atlantic halibut *Hippoglossus hippoglossus* in Norway. *Diseases of Aquatic Organisms* 43:183–189.

Duym, T. 1996. Submersible halibut cage project: interim report. In: Polk, M. (ed.) *Open Ocean Aquaculture. Proceedings of an International Conference*. Portland, Maine. New Hampshire/Maine Sea Grant College Program Rpt. # UNHMP-CP-SG-96–9, pp. 383–388.

Engelsen, R., Asche, F., Skjennum, F., and Adoff, G. 2004. New species in aquaculture: some basic economic aspects. In: Moksness, E., Kjorsvik, E., and Olsen, Y. (eds) *Culture of Cold Water Marine Fish*. Blackwell Publishing, Oxford, UK, pp. 487–515.

Forster, J. 1999. Halibut farming. Its development and likely impact on the market for wild Alaska halibut. Report written for the State of Alaska, Alaska Department of Commerce and Economic Development, 36 pp.

Gara, B., Shields, R.J., and McEvoy, L. 1998. Feeding strategies to achieve correct metamorphosis of Atlantic halibut, *Hippoglossus hippoglossus* L. using enriched Artemia. *Aquaculture Research* 29:935–948.

Gardner Pinfold 2003. Maine aquaculture – viability of selected species and culture systems: a report prepared for the Maine Department of Marine Resources.

Greaves, K., and Tuene, S. 2001. The form and context of aggressive behaviour in farmed Atlantic halibut *Hippoglossus hippoglossus* L. *Aquaculture* 193:139–147.

Grotmol, S., Dahl-Paulsen, E., and Totland, G.K. 2003. Hatchability of eggs from Atlantic cod, turbot and Atlantic halibut after disinfection with ozonated seawater. *Aquaculture* 221:245–254.

Grotmol, S., and Totland, G.K. 2000. Surface disinfection of Atlantic halibut (*Hippoglossus hippoglossus*) eggs with ozonated sea-water inactivates nodavirus and increases survival of the larvae. *Diseases of Aquatic Organisms* 39:89–96.

Grotmol, S., Totland, G.K., Thorud, K., and Hjeltnes, B.K. 1997. Vacuolating encephalopathy and retinopathy associated with a nodavirus-like agent: a probable cause of mass mortality of cultured larval and juvenile Atlantic halibut *Hippoglossus hippoglossus*. *Diseases Aquatic Organisms* 29:85–97.

Guldbransen, J., Lein, I., and Holmefjord, I. 1996. Effects of light administration and algae on first feeding of Atlantic halibut larvae, *Hippoglossus hippoglossus* (L.). *Aquaculture Research* 27:101–106.

Hamre, K., and Harboe, T. 2008. Artemia enriched with high n-3 HUFA may give a large improvement in performance of Atlantic halibut (*Hippoglossus hippoglossus* L.) larvae. *Aquaculture* 277:239–243.

Hamre, K., Holen, E., and Moren, M. 2007. Pigmentation and eye migration in Atlantic halibut (*Hippoglossus hippoglossus* L.) larvae: new findings and hypotheses. *Aquaculture Nutrition* 13:65–80.

Hamre, K., Øfsti, A., Næss, T., Nortvedt, R., and Holm, J.C. 2003. Macronutrient composition of formulated diets for Atlantic halibut (*Hippoglossus hippoglossus*, L.) juveniles. *Aquaculture* 227:233–244.

Harboe, T., Huse, I., and Øie, G. 1994a. Effects of egg disinfection on yolk-sac and first feeding stages of halibut (*Hippoglossus hippoglossus* L.) larvae. *Aquaculture* 119:157–165.

Harboe, T., Naess, T., Naas, K.E., Rabben, H., and Skjolddal, L.H. 1990. Age of Atlantic halibut larvae (*Hippoglossus hippoglossus*) at first feeding. International Council for the Exploration of the Sea. C.M. 1990. F:53.

Harboe, T., Tuene, S., Mangor-Jensen, A., Rabbien, H., and Huse, I. 1994b. Design and operation of an incubator for yolk-sac larvae of Atlantic halibut. *The Progressive Fish Culturist* 56:188–193.

Hatlen, B., Grisdale-Helland, B., and Helland, S.J. 2005. Growth, feed utilization and body composition in two size groups of Atlantic halibut (*Hippoglossus hippoglossus*) fed diets differing in protein and carbohydrate content. *Aquaculture* 249:401–408.

Haug, T. 1990. The biology of the Atlantic halibut, *Hippoglossus hippoglossus* L., 1758. *Advances in Marine Biology* 26:1–70.

Haug, T., and Gulliksen, B. 1988. Fecundity and oocyte sizes in ovaries of female Atlantic halibut, *Hippoglossus hippoglossus* (L.). *Sarsia* 73:259–261.

Helvik, J.V., and Walther, B.T. 1993. Environmental parameters affecting induction of hatching in halibut (*Hippoglossus hippoglossus*) embryos. *Marine Biology* 116:39–45.

Hendry, C.I., Martin-Robichaud, D.J., and Benfey, T.J. 2003. Hormonal sex reversal of Atlantic halibut (*Hippoglossus hippoglossus* L.). *Aquaculture* 219:769–781.

Hjertnes, T., and Opstvedt, J. 1990. Effects of dietary protein levels on growth in juvenile halibut (*Hippoglossus hippoglossus*, L.). In: Takeda, M., Watanabe, T. (eds) *The Current Status of Fish Nutrition in Aquaculture*. Proceedings of the 3rd International Symposium on Feeding and Nutrition of Fish, Toba, Japan.

Holmefjord, I. 1991. Timing of stripping relative to spawning rhythms of individual females of Atlantic halibut (*Hippoglossus hippoglossus* L.). In: Lavens, P., Sorgeloos, P., Jaspers, E., Ollevier, F. (eds) *Larvi '91. Fish and Crustacean Larviculture Symposium*. European Aquaculture Society Special Publication No. 15, Ghent, Belgium, pp. 203–204.

Holmefjord, I. 1996. Spawning of Atlantic halibut in captivity. PhD thesis, University of Bergen, Norway.

Holmefjord, I., Gulbrandsen, J., Lein, I., Reftsie, T., Lèger, P., Harboe, T., Huse, I., Sorgeloos, P., Bolla, S., Reitan, K.I., Vadstein, O., Øie, G., and Danielsberg, A. 1993. An intensive approach to Atlantic halibut fry production. *Journal of the World Aquaculture Society* 24:275–284.

Holmefjord, I., and Lein, I. 1990. Natural spawning of Atlantic halibut (*Hippoglossus hippoglossus* L.) in captivity. International Council for the Exploration of the Seas, C.M. 1990/F:74, 5 pp.

Holmefjord, I., and Refstie, T. 1997. Induction of triploidy in Atlantic halibut by temperature shocks. *Aquaculture International* 5:169–173.

Howell, W.H., and Chambers, M.D. 2005. Growth performance and survival of Atlantic halibut (*Hippoglossus hippoglossus*) grown in submerged net pens. *Bulletin of the Aquaculture Association of Canada* 9:35–37.

Imsland, A.K., Gustavsson, A., Gunnarsson, S., Foss, A., Arnason, J., Arnarson, I., Jonsson, A.F., Smaradottir, H., and Thorarensen, H. 2008. Effects of reduced salinities on growth, feed conversion efficiency and blood physiology of juvenile Atlantic halibut (*Hippoglossus hippoglossus* L.). *Aquaculture* 274:254–259.

Jackson, T.R., Martin-Robichaud D.J., and Reith, M.E. 2003. Application of DNA markers to the management of Atlantic halibut (*Hippoglossus hippoglossus*) broodstock. *Aquaculture* 220:245–259.

Jákupsstovu, S.H., and Haug, T. 1988. Growth, sexual maturation and spawning season of Atlantic halibut, *Hippoglossus hippoglossus*, in Faroese waters. *Fisheries Research* 6:201–215.

Jelmert, A., and Rabben, H. 1987. Upwelling incubators for eggs of the Atlantic halibut (*Hippoglossus hippoglossus* L.). International Council for the Exploration of the Sea. C.M. 1987. F:20, 8 pp.

Kanwit, J.K. 2007. Tagging results from the 2000–2004 Federal experimental fishery for Atlantic halibut (*Hippoglossus hippoglossus*) in the Eastern Gulf of Maine. *Journal of Northwest Atlantic Fisheries Science* 38:37–42.

Kjorsvik, E., Haug, T., and Tjelmsland, J. 1987. Spawning season of the Atlantic halibut (*Hippoglossus hippoglossus*) in northern Norway. *Journal du Conseil Internationale pour Exploration de la Mer* 43:285–293.

Kobayashi, T., Pakarinen, P., Torgersen, J., Huhtaniemi, I., and Andersen, Ø. 2008. The gonadotropin receptors FSH-R and LH-R of Atlantic halibut (*Hippoglossus hippoglossus*)-2. Differential follicle expression and asynchronous oogenesis. *General and Comparative Endocrinology* 156:595–602.

Kramer, D.E., and Paust, B.C. 1985. *Care of Halibut Aboard the Fishing Vessel*. Marine Advisory Bulletin No. 18. University of Alaska Sea Grant College Program, University of Alaska, Fairbanks, 30 pp.

Kristjansson, B.A. 1995. Egg incubation of Atlantic halibut (*Hippoglossus hippoglossus* L.): bacterial loading and the use of peracetic acid as an egg surface disinfectant. MSc thesis, Institute of Aquaculture, University of Stirling, UK, 63 pp.

Kvenseth, A.M., Pittman, K., and Helvik, J.V. 1996. Eye development in Atlantic halibut (*Hippoglossus hippoglossus*): differentiation and development of the retina from early yolk sac stages through metamorphosis. *Canadian Journal of Fisheries and Aquatic Sciences* 53:2524–2532.

Lein, I., and Holmefjord, I. 1992. Age at first feeding of Atlantic halibut larvae. *Aquaculture* 105:157–164.

Lein, I., Holmefjord, I., and Rye, M. 1997a. Effects of temperature on yolk sac larvae of Atlantic halibut (*Hippoglossus hippoglossus* L.). *Aquaculture* 157:121–133.

Lein, I., Tveite, S., Gjerde, B., and Holmefjord, I. 1997b. Effects of salinity on yolk sac larvae of Atlantic halibut (*Hippoglossus hippoglossus* L.). *Aquaculture* 156:295–307.

Lonning, S., Kjorsvik, E., Haug, T., and Gulliksen, B. 1982. The early development of the halibut, *Hippoglossus hippoglossus* (L.), compared with other marine teleosts. *Sarsia* 67:85–91.

MacIntyre, A.D. 1953. The food of halibut from north Atlantic fishing grounds. *Marine Research* 3:1–20.

Mangor-Jensen, A., Harboe, T., Shields, R.J., Gara, B., and Naas, K.E. 1998. Atlantic halibut, *Hippoglossus hippoglossus* L., larvae cultivation literature, including a bibliography. *Aquaculture Research* 29:857–886.

Martinez-Cordero, F.J., Beveridge, M.C.M., Muir, J.F., Mitchell, D., and Gillespie, M. 1994. A note on the behaviour of adult Atlantic halibut, *Hippoglossus hippoglossus* (L.) in cages. *Aquaculture and Fisheries Management* 25:475–481.

Martin-Robichaud, D.J., Powell, J., and Wade, J., 2000. Gonadotropin-releasing hormone affects sperm production of Atlantic halibut (*Hippoglossus hippoglossus*). *Bulletin of the Aquaculture Association of Canada* 4:45–48.

Martin-Robichaud, D.J., and Rommens, M. 2001. Assessment of sex and evaluation of ovarian maturation of fish using ultrasonography. *Aquaculture Research* 32:113–120.

Martins, D., Valente, L., and Lall, S. 2007. Effects of dietary lipid level on growth and lipid utilization by juvenile Atlantic halibut (*Hippoglossus hippoglossus*, L.). *Aquaculture* 263:150–158.

Mazorra, C., Bruce, M., Bell, J.G., Davie, A., Alorend, E., Jordan, N., Rees, J., Papanikos, N., Porter, M., and Bromage, N. 2003. Dietary lipid enhancement of broodstock reproductive performance and egg and larval quality in Atlantic halibut (*Hippoglossus hippoglossus*). *Aquaculture* 227:21–33.

McCallum, T. 2000. Saltwater recirculation for land-based farming of Atlantic halibut (*Hippoglosus hippoglossus*). MSc thesis, University of New Brunswick, Canada.

McCracken, F.D. 1958. On the biology and fishery of the Canadian Atlantic halibut, *Hippoglossus hippoglossus* L. *Journal of Fisheries Research Board of Canada* 15:1269–1311.

McEvoy, L.A., Næss, T., Bell, J.G., and Lie, Ø. 1998. Lipid and fatty acid composition of normal and malpigmented Atlantic halibut (*Hippoglossus hippoglossus*) fed enriched Artemia: a comparison with fry fed wild copepods. *Aquaculture* 163:237–250.

Methven, D.A., Crim, L.W., Norberg, B., Brown, J.A., Goff, G.P. and I. Huse. 1992. Seasonal reproduction and plasma levels of sex steroids and vitellogenin in Atlantic halibut (*Hippoglossus hippoglossus*). Canadian Journal of Fisheries and Aquatic Sciences 49:754–759.

Naas, K.E., and Mangor-Jensen, A. 1990. Positive phototaxis during late yolk-sac stage of Atlantic halibut larvae (*Hippoglossus hippoglossus* L.). *Sarsia* 75:243–246.

Naas, K.E., Naess, T., and Harboe, T. 1992. Enhanced first feeding of halibut larvae (*Hippoglossus hippoglossus* L.) in green water. *Aquaculture* 105:143–156.

Naess, T., Germain-Henry, M., and Naas, K.E. 1995. First feeding of Atlantic halibut (*Hippoglossus hippoglossus*) using different combinations of Artemia and wild zooplankton. *Aquaculture* 130:235–250.

Naess, T., Harboe, T., Mangor-Jensen, A., Naas, K.E., and Norberg, B. 1996. Successful first feeding of Atlantic halibut larvae from photoperiod-manipulated broodstock. *The Progressive Fish Culturist* 58:212–214.

Neilson, J.D., Kearney, J.F., Perley, P., and Sampson, H. 1993. Reproductive biology of Atlantic halibut (*Hippglossus hippoglossus*) in Canadian waters. *Canadian Journal of Fisheries and Aquatic Sciences* 50:551–563.

Norberg, B., Valkner, V., Huse, J., Karlsen, I., and Lerøy Grung, G. 1991. Ovulatory rhythms and egg viability in the Atlantic halibut (*Hippoglossus hippoglossus*). *Aquaculture* 97:365–371.

Nortvedt, R., and Tuene, S. 1998. Body composition and sensory assessment of three weight groups of Atlantic halibut *Hippoglossus hippoglossus* fed three pellet sizes and three dietary fat levels. *Aquaculture* 161:295–313.

Olsen, Y., Evjemo, J.O., and Olsen, A. 1999. Status of the cultivation technology for production of Atlantic halibut (*Hippoglossus hippoglossus*) juveniles in Norway/Europe. *Aquaculture* 176:3–13.

Ottesen, O.H., Noga, E.J., and Sanda, W. 2007. Effect of substrate on progression and healing of skin erosions and epidermal papillomas of Atlantic halibut, *Hippoglossus hippoglossus* (L.). *Journal of Fish Diseases* 30:43–53.

Ottesen, O.H., and Strand, H.K. 1996. Growth, development, and skin abnormalities of halibut (*Hippoglossus hippoglossus* L.) juveniles kept on different substrates. *Aquaculture* 146:17–25.

Penney, R. 1999. An economic analysis of land-based vs sea cage growout of Atlantic halibut *Hippoglossus hippoglossus*. MSc thesis. Simon Fraser University, 86 pp.

Pittman, K., Bergh, O., Opstad, I., Skiftesvik, A.B., Skjoldal, L., and Strand, H. 1990a. Development of eggs and yolk sac larvae of halibut (*Hippoglossus hippoglossus* L.). *Journal of Applied Icthyology* 6:142–160.

Pittman, K., Skiftesvik, A.B., and Berg, L. 1990b. Morphological and behavioural development of halibut, *Hippoglossus hippoglossus* (L.) larvae. *Journal of Fish Biology* 37:455–472.

Rabben, H., Nilsen, T.O., Huse, I., and Jelmert, A. 1986. Production experiment of halibut fry in large enclosed water columns. Council Meeting of the International Council for the Exploration of the Sea F19, 10 pp.

Rana, K., Edwardes, S., and Shields, R. 1995. Potential application of low temperature preservation of Atlantic halibut *Hippoglossus hippoglossus* L. and salmon Salmo salar spermatozoa for seed production. In: Lavens, P., Jaspers, E., Roelants, I. (eds) *Larvi '95. Fish and Shellfish Larviculture Symposium. European Aquaculture Society Special Publication No. 24*. Ghent, Belgium, pp. 53–56.

Riis-Vestergaard, J. 1982. Water and salt balance of halibut eggs and larvae (*Hippoglossus hippoglossus*). *Marine Biology* 70:135–139.

Rollefsen, G. 1934. The eggs and larvae of halibut (*Hippoglossus vulgaris*). *Det Kongelige Norske Videnskabers Selskab Forhandlinger* 7(7): 20–23.

Sargent, J., Bell, J.G., McEvoy, L.A., Tocher, D., and Estevez, A. 1999. Recent developments in essential fatty acid nutrition of fish. *Aquaculture* 177:191–199.

Shields, R.J., Brown, N.P., and Bromage, N.R. 1997. Blastomere morphology as a predictive measure of fish egg viability. *Aquaculture* 155:1–12.

Shields, R.J., Davenport, J., Young, C., and Smith, P.L. 1993. Oocyte maturation and ovulation in the Atlantic halibut, *Hippoglossus hippoglossus* (L.), examined using ultrasonography. *Aquaculture and Fisheries Management* 24:181–186.

Smith, P., Bromage, N., Shields, R., Ford, L., Gamble, J., Gillespie, M., Dye, J., Young, C., and Bruce, M. 1991. Photoperiod controls spawning time in the Atlantic halibut (*Hippoglossus hippoglossus*, L.). In: Scott, A.P., Sumpter, J.P., Klime, D.E., and Rolfe, M.S. (eds) *Proceedings of the Fourth International Symposium on the Reproductive Physiology of Fish*. University of East Anglia, Norwich. Fishsymp 91, Sheffield, p. 172.

Smith, P.L. 1987. The establishment of a potential broodstock at Ardtoe 1983–1987. Seafish Industry Authority Technical Report 137, 31 pp.

Solbakken, J.S., Berntssen, M.H.G., Norberg, B., Pittman, K., and Hamre, K. 2002. Differential iodine and thyroid hormone levels between Atlantic halibut (*Hippoglossus hippoglossus* L.) larvae fed wild zooplankton or Artemia from first exogenous feeding until post metamorphosis. *Journal of Fish Biology* 61:1345–1362.

Solbakken, J., and Pittman, K. 2004. Photoperiodic modulation of metamorphosis in Atlantic halibut (*Hippoglossus hippoglossus* L.). *Aquaculture* 232:613–625.

Stobo, W.T., Neilson, J.D., and Simpson, P.G. 1988. Movements of Atlantic halibut (*Hippoglossus hippoglossus*) in the Canadian North Atlantic. *Canadian Journal of Fisheries and Aquatic Sciences* 45:484–491.

Van Der Meeren, T., Harboe, T., Holm, J.C., and Solbakkena, R. 1998. A new cleaning system for rearing tanks in larval fish culture. International Council for the Exploration of the Sea. C.M. 1998. L:13, 11 pp.

Van Der Meeren, T., Olsen, R.E., Hamre, K., and Fyhn, H.J. 2008. Biochemical composition of copepods for evaluation of feed quality in production of juvenile marine fish. *Aquaculture* 274:375–397.

Vermeirssen, E.L.M., Mazorra de Quero, C., Shields, R., Norberg, B., Kime, D.E., and Scott, A.P. 2004. Fertility and motility of sperm from Atlantic halibut (*Hippoglossus*

hippoglossus) in relation to dose and timing of gonadotrophin-releasing hormone agonist implant. *Aquaculture* 230:547–567.

Vermeirssen, E.L.M., Shields, R., Mazorra de Quero, C., and Scott, A.P., 1999. Gonadotropin-releasing hormone agonist raises plasma concentrations progestogens and enhances milt fluidity in male Atlantic halibut (*Hippoglossus hippoglossus*). In: Norberg, B., Kjesbu, O.S., Taranger, G.L., Andersson, E., and Stefansson, S.O. (eds) *Proceedings of the Sixth International Symposium on the Reproductive Physiology of Fish*, Bergen 2000, Bergen, pp. 399–401.

Verner-Jeffreys, D.W., Shields, R.J., Bricknell, I.R., and Birkbeck, T.H. 2003. Changes in the gut-associated microflora during the development of Atlantic halibut (*Hippoglossus hippoglossus* L.) larvae in three British hatcheries. *Aquaculture* 219:21–42.

Yacoob, S.Y., and Browman, H.I. 2007. Olfactory and gustatory sensitivity to some feed-related chemicals in the Atlantic halibut (*Hippoglossus hippoglossus*). *Aquaculture* 263:303–309.

Chapter 2

Culture of Chilean flounder

Alfonso Silva

The production of marine fishes through their cultivation has experienced significant growth recently in Europe and Asia. Chile has participated in this growth by adapting technology for the cultivation of the turbot (*Scopthalmus maximus*) hirame (*Paralichthys olivaceus*) and by conducting research on the cultivation of the two native flounders (*P. microps* and *P. adspersus*). Today, after more than 10 years of efforts developing technology for the culture of flatfish, Chile's flatfish industry has become an important contributor to the nation's aquaculture production. At the moment, turbot is commercially cultured and, since 1998, between 268 and 426 tons of turbot have been produced annually. The two Chilean flounders, *P. microps* and *P. adspersus*, have been produced at pilot-scale level but are at the beginning of commercial cultivation. At the same time, the Halibut *(Hippoglossus hippoglossus)* has been introduced to evaluate their cultivation feasibility in southern Chile. Following, the current state of the cultivation of Chilean flounder *(P. adspersus)* is described as well as the technological aspects of its cultivation (Silva and Vélez 1998; Alvial and Manríquez 1999).

2.1 Life history and biology

2.1.1 Taxonomy

The flounder is an endemic resource off the coast of Chile among which the families Bothidae and Paralichthyidae (Zúñiga 1988) are present. From the latter, whose genus *Paralichthys* is composed of 17 species distributed along both coasts of America (Ginsburg 1952), 8 species have been described for Chile. The two flounder with the highest economic relevance are *P. adspersus*, also denominated Chilean flounder or "three stains flounder" (Figure 2.1) and *P. microps* or "small eyes" flounder.

Figure 2.1 Adult Chilean flounder (*Paralichthys adspersus*) from Fish Culture Laboratory, Department of Aquaculture, North Catholic University (UCN), Coquimbo, Chile.

From the taxonomic point of view, the Chilean flounder is described in the following way:

Class: Osteichthyes
Subclass: Teleostei
Superorder: Acantopterygii
Order: Pleuronectiformes
Family: Paralichtyidae
Genus: *Paralichthys*
Species: *Paralichthys adspersus* (Steindachner, 1867)

The taxonomic separation between *P. adspersus* and *P. microps*, although difficult because of their similar morphology, can be based on three characteristics:

(a) Origin of the dorsal fin; in *P. microps* the origin is located above the eye and in *P. adspersus* above or behind the margin of the eye (Ginsburg 1952).
(b) Number of branchialthorns; in *P. adspersus* the superior branch of the first branchial arch (6–7) differs from the one presented by *P. microps* (9–10) (Chirichigno 1974).
(c) Relative size of the nostril excurrente; the nostril of *P. microps* has visible higher diameter than the one from *P. adspersus* (Zúñiga 1988).

2.1.2 Natural range

The flounder *P. adspersus* is distributed from the town of Paita (North of Peru) to the Gulf of Arauco (Chile) including the island Juan Fernández (Pequeño and Plaza 1987; Pequeño 1989; Siefeld *et al.* 2003).

Their common habitat are gulfs and shallow bays, with soft sand bottom, similar to other flounder species like *P. dentatus* and *P. californicus*, basically looking for protection against predators, for more appropriate temperatures and food abundance (Able *et al.* 1990; Kramer 1991; Acuña and Cid 1995).

2.1.3 Fishing companies

Approximately 99% of the national flounder fishery in Chile is done by artisanal fishermen for local consumption and possibly export, and does not distinguish between the two species *(P. microps* and *P. adspersus*). The capture fishery landings show a significant decrease during the last decade, from 821 tons in 1990 to only 76 tons in 2006.

At the moment, commercial landings are in the north region but the majority of landings are in the south center of the country (Region VIII) with 68.4% of the national landings. In most months, female flounder constitute the majority of the commercial catch (Acuña and Cid 1995).

2.1.4 Reproduction

Females *P. adspersus* develop large ovaries at maturation which extend from the abdomen back to the caudal region of their bodies. Chilean flounder spawn multiple batches of small pelagic eggs during the spawning season from August to December (Acuña and Cid 1995) when temperatures oscillate between 10.3 and 16.8°C in the south area of Concepción Bay (Ahumada and Chuecas 1979); and between 13 and 17°C in the center north area of Coquimbo Bay (Olivares 1989).

P. adspersus reach first maturity at 21 months of age (24 cm, 220 g) (Zúñiga 1988). Mature oocytes reach a diameter between 0.66 and 0.80 mm. The total average fecundity is 2,125,000 eggs, with an average of 1,500 eggs per gram of fish (Angeles 1995).

Physical differences between sexes are only evident during the process of sexual maturation, when females show an easily identifiable swollen stomach and the males show presence of sperm when abdominal pressure is applied. However, Angeles (1995) reported the presence of a genital orifice in females above the mid line and behind the anus, which is nonexistent in males and distinguish them by sex. The author also reported clear sexual dimorphism with respect to growth; females reach larger size than males.

2.1.5 Feeding

Flounders are fundamentally marine carnivores that consume benthic and actively swimming pelagic prey. Therefore, the natural feed of *P. adspersus* is composed basically of fish, crustaceans, and mollusks, the importance of each item prey differs depending on geographic location and according to the seasonal fluctuations in the abundance of the organisms (Bahamonde 1954; Klimova and Ivankova 1977; Silva and Stuardo 1985; Zúñiga 1988; Gonzalez and Chong 1994; Kong *et al.* 1995). Zúñiga (1988) indicated that in the central area of the country, *P. adspersus* preferably consumes anchovy (*Engraulis ringens*) and mysid shrimp (*Metamysidopsis* sp.). He also pointed out a marked difference in the diet between juvenile and adults, from the presence of numerous small epifauna preys in juvenile to few big pelagic preys in adults. On the other hand, Kong *et al.* (1995) pointed out that in the north, *P. adspersus* prey mainly on mid-water fish (*E. ringens*) and occasionally on benthic crustaceans (*Emerita analoga*).

In culture, *P. adsperdsus* readily feeds on moist or dry pellets. However, when compared with other species, *P. adsperdsus* feed slowly from the water column and from the bottom, and presents different feeding patterns depending on the kind of food and food's movement in the tanks (Castro 1995). Likewise, it has been detected that their consumption varies depending on fish size and season. Thus, daily food consumption declines from 11 to 9% of biomass/day between 2 and 5 g to 2.7 and 1.4% of biomass/day from 46 grams until market size (600 g) (Silva *et al.* 2001).

Once sexual maturity is reached, food consumption increases during the months prior to spawning, followed by a decrease in consumption at the time of spawning (Silva 2001).

2.1.6 Growth

Few studies exist about the natural and/or artificial growth of *P. adspersus*. Silva and Flores (1994) proposed the following von Bertalanffy growth equation for length for *P. adspersus* by using 182 captured wild flounders maintained in captivity in Coquimbo, Chile, during 336 days and fed with moist pellets.

$$\text{Lt} = 54.52\,(1 - e^{0.2725(t+0.1104)})$$

The same authors projected that under such culture conditions, *P. adspersus* would reach 500 g of weight in 1,030 days of cultivation, showing instantaneous rates of growth of 1.5 g/day in March for fishes of 5–10 cm and a minimum of 0.09 g/day in September for fishes of 15–20 cm.

Angeles (1995) presented the von Bertalanffy growth equations for length and weight for both sexes, calculated from 150 *P. adspersus* collected in the ports of Ancón, Callao, Chorrillos, and Pucusana, Peru:

$$\text{Length growth}: \quad \begin{aligned} &\text{Lt} = 101{,}169\,(1 - e^{-0.139(t+0.584)}) \text{ for females} \\ &\text{Lt} = 60{,}539\,(1 - e^{-0.253(t+0.310)}) \text{ for males} \end{aligned}$$

$$\text{Weight growth}: \quad \begin{aligned} &\text{Wt} = 16{,}412.72\,(1 - e^{-0.139\,(t+0.584)})^{3.27} \text{ for females} \\ &\text{Wt} = 3{,}145.54\,(1 - e^{-0.253\,(t+0.310)})^{3.27} \text{ for males} \end{aligned}$$

Using these equations, length and weights, calculated by age (Table 2.1), show a higher growth in females than in males, as well as the time of 3 years in arriving to a market size of 770 g for females.

Chong and González (1995) reported for flounders collected in Concepción (south of Chile) that *P. adspersus* would reach its commercial size of 1 kg at younger ages than their national congeners, which would imply advantages in their eventual cultivation compared to the other species of national flounders.

Table 2.1 Length and weight per age of *P. adspersus* according to von Bertalanffy equation.

	Female		Male	
Edad (Years)	**Length (cm)**	**Weight (g)**	**Length (cm)**	**Weight (g)**
1	20.08	81.76	17.10	50.18
2	30.65	326.30	26.82	218.82
3	39.85	770.08	34.37	492.45
4	47.84	1,401.56	40.23	824.20
5	54.79	2,186.16	44.77	1,170.20
6	60.84	3,080.91	48.30	1,500.18
7	66.09	4,043.20	51.04	1,797.22
8	70.67	5,035.34	53.17	2,054.30
9	74.64	6,026.42	54.82	2,270.74
10	78.10	6,992.64	56.10	2,449.34

Redrawn from Angeles 1995.

2.2 Broodstock husbandry

2.2.1 Acquisition of broodstock

Chilean flounder broodstock are obtained from two sources: capture of juvenile or adults by means of bottom trawling or gillnetting, or from research laboratories dedicated to flounder culture. When using bottom trawls, different fishing strategies are employed from those usually used by fishermen. In that sense, it is recommended to use shorter haul times than those usually used by fishermen. Once captured, the fishes should be selected among those that show the least damage before placing them in the transfer tanks. The transfer tanks should be covered and supplied with direct oxygenation in order to maintain the dissolved oxygen (DO) above 7 ppm. Bags of ice are used to maintain low temperatures below 13°C and transfer densities should not exceed 30 kg/m^3. These simple precautions during the transfer assure survivals between 60 and 80% of the transferred fishes and minimize mortalities caused by stress.

2.2.2 System design and requirements

After arriving at the hatchery, the fish are put in half-covered quarantine tanks no smaller than 3 m^3 with circulating water and constant aeration. The fish are subjected to antiparasite treatments (Formalin 50–100 ppm) and antibiotics (Oxolinic acid 10 ppm; Oxytetracycline 50–70 ppm) to avoid infections. Later, they are sampled, sexed, and tagged for their definitive selection. Feeding with frozen chopped fish usually begins 4–5 days after their arrival since they don't consume food during the first few days.

After the quarantine period (30–40 days), the fish can be transferred to the final reproduction tanks. These are 10 m^3 circular tanks or larger, receiving filtered (50 μ) seawater (33 ppt) and constant aeration. The fish are stocked at two males per female and at maturation densities of 5 kg/m^3 or under

2 kg/m^3 for spontaneous spawns. During the first year of captivity, wild flounders usually don't produce spontaneous spawns; nevertheless, females with swollen and turgid abdomens are encountered as well as males with the presence of sperm. Usually the wild flounder require two complete years of captivity before the first spontaneous spawn is observed (Silva and Castelló 2005).

2.2.3 Diet and nutrition

Feeding consists of either frozen chopped fish (*Trachurus murphyi, Sardinops sagax*) and moist pellet (a mix of fresh fish, fish-meal, fish-oil, and vitamins), or dry pellets fed at 1–2% of the biomass, trying to keep within the nutritional ranges of 50% protein and 12% lipid. The food is either given daily or four times per week according to variations in seasonal consumption, since the species shows marked consumption differences before, during, and after the spawning season (Silva 2001). Four months prior to spawning and to assure reproductive conditioning, feeding is usually reinforced by adding to the diet a premixture of C, B1, and E vitamins (500–700 mg/kg in feed) and an additional source of fatty acids (EPA and DHA) usually DHA Selco, or other commercial products that possess these characteristics.

2.2.4 Controlled spawning

Controlled spawning is achieved by injecting females with GnRHa (10 μg/kg) at different stages of maturity. This procedure is effective to induce spawning in females during early maturation when average oocyte diameter is between 320 and 500 μm. In later developmental stages, this procedure is less effective (Manterola 2006). Currently, the natural maturation and spontaneous spawning of flounder broodstock is routinely achieved with good results. Broodstock from 3 to 4 years of age (700–1,500 g) are maintained under natural light and temperature conditions in tanks between 6 and 10 m^3, with flow-through seawater and constant aeration, at a ratio of two males per female and densities from 1 to 2 kg/m^3 (Silva 1996).

Prior to spawning (12–24 hours), mature females show an enlarged abdomen (Figure 2.2) and they are constantly accompanied by one or two males swimming in tanks. Although spontaneous spawnings take place in the morning and in the afternoon, it is more common to find eggs in the first stage of cell division in the morning.

Spontaneous spawns (24–48 hours duration) are frequently separated by 4–7 days of little or no egg production. The spawning season begins in mid-August (end of winter) and lasts for approximately five months, until December. There is a latency period between January and July when spawning ceases or becomes intermittent. However, during the year-round control of environmental conditions, the peak egg production and viability remains stable between September and October (spring) and when the temperatures fluctuate between 14 and 15.5°C. Above this range, both the production and the viability of the

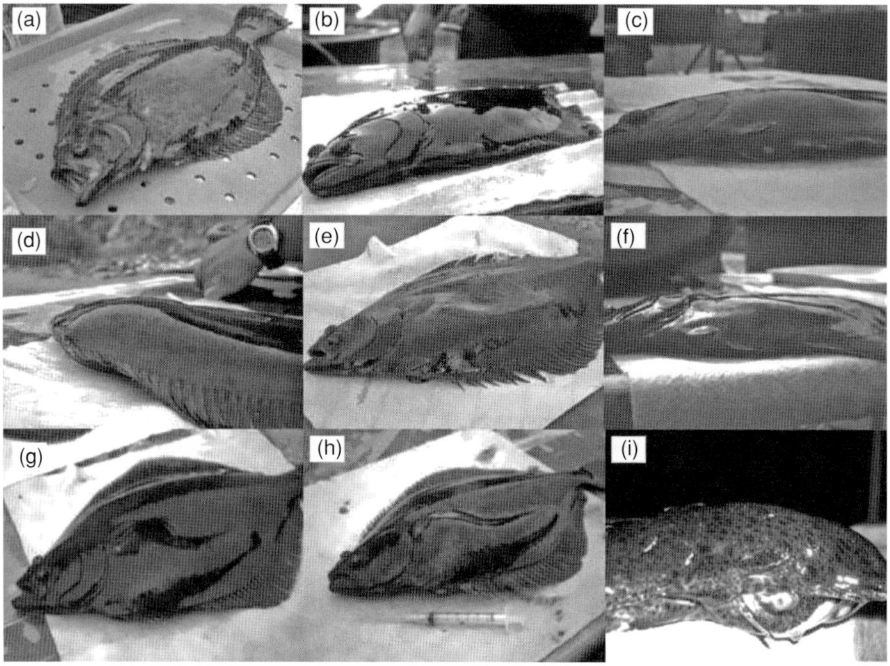

Figure 2.2 Different maturation states (enlarge abdomen) of *P. adspersus*. (a) without enlarge; (b) and (c) light enlarge; (d)–(f) very defined enlarge; (g) and (h) marked enlarge; and (i) maximum enlarge, extensive genital pore. (adapted from Manterola 2006).

eggs decline. The minimum and maximum temperature values between which broodstock spawn are from 12.7 to 19.7°C.

Female *P. adspersus* can annually produce an average of 2×10^6 eggs/kg of bw, from which between 30 and 50% are viable (floating) with fertilization success between 0 and 100%. Although the period can be extensive, the highest percentage of viable eggs is produced in a period of two months when temperatures range from 14 to 15°C. The nonviable eggs are generally opaque, with irregular shape and surface irregularities, abnormal distribution of the yolk, and multiple oil droplets. Eggs are checked throughout the entire spawning season, but in our experience the number of nonviable eggs tends to increase when temperatures exceed 16°C.

2.2.5 Collection and incubation of eggs

Eggs can be collected in two ways, by means of dry stripping of mature females, in which case the eggs are received and fertilized with sperm of the males in a disinfected dry container filled with seawater and left to rest for 10–15 minutes (Figure 2.3), or by means of spontaneous tank spawns in which case the eggs are obtained at the tank outlet with a 500–600 μm mesh collector bag. The eggs of *P. adspersus* are buoyant, transparent, contain a single oil globule, and reach an average diameter of 0.8 mm. Once the eggs are collected, they are washed

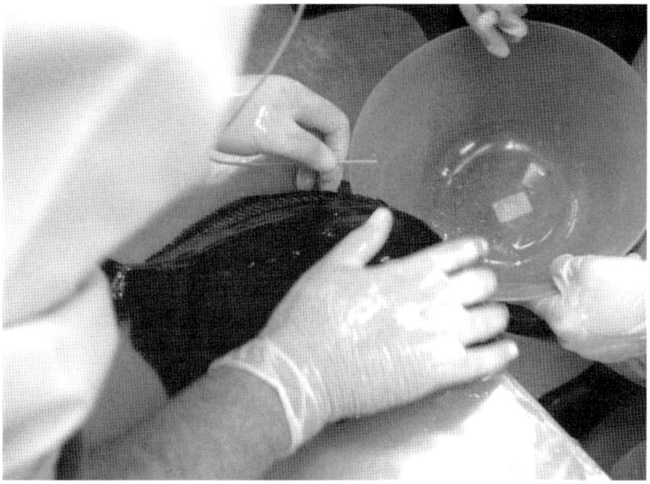

Figure 2.3 Artificial spawn of *P. adspersus* female by stripped. (Photographs from Fish Culture Laboratory of UCN.)

with UV-sterilized water and carefully disinfected (glutaraldehyde 100 ppm). Then they are put in a conical transparent tank (100–200 liters) to allow the separation of viable (floating) and nonviable (sinking) eggs. The percentage of floating eggs normally varies among spawns from 0 to 98%. After separation and enumeration (count, fertilization, development, quality), the floating eggs are transferred to a 500- to 1,000-liter incubator filled with sterilized (UV) and filtered seawater (1 μm). The incubation densities range from 500 to 1,000 eggs/L with a water exchange of 50 to 100% daily with seawater. Moderate density and water exchange are best. Dead eggs should be removed daily to prevent bacterial growth. Time to hatch depends on temperature (Figure 2.4). At 13°C, 50% hatch is achieved by 80 hours; at 16°C, time to hatch is 60 hours and at

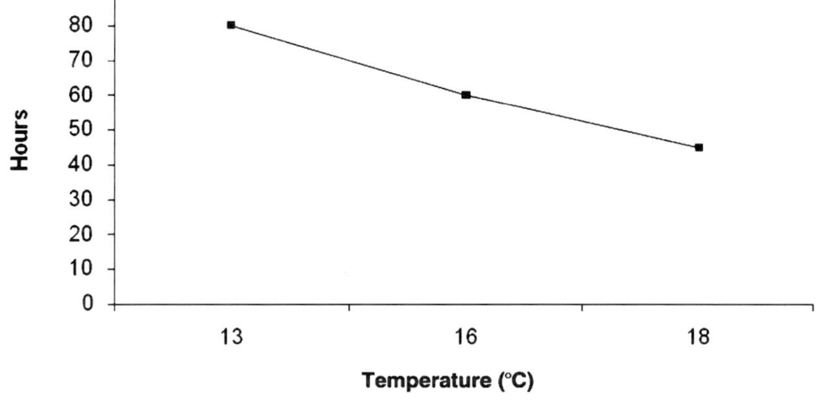

Figure 2.4 Time (hours) to hatch at different temperatures for Chilean flounder *P. adspersus* eggs. (Data from Fish Culture Laboratory of UCN.)

18°C hatch time is only 45 hours (Silva 1988; Silva *et al.* 1994; Silva and Vélez 1998).

Hatch percentages vary (30–90%) and depend on the quality of spawns, which in turn depends on the nutritional conditioning of the broodstock and a good separation of viable and nonviable eggs before incubation.

2.3 Larval culture

At hatching, yolk-sac larvae measure between 1.7 and 2.0 mm total length (TL). Newly hatched larvae have not completed the development of the eyes or the digestive tract, and their survival depends exclusively on the prominent yolk sac (Figure 2.5). In our laboratory, yolk-sac larvae are stocked at 30–100 larvae/L into 500–1,000 liters conical fiberglass tanks supplied with micron-filtered and UV sterilized water in a flow-through system exchanged at 25–50% of tank volume daily. After 4–5 days, size averages 3.7 mm TL, the larvae has totally consumed the yolk-sac and has completed the development of its eyes and shows a functional digestive tract. During this stage, survival is 80–90% if the appropriate hygienic conditions are maintained (Silva 2001).

The larval culture is done in 2 m^3 circular tanks at densities between 20 and 30 larvae per liter. The daily exchange of micro filtered and sterilized seawater is increased from 0 to 100% between days 4 and 20 of cultivation. Live microalgae cultures of *Isochrysis* and *Nannochloropsis* are added daily to larval rearing tanks (150,000–200,000 cells/mL). In this first phase, live rotifers (5–10 ind./mL twice daily) are enriched by feeding them with a high-density mixture of microalgae (80% *Isochrysis* and 20% *Nannochloropsis*) or with commercial enrichment diets (e.g., Algamac, DHA Selco) until day 15–20. At 15 dph, feeding begins with *Artemia* nauplii at 0.5–1 nauplius/mL accompanied by a decrease of rotifers until day 20. By this time, larvae are able to consume enriched

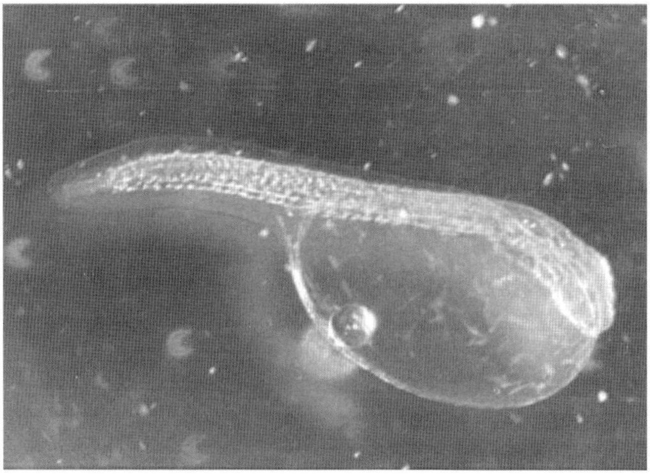

Figure 2.5 Newly hatched *Paralichthys adspersus* larva (the standard length is 1.9 mm). (Photographs from Fish Culture Laboratory of UCN.)

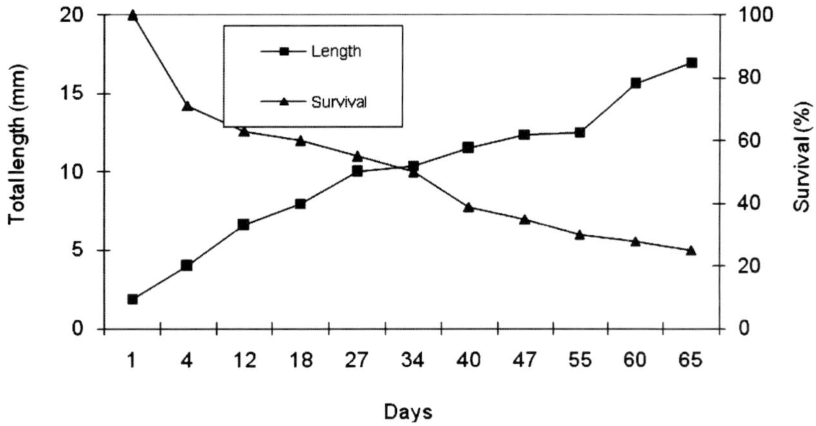

Figure 2.6 Growth and survival of larval *Paralichthys adspersus* during 65 days of culture. (Redrawn from Silva 2001.)

Artemia meta nauplii at an average of 1–3 art/mL and are cofed with commercial feeds (100–400 μm) until day 60. At this time, the larvae have completed metamorphosis between 15 mm (67%) and 20 mm (33%) TL, and become benthic juveniles. Survival by this stage fluctuates between 10 and 25% (Figure 2.6).

Studies carried out on this stage indicate that growth and larval survival depend on factors primarily related to nutritional quality of prey, temperature, and water quality of the culture medium. Silva (1999) reported that the use of microalgae as an enrichment for rotifers and as a part of the culture medium significantly increased growth, survival, development, and handling of the larvae of Chilean flounder during the first stage of the culture. This was due to the enrichment's healthy effect on the culture medium and in maintaining the nutritional quality of the live prey.

With respect to the nutritional needs of larvae, Wilson *et al.* (1999) reported on the changes experienced by lipids in eggs and prelarvae of Chilean flounder, showing that arachidonic acid remained constant during development, though in unfed larvae highly unsaturated fatty acids (HUFA) such as eicosapentaenoic acid (EPA) and docosahexaenoic acid (DHA) decreased, with DHA decreasing at a lower rate than EPA. This indicated the importance of providing adequate amounts of such fatty acids in the larvae's live food to assure their optimal development. Similarly, recent research on the different levels of n-3 HUFA in the larval development of Chilean flounder has found that levels of 0.7–1% of n-3 HUFA in rotifers provided improved results in terms of growth rate, survival, and larval quality (Silva 2001). Also, ratios of DHA/EPA of 2:1 in enrichment diets were the best to maintain optimal results of the same parameters (unpublished data).

Other experiments related to the determination of optimal temperatures for larval culture (16°, 18°, and 20°C) showed that better survival was obtained at 18°C than at 16° and 20°C, though the best growth was obtained at 20°C (Silva 2001).

Regarding the feeding strategies used in the larval stage of *P. adspersus*, it has been determined that the early replacement of *Artemia* and the cofeeding starting from day 20 of cultivation up to day 40, improves the growth and survival and improves weaning success (Piaget *et al.* 2007a).

Studies done to determine the effect of immunostimulants indicate that the application of 5 mg/L of β-glucans (βG) and mannan-oligosaccharides (βG MOS) in the cultivation water increases the survival and the growth of the larvae versus the control, whereas 15 mg/L of βG MOS has a negative effect on both production parameters. This effect increases if it is applied to larvae that have newly absorbed the yolk sac. The histological analysis of the intestinal epithelium of the larvae suggests that the βG MOS promotes the Monocytes manifestation (macrophage cells precursors) associated with the nonspecific immune system of the fish (Piaget *et al.* 2007c).

2.4 Weaning and nursery culture and grow out

The process of weaning (40–60 dph) basically consists of progressive replacement of live prey (*Artemia*) with formulated feeds of different sizes (0.2–1.0 mm) technically similar to diets for bigger fish over a determined period of time (10–15 days). This is normally carried out in 400–1,000 liters semicircular or rectangular, flat-bottom tanks with a water column height of 50 cm and fish densities of 1,000–3,000 per m^2 (Silva and Vélez 1998; Embry 1999). Contrary to other flatfish, it is important to mention the necessity to not permit postlarvae to contact the bottom of the tanks during the weaning process; for this reason, floating mesh cages are used within the tanks. This strategy improves the cleaning of the tanks and significantly increases survival. Duration of use depends on the weaning strategy process at this stage, which is recommended until the end of the nursery stage (150–180 dph).

Although weaning may be initiated before the larvae finish metamorphosis, best results in terms of survival (71%) and growth rate (3.3% per day) are obtained with metamorphosed benthic juveniles nearly 50–60 days old. Embry (1999) reported that initial density of fishes during the period of weaning has a greater effect on survival, leading to the conclusion that the best density at which to carry out weaning within the range tested was 1,000 per m^2 (Figure 2.7).

The same author suggested that the duration of the weaning period significantly affected larval growth (2.4–2.7% per day) and survival (7–25%), being much more important than starting density. He recommended increasing the length of the weaning period and always maintaining the availability of live food, while progressively decreasing its concentration during this period.

Once the fishes are adapted to the formulated diet and they are all on the bottom of the cultivation tanks (60–90 days), the nursery stage begins with the main objective of producing a flounder of size (8–10 g) and quality adequate to begin the stage of ongrowth of Chilean flounder.

Table 2.2 shows growth in length and weight of the Chilean flounder (*P. adspersus*) cultivated in tanks in nursery stage according to the cultivation protocol used in the Laboratory of Fish Cultivation of Universidad Católica del

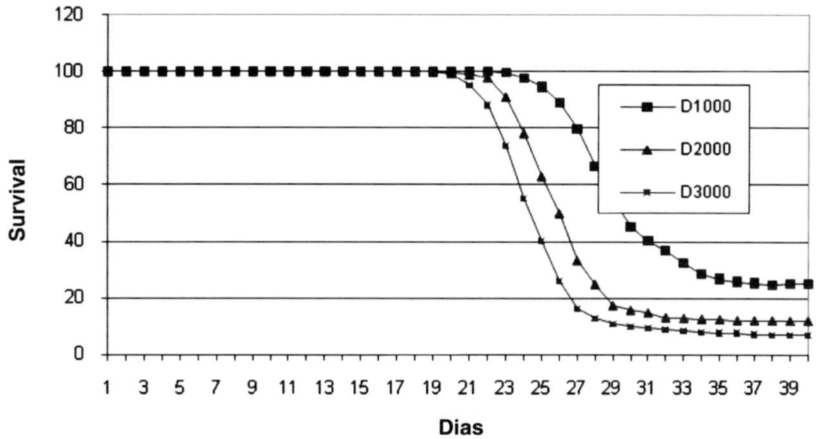

Figure 2.7 Effect of density in early replacement of *Artemia* with formulated diets in *P. adspersus*. (Redrawn from Embry 1995.)

Norte, Coquimbo. The fishes are maintained in semicircular tanks and in mesh cages with natural illumination and subjected to daytime automatic feeding. Cultivation temperatures for the stage fluctuate between 15 and 18°C.

Growth in this stage and during early juvenile production is certainly lower than other species of cultured flatfish, such as turbot, which reaches 9–10 g of weight in 90–100 days (Stoss *et al.* 2004)

2.5 Growout

Work related to the growout of juvenile Chilean flounder, either captured from the wild or cultured and maintained in culture facilities, indicates that this species may be grown in tanks and cages from juveniles to commercial size without major difficulties encountered in their growth, survival, or management (Silva and Flores 1994; Rolando and Ramirez 1998; Kelly *et al.* 1999; Silva *et al.* 2001).

Silva and Flores (1994) cultured three groups of wild-caught Chilean flounder (5–10 cm; 15–20 cm; 20–24 cm) over a period of 335 days in flow-through seawater tanks. The fish were fed to satiation four times per week with moist

Table 2.2 Results of nursery phase of Chilean flounder (*P. adspersus*) between 15 and 18°C, according to the cultivation protocol used in the Laboratory of Fish Cultivation of North Catholic University, Coquimbo.

Age (dph)	Length (cm)	Weight (g)	Feed particle size (mm)	Density (kg/m^2)
90	3.1	0.4	0.7–1.0	0.3
120	6.3	3.3	1.0–1.5	0.3
150	7.4	5.2	1.5–2.5	0.5
180	8.9	9.5	2.0–3.0	1
210	9.6	11.5	3.0–4.0	1.1

pellets, obtaining growth rates per group of 0.79, 0.49, and 0.19% per day; under those conditions, it was suggested that Chilean flounder could be grown to 500 g in 1,030 days. The same authors working with hatchery-cultured fish of smaller size (2–8 cm) fed with diets formulated for salmon and cultured under similar conditions to those mentioned above, reported growth rates between 1.7 and 1.5% per day (Silva and Flores 1998). These results were higher than previously observed for flounder of this size and coincided with data reported for the same species cultured in similar conditions over 120 days (Rolando and Ramirez 1998).

Silva *et al.* (2001) examined the growout of Chilean flounder fed with extruded diets formulated for turbot. The authors showed no difference in growth rates among different groups of fishes (large, medium, and small) that had originated from the same spawning and concluded that within a temperature range of 14.9–17.3°C, first market size (250–550 g) of this species should be achieved in 19–20 months and 1 kg in 42 months.

Improved results were obtained in 2005–2007 in our laboratory with fingerling *P. adspersus* stocked in tanks at densities from 18 to 25 kg/m^2 and fed with moist pellets and extruded commercial pelleted diets (crude protein 50–55%). Fish grew to 0.3–0.5 kg in 20–25 months and 1 kg in 35–37 months (Figure 2.8).

With regard to disease, different pathologies have been observed in broodstock, caused principally by *Vibrio* and *Pseudomonas* spp. and characterized by bloody eye inflammation, hemorrhages on the mouth and gills, and disintegration of the lower jaw. Major participation of *V. splendidus* and to a lesser extent *V. anguillarum* has been observed, coinciding with pathology observed in other flatfishes such as the turbot (Miranda and Rojas 1993, 1996).

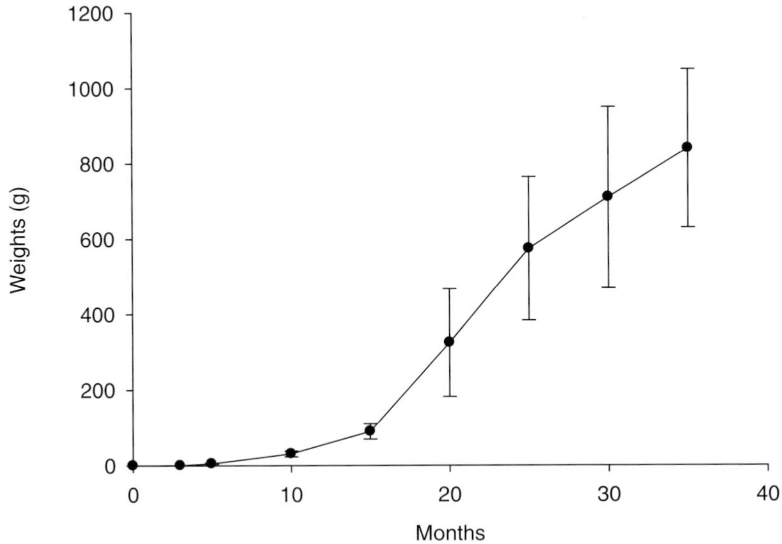

Figure 2.8 Growth in weight of Chilean flounder *P. adspersus* between 15 and 18°C, according to the cultivation protocol used in the Laboratory of Fish Cultivation of North Catholic University, Coquimbo (2005–2007). Vertical bars indicate standard deviation (unpublished results, 2007).

2.6 Needs for future research

Research is needed on the nutritional requirements of juvenile and adult *P. adspersus*, in order to design species-specific diets. Piaget *et al.* (2007b) carried out experiments to determine the optimum protein level for the cultivation of juveniles using flounder (100 g avg. wt) cultivated in semicircular tanks with flow-through seawater, at densities of 5 kg/m^2, and using diets containing 50% protein (control) and three experimental diets (44, 53, and 55% protein). The results indicate that the optimum protein level for growth of juvenile of Chilean flounder is 50–55%.

In technical terms, it is necessary to test and evaluate new production systems that optimize growout of Chilean flounder, such as the use of cages, recirculation systems, and tanks with laminar flow. Parallel studies need to be conducted on identification of pathogenic agents leading to diseases in adult flounder, with the objective of defining protocols for their detection and for the treatment of recurrent diseases. This would minimize the effects of disease at this stage, which is a recurrent problem in commercial flounder culture.

Literature cited

Able, K., Matheson, R.E., Morse, W.W., Fahay, M.P., and Sheperd, G. 1990. Pattern of summer flounder *Paralichthys dentatus* early life history in the mid-Atlantic bight and New Jersey estuaries. *Fisheries Bulletin* 88:1–12.

Acuña, E., and Cid, L. 1995. On the ecology of two sympatric flounders of the genus *Paralichthys* in the bay of Coquimbo. *Netherlands Journal of Sea Research* 34(1–3):7–18.

Ahumada, R., and Chuecas, L. 1979. *Algunas características hidrográficas de la Bahía de Concepción (36° 40′S; 73° 03′W) y áreas adyacentes*, 1–56. Miscelánea, Gayana, Chile.

Alvial, A., and Manríquez, J. 1999. Diversification of ftatfish culture in Chile. *Aquaculture* 176:65–73.

Angeles, B. 1995. Diformismo sexual, crecimiento y fecundidad del lenguado común (*Paralichthys adspersus*) de la costa central del Perú. Tesis presentada para optar al titulo de Ingeniero Pesquero. Facultad de Pesquerias. Universidad Nacional Agraria La Molina. Lima, Perú.

Bahamonde, N. 1954. Alimentación de los lenguados (*Paralichthys microps* Steindachner e *Hoppoglossina macrops* Günther). *Investigaciones Zoológicas Chilenas* 2:72–74.

Castro, J. 1995. Alimentación artificial de lenguados del género *Paralichthys*. Informe Final Práctica Profesional. *Carrera de Tecnología en Recursos del Mar*. Pontificia Universidad Católica de Chile, Talcahuano, Chile.

Chirichigno, N. 1974. Clave para identificar los peces marinos del Perú. *Informe Instituto del Mar del Perú* 44:1–387.

Chong, J., and González, P. 1995. Ciclo reproductivo del lenguado de ojos chicos *Paralichthys microps* (Gunther, 1881) (Pleuronectiformes, *Paralichthydae*) frente al litoral de Concepción, Chile. *Biología Pesquera* 24:39–50.

Embry, D. 1999. Deshabituación de juveniles de lenguado *Paralichthys adspersus* (Steindachner, 1867). Efecto de la densidad de arranque y la técnica de deshabituación. *Memoria para obtener el título de Ingeniero en Acuicultura*. Departamento de

Acuicultura. Facultad de Ciencias del Mar. Universidad Católica del Norte, Sede Coquimbo, Chile.

Ginsburg, I. 1952. Flounders of the genus Paralichthys and related genera in American Waters. *U.S. Fish and Wildlife Service Fish Bulletin* 52(71):1–51.

Gonzalez, P., and Chong, J. 1994. Examen de los contenidos gástricos de *Paralichthys microps* (Gunther 1881) de la Bahía Concepción. XIV Jornadas Ciencias del Mar, Universidad Austral de Chile. Puerto Montt. Abstract Book: 211.

Kelly, R., Ramirez, D., Comte, S., Adam, F., and Solari, M. 1999. Cultivo experimental de lenguado chileno (*Paralichthys adspersus*) en jaulas sumergibles. Desarrollo de un protocolo operacional. XIX Congreso de Ciencias del Mar. Universidad de Antofagasta, Antofagasta, Chile. Abstract Book:131.

Klimova, V., and Ivankova, Z. 1977. The effect of changes in bottom population from Peter the Great Bay on feeding and growth rates in some flatfishes. *Oceanology* 17:896–900.

Kong, I., Clarke, M., and Escribano, R. 1995. Alimentación de *Paralichthys adspersus* (Steindachner, 1867) en la zona norte de Chile. *Osteichthyes: Paralichthyidae. Revista Biologıia Marina Valparaiso* 30:29–44.

Kramer, S.H. 1991. Growth, mortality and movements of juveniles California halibut, *Paralichthys*, in shallow coastal and bay habitats on San Diego County, California. *Fishery Bulletin* 89:195–207.

Manterola, R. 2006. Respuesta endocrina y ovulatoria en hembras de lenguado chileno (*Paralichthys adspersus*) post inducción hormonal con GnRHa. *Tesis para optar al grado de Magister en Ciencias de la Acuicultura*, p. 68. Facultad de Ciencias Agronómicas, Ciencias Veterinarias y Pecuarias e Instituto de Nutrición y Tecnología de los Alimentos. Universidad de Chile.

Miranda, C., and Rojas, R. 1993. Prevalencia de patologías oportunistas en el cultivo experimental del lenguado *Paralichthys adspersus*. *Anales Microbiología* 1:51–54.

Miranda, C., and Rojas, R. 1996. Vibriosis en el lenguado *Paralichthys adspersus* (1867) en cautiverio. *Revista Biología Marina* 31 Steindachner (1):1–9.

Olivares, J. 1989. Aspectos hidrográficos de la Bahía Coquimbo. *Biología Pesquera* 18:97–108.

Pequeño, G. 1989. Lista de peces de Chile. Revisada y comentada. *Revista de Biologia Marina* 24:1–132.

Pequeño, G., and Plaza, R. 1987. Descripción de *Paralichthys delfini* n. Sp., con notas sobre otros lenguados congenéricos de Chile. (Pleuronectiformes, Bothidae). Resúmenes de las VII Jornadas de Ciencias del Mar. Universidad de Concepción, Chile.

Piaget, N., Silva, A., Vega, A., and Toledo, P. 2007a. Optimización de la alimentación en el cultivo intensivo de larvas de lenguado (*Paralichthys adspersus*) usando microdietas. *Libro Resumenes 1er Congreso Nacional de Acuicultura*. Universidad Católica del Norte, Coquimbo, Chile, pp. 150–152.

Piaget, N., Toledo, P., Silva, A., and Vega, A. 2007b. Effects of dietary protein levels on the growth of *Paralichthys adspersus* flounder cultured in controlled conditions. *Libro Resumenes 1er Congreso Nacional de Acuicultura*. Universidad Católica del Norte, Coquimbo, Chile, pp. 253–255.

Piaget, N., Vega, A., Silva, A., and Toledo, P. 2007c. Efecto de la aplicación de β-glucanos y manano-oligosacáridos (βG MOS) en un sistema de cultivo intensivo de larvas de *Paralichthys adspersus* (Paralichthydae). *Investigaciones Marinas Valparaiso* 35(2):35–43.

Rolando, H., and Ramirez, D. 1998. Evaluación del crecimiento y parámetros alimenticios en juveniles de lenguado chileno *Paralichthys adspersus* (Steindachner, 1867).

XVIII Congreso de Ciencias del Mar. Universidad Arturo Prat, Iquique, Chile. Abstract Book: 116.

Siefeld, W., Vargas, M., and Kong, I. 2003. Primer registro de *Etropus ectenes* Jordan, 1889, *Bothus constellatus* Jordan & Goss, 1889, Achirus klunzingeri (Steindachner, 1880) y *Symphurus elongatus* (Gunther, 1868) (Piscis, Pleuronectiformes) en Chile, con comentarios sobre la distribución de los lenguados chilenos. *Investigaciones Marinas Valparaiso* 31:51–65.

Silva, A. 1988. Observaciones sobre el desarrollo del huevo y estadios larvarios de lenguado *Paralichthys microps* (Gunther 1881). *Revista Latinoamericana de Acuicultura* 35:18–25.

Silva, A. 1996. Conditioning and spawning of the flounder, *Paralichthys microps*, Gunther, 1881 in captivity. In: Gajardo G. and Coutteau P. (eds) *Improvement of the Commercial Production of Marine Aquaculture Species*. Proceeding of a workshop on fish and mollusc larviculture. Impresora Creces, Santiago, Chile, pp. 97–102.

Silva, A. 1999. Effect of the microalga *Isochrysis galbana* on the early larval culture of *Paralichthys adspersus*. *Ciencias Marinas* 25:267–276.

Silva, A. 2001. Advance in the culture research of small-eye flounder, *Paralichthys microps,* and Chilean flounder, *P. adspersus*, in Chile. *Journal of Applied Aquaculture* 11(1/2):147–164.

Silva, A., and Castelló, F. 2005. Técnicas de producción de huevos y larvas de peces marinos. In: Silva A. (ed.) *Cultivo de Peces Marinos*. Facultad de Ciencias del Mar, Universidad Católica del Norte, Coquimbo, Chile, pp. 159–184.

Silva, A., and Flores, H. 1994. Observations on the growth of the chilean flounder (*Paralichthys adspersus*, Steindachner, 1867) in captivity. In: Lavens, P., and Remmerswaal, R.A.M. (eds) *Turbot Culture: Problems and Prospects*. European Aquaculture Society, Special Publication No 22, Gent, Belgium, pp. 323–332.

Silva, A., and Flores, H. 1998. Observaciones sobre el crecimiento de juveniles de lenguado *Paralichthys adspersus* (Steindachner, 1867) cultivado en estanques. XVIII Congreso de Ciencias del Mar. Universidad Arturo Prat, Iquique, Chile. Abstract Book: 158.

Silva, A., Henriquez, C., and Munita, C. 1994. Desafío del lenguado: de cultivo experimental pasar a etapa piloto. *Aquanoticias Internacional* 22:42–51.

Silva, A., Oliva, M., and Castelló, F. 2001. Evaluación del crecimiento de juveniles de lenguado chileno (*Paralichthys adspersus*, Steindachner, 1867) cultivado en estanques. *Biología Pesquera* 29:21–30.

Silva, A., and Vélez, A. 1998. Development and challenges of turbot and flounder aquaculture in Chile. *World Aquaculture* 29(4):48–51.

Silva, M., and Stuardo, J. 1985. Alimentación y relaciones tróficas generales entre algunos peces demersales y el bentos de bahía de Coliumo (Provincia de Concepción, Chile). *Gayana Zoología* 49(3–4):77–102.

Stoss, J., Hamre, K., and Ottera, H. 2004. Weaning and nursery. In: Moksness, E., Kjorsvik, E., and Olsen, Y. (eds) *Culture of Cold-Water Marine Fish*. Blackwell Publishing, Iowa, pp. 337–362.

Wilson, R., Velez, A., and Avila, R. 1999. Cambios en el contenido de lípidos, clases de lípidos y ácidos grasos en huevos fertilizados, larva vitelínica, larvas prealimentación y en ayuno del lenguado *Paralichthys adspersus*. XIX Congreso de Ciencias del Mar. Universidad de Antofagasta, Antofagasta, Chile. Abstract Book: 213.

Zúñiga, H. 1988. Comparación morfológica y dietaria de *Paralichthys adspersus* (Steindachner, 1867) y *Paralichthys microps* (Gunther, 1881) en Bahía de Coquimbo. *Tesis para obtener el título de Biologo Marino*. Facultad de Cienciasd del Mar. Universidad Católica del Norte, Coquimbo, Chile, 144 pp.

Chapter 3
California halibut*

Douglas E. Conklin and Raul Piedrahita

As a group, flatfish are highly valued in the market. Flatfish fillets have an appealing firm texture along with a subtle mild seafood taste. In that fisheries are unable to meet market demand, continued development of commercial production of various flatfishes is to be expected (Sjoholt 2000).

The technology for commercial culture developed in the early 1980s with production of the common sole *Solea solea* followed by the turbot *Psetta maxima* in Europe (Brown 2002). At about the same time, Japan (Kikuchi 2001) and later Korea (Seikai 2002) began producing commercially significant amounts of hirame (the bastard halibut) *Paralichthys olivaceus*. Production increases were relatively slow until the last few years. As a result of various refinements in culture technology, there has been dramatic increases in the production of turbot and hirame in both Europe and Asia. Existence of these improved methods also encouraged the transfer of the culture technology to other countries having a robust tradition of aquaculture production, such as Korea, China (Seikai 2002), and Chile (Alvial and Manriquez 1999). Thus, while global production over the first two decades of commercial culture had only risen to a little more than 26,000 MT by 2000, the next 6 years saw an increase of more than 100,000 MT to a total of 126,579 MT in 2006 (Table 3.1) (FAO 2006). Much of the increased production is due to flatfish culture in China. While FAO statistics relating to Chinese production are presently limited only to the general designation of left or right eyed flounders, it is likely that these primarily are the endemic bastard halibut and the turbot which was introduced into China from Europe in the 1990s (Chen et al. 2003; Hong and Zhang 2003).

The widespread success of commercial flatfish culture has also stimulated interest in the United States in various endemic flatfish species as potential aquaculture candidates. On the eastern seaboard, extensive research has been done with the southern flounder *P. lethostigma*, which would appear to have commercial feasibility (Daniels and Watanabe 2002; Luckenbach *et al.* 2002; Watanabe

*In memory of Jean-Benoit Muguet 1977–2007.

Table 3.1 Flatfish aquaculture production by country (FAO 2006)

Country	Information on species	Metric tons
China	Lefteye flounders	63,490
China	Righteye flounders	5,196
Korea	Bastard halibut	43,852
Spain	Turbot	6,419
Spain	Senegalese sole	32
Spain	Soles	11
Japan	Bastard halibut	4,613
Norway	Atlantic halibut	1,185
France	Turbot	800
United Kingdom	Atlantic halibut	233
Chile	Flatfishes	215
Portugal	Turbot	185
Portugal	Common sole	9
Iceland	Atlantic halibut	110
Netherlands	Turbot	100
United Kingdom	Turbot	62
Germany	Turbot	60
Denmark	Turbot	7
	Total	126,579

et al. 2006). On the Pacific coast, the California halibut, *P. californicus*, supports important recreational and commercial fisheries throughout its range from the Oregon–Washington border in the United States to the southern part of Baja California in Mexico (Allen 1990; Kramer 1990). Commercial landings have declined since the 1920s to less than 500 MT/year, about a fourth of its maximum tonnage (Kramer 1990). Interest in locally produced foods and the general environmental concern regarding introduced species further suggests that commercial culture of this species is worth exploring. This chapter brings together available information on the culture of the California halibut particularly the result of studies done at the University of California (UC), Davis, both on campus as well as studies with larger animals transferred from the campus to larger systems at the University's marine laboratory. Addition insights are taken from early work with the species as well as other flatfish species.

3.1 Broodstock culture

Early work establishing basic parameters for holding broodstock indicated that adult California halibut were amenable to culture in outdoor tanks and would produce large numbers of fertilized eggs without hormonal intervention (Caddell *et al.* 1990). Several productive groups of captive wild adult fish have been established although extensive experimentation directed toward complete control of reproduction has yet to be carried out.

The first population was established in the mid-1980s with adults taken from the wild for culture at the newly established California Halibut Hatchery in Redondo Beach, California. The ultimate goal of the hatchery was to support a restocking program for the California Department of Fish and Game (Drawbridge and Kent 2001). Later, eggs and larvae were provided to a number of other institutions investigating various aspects of halibut culture and biology. Presently, the hatchery is part of SEA Lab, a coastal science education center managed by the Los Angeles Conservation Corps. Although the broodstock at Redondo Beach are still producing eggs, the focus of the SEA Lab staff has shifted primarily to education relating to marine and coastal issues and the restocking program is inactive (B. Scheiwe, personal communication). The spawning populations were successfully established in two large redwood outdoor tanks; the largest $\sim$150 m^3 and a smaller $\sim$37 m^3 tank (which has recently been taken out of service). The open-topped tanks covered with shade cloth and provided with flow-through ocean water manifest natural conditions with regard to water temperature and photoperiod regimes.

Additional captive spawning groups have been established in California at the Leon Raymond Hubbard, Jr., Marine Fish Hatchery operated by Hubbs-SeaWorld Research Institute at Carlsbad, California (M. Drawbridge, personal communication) and at the commercial facility of The Cultured Abalone Inc. located at Goleta, California (D. Bush, personal communication). Most recently, an additional large hatchery has been constructed in Ensenada, Mexico, by Centro de Investigación Científica y de Educación Superior de Ensenada (CICESE). The new hatchery, a joint effort between CICESE and local commercial interests, is envisioned to produce as many as 500,000 juveniles per year to support halibut farms in the local area (J. Lazo, personal communication).

All of the broodstock groups are exposed to essentially natural temperature and photoperiod regimes. Typically, these spawning groups consist of multiple males and females (10–30 adults) with male to female ratios ranging from 2:1 to 1:2. The fish are fed abundantly with various combinations of fresh or frozen natural prey items. In addition, gelatin-bound mixtures of commercial fish feeds and extra vitamins, such as stabilized vitamin C and thiamine, made into appropriately large-sized cubes are sometimes used in an effort to avoid any micronutrient deficiencies. Lipid supplements should be used with caution in that research with hirame has shown that essential fatty acids such as n-3 HUFAs (highly unsaturated fatty acid) and the n-6 HUFA arachidonic acid are needed for viable eggs and larvae but excessive levels can be counterproductive (Furuita *et al.* 2000, 2003). In that, established broodstock appear to feed readily on a variety of items, the development of formulated rations for broodstock would appear to be rather straightforward once specific nutrient requirements have been established. While identification of nutrient combinations supporting optimal reproductive outcomes in California halibut is unlikely to be carried out in the short-term, significant insights arising from research with other marine fish species will undoubtedly continue to be incorporated into commercial flatfish rations.

3.2 Spawning

While the exact roles of temperature and photoperiod on gonadal maturation and egg productions for the California halibut have yet to be identified, the general pattern can be outlined. In the wild and in captive broodstock subjected to natural conditions, egg production is strictly seasonal. At the California Halibut

Hatchery, females stop producing eggs when the water temperature is above 20°C for extended periods (J. Rounds, personal communication). Based on limited experimentation but consistent with what is known for other flatfish species (see review by Bromage *et al.* 2001), resumption of oogenesis probably requires a recovery period allowing for nutrient reserves to be reestablished and ovarian yolk deposition to occur during the fall and early winter, a period of decreasing temperature and photoperiod. In the case of the southern flounder, mature females averaging just over 1 kg required a recovery period of 5 months (Watanabe *et al.* 2006). The key exogenous cue inducing final maturation and spawning, as early as February, is thought to be the subsequent late winter increase in photoperiod. Based on their experience with captive broodstock, Caddell and co-authors (1990) suggested spawning is initiated when temperatures range between 15.0 and 16.5°C and day length exceeds 10.5 hours. These observations of spawning patterns are similar to those reported from field studies (Moser and Watson 1990). Peak spawning periods for the captive broodstock occur during the winter and spring with some spawning extending into the summer until the temperature becomes too warm. Manipulation of photoperiod and temperature in order to produce seedstock throughout the year is routinely used in the commercial culture of both turbot (Person-Le Ruyet *et al.* 1991) and hirame (Seikai 1998). While realizing the potential to control the timing of egg production in the California halibut is straightforward in theory, gathering the required information would require fairly extensive facilities as shown by the recent work of Watanabe and coauthors (2006) with the southern flounder. Presumably, hormonal implants could also be used to make spawning more predictable. Insertion of a commercially available implant (typically an analog of the Gonadotropin-Releasing Hormone) into fully mature southern flounder females results in spawning some 48 hours later. Ripeness or maturity of the females is determined using backlighting to ascertain gonadal development (Daniels and Watanabe 2002). Normally, stripping sexually mature flatfish by gently squeezing the body cavity to collect eggs and sperm is straightforward. Cryopreservation of sperm taken from turbot males has recently been shown to be effective (Chen *et al.* 2003) allowing for greater flexibility with regard to the need to hold male broodstock.

The pelagic eggs of the California halibut are broadcast by the females for subsequent external fertilization by sperm from the male. In nature, spawning is thought to occur along the coast outside the various bays from San Francisco Bay in California to Magdalena Bay in Baja California (Allen 1990). At the California Halibut Hatchery, spawning has been observed to occur near the water surface (J. Rounds, personal communication). The subsequent establishment of spawning populations at Carlsbad and Goleta, California, in relatively shallow tanks (~1 m) has shown that the unusually deep tanks (2–3 m) in use at the California Halibut Hatchery are not essential for successful spawning. In that, the California halibut are multiple spawners, even a small broodstock population can produce around 50 million eggs per season. The number of eggs produced daily depends on the number and size of the females. Spawning generally is in the late afternoon or evening. Eggs, typically, are collected from the tank outflow using a fine mesh basket. After decanting to separate unfertilized and nonviable eggs which sink to the bottom of the container, the remaining viable oblong eggs (~0.7 mm in length) are concentrated with a 400–500 μm screen, rinsed with filtered seawater, and either placed directly in larval rearing containers, or, when necessary, plastic bags (~1,000 eggs/L) for transport to off-site rearing facilities. Bags containing eggs to be shipped are provided with an oxygenated headspace, roughly a third of the bag volume, and transported in insulated containers to provide a constant low transport temperature. Eggs of the California halibut have been found to be particularly sensitive to handling between 3 and 17 hours postfertilization (Bush *et al.* 2002), so manipulation or shipment should be carried out to avoid this vulnerable period. No problems in development were noted in the range between 12 and 20°C; however, development stopped at the 32-cell stage when eggs were held at 8°C and was abnormal at 24°C (Gadomski and Caddell 1991).

3.3 Larval rearing

Newly hatched larvae are about 2 mm in length with the mouth opening a day or two after hatching when cultured at 18°C. First feeding coincides with the completion of eye pigmentation on day 3 by which time the digestive tract has become regionally differentiated but lacks a functional stomach (Gisbert *et al.* 2002). In that, any delay in feeding retards larval development, food is ideally provided during the second day after hatching to ensure that rapidly developing larvae have access to food when they are ready. Larvae are somewhat flexible with regard to food intake, in that, the point of no return for starvation at 18°C does not occur until between 6 and 8 days posthatch (dph) (Gisbert *et al.* 2004a). Increasing culture temperatures hastens the time to starvation (Gadomski and Caddell 1991).

Early California halibut larvae are fed live rotifers *Brachionus plicatilis* in static water conditions (Muguet *et al.* 2005). Water exchange is limited to the replacement of volume lost during cleaning. Gentle aeration is used to maintain oxygen levels. The rotifers are reared using baker's yeast and enriched with RotiMac, a commercial rotifer growout diet (Aquafauna Bio Marine, Inc., Hawthorne, California). A number of other enrichments have been tried but the most successful strategy has proved to be a greenwater approach of adding either live or preserved algae *Isochrysis* (Reed Mariculture, San Jose, California). Larval rearing tanks are illuminated from above with fluorescent lights using a 16-hour light:8-hour dark photoperiod. The addition of algae to the culture water causes a reduction in the amount of light reaching the larvae as well as providing additional food for the rotifers. It does not appear that the fish larvae feed directly on the algal cells. The greenwater regime results in a decrease in the number of unpigmented fish. The lack of pigmentation has been a significant problem in flatfish culture often associated with poor larval nutrition (for review see Bolker and Hill 2000). In the case of California halibut larvae reared in greenwater, even juveniles that were poorly pigmented following metamorphosis, tend to become fully pigmented on the eyed side with time. Larvae reared in greenwater also are larger in size at 17 and 45 dph. Subsequent weaning of the juveniles is enhanced most likely because of the larger size of the fish. Most dramatic is the impact on survival from less than 10% in clear water trials to over 50% at 38 dph with algae (Muguet *et al.* 2005). There is little difference in response of the larvae between the live and the preserved but intact algal cells.

The specific role of algae is unclear although beneficial effects of greenwater culture for marine fish larvae have been noted for some time (May 1971). Among a number of suggestions (Muller-Feuga *et al.* 2003) is that the greenwater culture approach has an environmental effect. The presence of high concentrations of algal cells alter lighting conditions providing a milieu that beneficially changes behavior or enhances prey capture as was found in the halibut *Hippoglossus hippoglossus* larvae (Naas *et al.* 1992). Another potential alternative is either a direct or indirect nutritional benefit in using a greenwater approach (Reitan *et al.* 1997). A number of studies have shown that the fatty acid profile of rotifers and *Artemia* is less than ideal with respect to fatty acids required by marine fish and thus these live feed organisms are typically enriched with an array of fatty acids (see Bell *et al.* 2003 and Koven 2003 for review). Specific requirements have been shown for flatfish larvae for arachidonic (20:4n-6) (Villalta *et al.* 2005a), eicosapentaenoic acid (20:5n-3) (Dickey-Collas and Geffen 1992; Izquierdo *et al.* 1992; Villalta *et al.* 2005b), as well as docosahexaenoic acid (22:6n-3) in the case of the turbot (Bell *et al.* 1985a, 1985b) and the yellowtail flounder (Copeman and Parrish 2002; Copeman *et al.* 2002). In the case of the Atlantic cod, *Gadus morhua*, addition of algal cells to the culture water was associated with changes in the fatty acid composition of phospholipids and triacylglycerols (Van Der Meeren *et al.* 2007). Of course, a host of other nutritional benefits are also possible with the addition of algae. Similar to the situation with regard to fatty acids, it is known that the amino acid profile found in rotifers and *Artemia* differs appreciably from that found in the natural prey items of marine fish (Ronnestad *et al.* 1999). Amino acid retention of postlarval Senegalese sole *Solea senegalensis* was found to be increased by balancing the amino acid profile of *Artemia* metanauplii with supplements (Aragão *et al.* 2004). At the moment, it is premature to focus on a single hypothesis to explain the value of greenwater culture techniques. It is likely that the effect varies by species. Use of algal additions in the larval culture water was more important for gilthead seabream *Sparus aurata* larvae than it was for larvae of the Senegalese sole (Rocha *et al.* 2008).

Around 17 dph, the California halibut larvae have grown large enough to start feeding on *Artemia* nauplii in place of rotifers. The nauplii and metanauplii for larger larvae are enriched with Selco, a commercial emulsified lipid preparation (Artemia Systems N.V., Ghent, Belgium). It is around this time that the stomach becomes distinct, gastric glands develop, and acid proteolytic activity is noted (Gisbert *et al.* 2004b; Alvarez-González *et al.* 2005; Zacarias-Soto *et al.* 2006). Gradual weaning of the larvae to manufactured microdiets from an enriched *Artemia* feeding regime can be started as early as 20 dph (Lazo *et al.* 2004); however, weaning is progressively more effective in terms of survival and growth if started later. Larger juveniles with a survival rate greater than 80% resulted when weaning the fish at 46 dph in comparison to weaning attempts at earlier

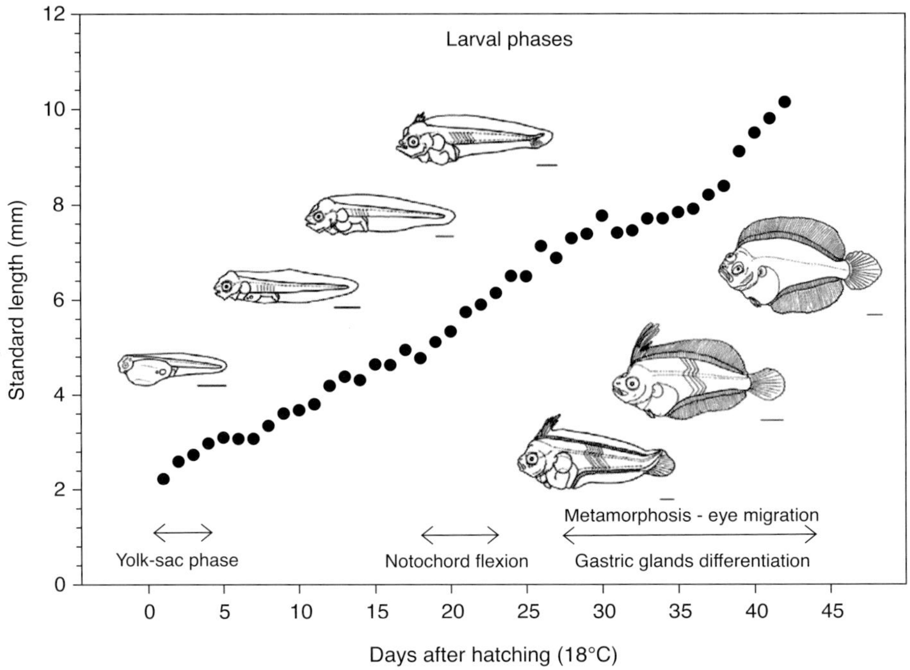

Figure 3.1 Growth in standard length of California halibut larvae from hatching to completion of metamorphosis (data redrawn from Gisbert *et al.* 2002).

ages of 16, 26, or 36 dph (Muguet *et al.* 2007). The complete transition to formulated diets from the live feeds takes place once the fish have undergone metamorphosis. A commercial feed, Otohime, manufactured in Japan (available from Reed Mariculture, San Jose, California) and formulated to meet the needs of marine fish proved to be an effective weaning diet. The diet is available in graded sizes so as the larvae growing toward metamorphosis feed particles of larger size are used.

Metamorphosis or the change from symmetric larvae inhabiting the water column to asymmetric juvenile flatfish favoring the benthic habitat is a unique characteristic of the group. Rapidly growing California halibut larvae (Figure 3.1) complete metamorphosis around 42 dph at 18°C (Gisbert *et al.* 2002).

Differing from most flatfish species, the California halibut does not show a preference for the side of eye development. Following eye migration at the time of metamorphosis, hatchery-reared fish could be either left or right eyed (Gisbert *et al.* 2002), a phenomenon also observed for wild fish of this species (Gadomski *et al.* 1990. While metamorphosis can be a difficult period in flatfish culture (Power *et al.* 2008), the surviving juveniles are quite hardy and mortality becomes negligible. Weaned juveniles are quite tolerant to salinity changes with growth being unaffected at salinities ranging from 5 to 30 ppt (unpublished data). Older juveniles, however, may not be as adaptable (Madon 2002).

3.4 Juvenile culture

A series of experiments relating to juvenile growth were carried out on the campus of the University of California, Davis, using a recirculation system (Figure 3.2).

The system, described in detail elsewhere (Merino 2004), had a total volume of approximately 3.0 m^3 and included four raceways (2.4 m long, 28 cm wide, with a water depth up to 22 cm). The system was filled with seawater trucked in from the coast. The seawater was chlorinated/dechlorinated before use. After the system was filled, additional seawater was stored in an outdoor tank until needed to replace losses from cleaning and maintenance of the system. A minimal amount (under 1% of the system volume) of seawater was exchanged on a daily basis. Salinity was maintained between 28 and 32 ppt by the addition of dechlorinated tap water to offset evaporation. Water from the raceways flowed first through a felt bag filter with a nominal retention size of 50 μm (Model FB50, Aquatic Ecosystems), then to a moving bed biological water treatment unit (0.45 m^3) filled with a combination of Kaldness™ 10 mm media; Rauschert™ Bioflow™ 9 mm media and Biolox™ 10 media (material density 1.05 g/cm^3) and then on to the main reservoir where a combination chiller/heater unit (Frigidunits™ D1–100, 2000 W) was calibrated to maintain a constant temperature in the system. Water was pumped (Jacuzzi™ S1KTM) from this main reservoir through cartridge filters (Hayward Star Clear™ 320L26) and a UV unit (Rainbow™ QL-25) up to a constant head tank that supplied seawater back to the raceways. The cartridge filter and UV unit were removed from the system after some months of operation with no apparent impact on fish health or water quality. Various other rearing and experimental units could be added into the main circuit of the system.

Additional systems for rearing larger juveniles were located at the Bodega Marine Laboratory (BML), Bodega Bay, California. Although larger and using both round, square, and rectangular tanks instead of the raceways, functionally, the systems on campus and those at BML were similar. Seawater was circulated from the rearing tanks to the other modules of the system for filtration, biological

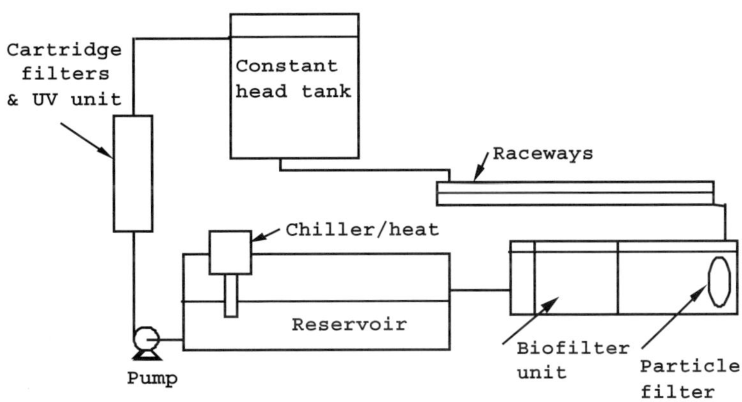

Figure 3.2 Schematic diagram of the recirculation system at UC Davis (Merino 2004).

treatment of the wastes, and temperature control. The one significant difference in the BML systems reflected the availability of piped seawater. Consequently, the marine laboratory systems were operated in a semi-recirculated fashion with a constant input of seawater.

The BML system, a prototype recirculation system for commercial culture (Figure 3.3) is described in detail elsewhere (De Vellis 2006) and had a total volume of approximately 8.9 m^3.

The system included three rectangular fiberglass tanks (3 m long × 1.5 m wide × 0.8 m deep) all with rounded corners and operated with about 0.3 m water depth. One of the rectangular tanks was partitioned into two separate square tanks (1.5 m sides). Each rectangular tank had two drains located 0.75 m from the end and sidewalls, while the square tanks had one center drain each. A perforated standpipe was installed in each of the drains. In addition, a PVC partition (0.64 cm thick PVC sheet) was placed between the standpipes in the rectangular tanks, creating a racetrack-type configuration. Water was introduced into the rectangular tanks through two inlets placed in opposite corners. A single inlet was used for the square tanks. Influent water was oxygenated in a short (about 0.75 m) column located within the fish tank and with an enriched oxygen atmosphere. The effluent from each of the fish tanks went to a separate fiberglass swirl separator (cylindro-conical about 0.91 m diameter, 0.69 cm deep in the cylindrical section, and 0.38 cm deep in the conical section) fabricated in-house. Effluent from the swirl separators flowed into a drum filter with a 60 μm screen (PRAqua Supplies Ltd, Model RFM 2014) before entering a moving bed biofilter filled with Rauschert's Biolox™ 10 media (material density 1.05 g/cm^3). The media in the biofilter were retained in three chambers (each about 2.5 m long × 0.5 m wide × 0.7 m deep) with the water flowing sequentially past each zone. The biofilter was filled to about 40% of bulk volume with the media that was fluidized by means of airstones placed on the bottom of the biofilter tank. Effluent from the biofilter was heated (with two Process Technology model ETA1.8117-PT-1) prior to being pumped back to the fish tanks (with two Jacuzzi™ Stingray pumps). Total system water flow rate was about 240 L/min with a makeup rate of approximately 65% of system volume per day. Make up water was filtered with a bag filter (50 μm felt, Model FB50, Aquatic Ecosystems) and disinfected with UV light (Advance Mark III Energy Saver Model R-140-TP-PC).

A number of studies were carried out in these systems to examine the impact of rearing halibut under conditions that would approximate commercial conditions. Culture temperature in the systems was maintained at 21–22°C. Salinity was maintained at 30 g/L or in the case of the marine laboratory systems, which reflected ambient salinity of the coastal water (~34 g/L). Light was provided to all systems by overhead fluorescent tubes with a photoperiod regime of 16:8 (light:dark). Survival was routinely noted and growth was periodically assessed both by blotted wet weight and image analysis of photographs of each individual fish to collect information on fish morphometics (Merino 2004).

A host of factors, such as environmental temperature, the amount of feed, etc., impact on the survival and growth of individuals during the growout phase of fish culture. The interaction of these factors in combination with the biological characteristics of each species must be optimized to maximize production. Key

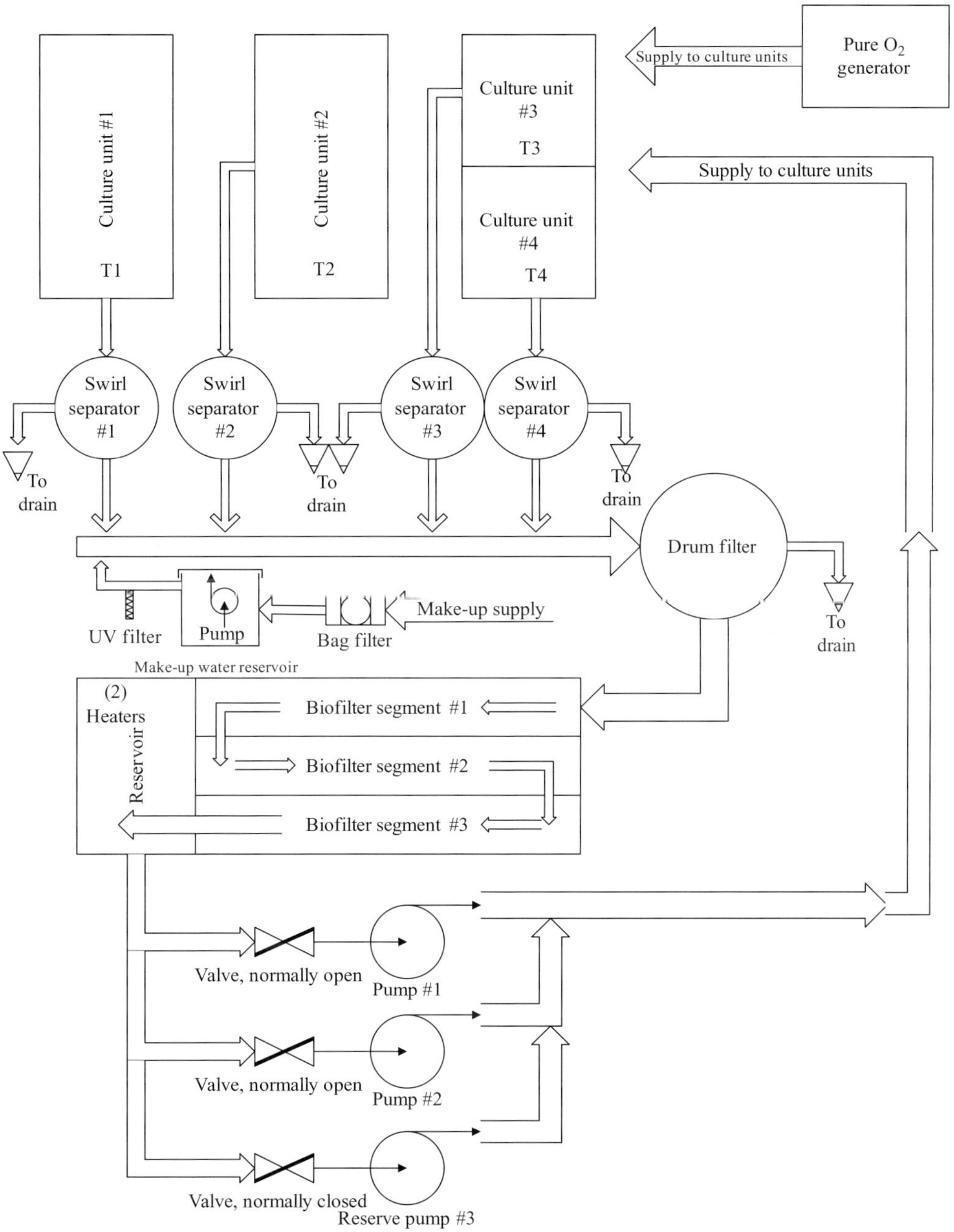

Figure 3.3 Schematic diagram of the recirculation system at BML (De Vellis 2006).

characteristics of cultured flatfish are their preference for the tank bottom and their relative low level of activity which impacts tank design, typically shallow raceways (Øiestad 1999) or round tanks and system inputs such as oxygen. A number of experiments were carried out at UC Davis and BML facilities with California halibut specifically to gather the information that would pertain to commercial culture.

3.5 Density

Culturing fish at high densities is important to maximize the utilization of economic resources such as tank space and water. Reflecting the biological characteristics of flatfish, tanks tend to be comparatively shallow with the focus of the culturist on the surface area of the bottom (Øiestad 1999). One important difference with respect to flatfish in the wild and in culture is that growout systems tend to have bare bottoms lacking the sandy environment that flatfish in the wild use to conceal themselves. Interestingly, in the case of the California halibut, it was found that when only a few individuals were in a bare tank, they tended to aggregate often laying on top of each other rather than distributing themselves across the tank bottom. This suggested, as for other flatfish species (Jeon *et al.* 1993; Björnsson 1994; King *et al.* 1998) that California halibut were gregarious and could be cultured at high densities. Juveniles at just over 10 grams in size were grown in shallow (~6 cm) raceway tanks of varying width stocked with appropriate number of fish to achieve the desired stocking density. In comparing growth in groups of fish initially stocked at 100, 200, and 300% of the coverage area (PCA = percent ratio of total fish ventral area to total tank bottom area), it was found that the best growth was obtained for the 100% PCA group (Merino *et al.* 2007a). While an optimum density was not defined, it does indicate that like other flatfish, California halibut can be grown in shallow tanks at high densities. Defining optimal density for growth is complicated by the possible influence of other factors related to either the biology of the species, system characteristics, or a combination of both that may impact on growth. An increased growth variation was noted in turbot when held at higher than optimum stocking densities (Irwin *et al.* 1999). When individually tagged (passive implant transponders) Atlantic halibut *H. hippoglossus* were grown at the highest of the three densities (112% PCA), slow-growing individuals were those recorded as spending significant time swimming at the surface (Kristiansen *et al.* 2004).

Typically, water flowing across or around cultured fish is used in an attempt to bring fresh oxygenated water to the fish and move particulate and soluble wastes away. Conversely, fish have to move against the flow to maintain position within the tank. While moderate exercise through swimming can improve growth rates in actively swimming fish species (Davison 1997; Jorgensen and Jobling 1993), ideally for demersal fish, current flow should not perturb sedentary behavior. Although survival of small juvenile (~1.5 g average weight) California halibut was unaffected by water velocities of up to 1.5 body lengths per second (bl/s), feed efficiency and growth were better at lower velocities, 0.5 and 1.0 bl/s (Merino *et al.* 2007b). Maximum growth was achieved at 1.0 bl/s which is similar to what was reported for Japanese flounder juveniles (~6 g average weight) (Ogata and Oku 2000). At flow rates higher than that promoting maximum growth, increased tail beating, presumably an energy expending behavior to maintain position, was noted in both the California and the Japanese flounder juveniles. Too little water exchange can lead to a reduction in oxygen levels and a buildup of metabolic byproducts. This may be acerbated in areas in the immediate vicinity of the fish if water flow and lack of activity by sedentary fish is insufficient to promote mixing of the water column. Recent studies of California halibut rearing systems through the measurement of the vertical distribution of dissolved oxygen indicate that effluent measurements are likely to underestimate detrimental environmental conditions faced by the fish at the bottom of tanks (Reig *et al.* 2007). Both low dissolved oxygen and a buildup of ammonia and other metabolites are of concern with respect to growth (Taylor and Miller 2001; Pinto *et al.* 2007). Although ammonia excretion in flatfish is relatively low, levels reflecting feeding activity vary significantly during the day (Kikuchi *et al.* 1991; Dosdat *et al.* 1995; Verbeeten *et al.* 1999). Even though feeding was spread out over a 12-hour period during the daily light phase, ammonia excretion rates of California halibut still peaked about 4–6 hours and 12–14 hours after the feeders activated in the morning (Merino *et al.* 2007c). Improved information to sustain the improved tank design and flow patterns (Cripps and Poxton 1992) will be advantageous to the further development of commercial flatfish culture including the California halibut.

Dietary protein requirements of flatfish tend to be relatively high, $\geq$45% of the diet (Bromley 1980; Guillaume *et al.* 1991; Helland and Grisdale-Helland 1998; Daniels and Gallagher 2000; Lee *et al.* 2002; Hebb *et al.* 2003; Kim *et al.* 2003). Carnivorous fish use amino acids from protein not only for tissue synthesis and maintenance but also as a preferential source of energy (Cowey and Sargent 1989; Wilson 1989). As protein typically is the most expensive component of feeds, one of the goals of fish feed formulation is to provide appropriate energy sources so as to ensure that a minimum of dietary protein is used for energy purposes. In that, flatfish like other carnivorous fish utilize carbohydrates poorly, lipids are the preferred dietary energy source. Attempts to promote protein sparing by increasing the amount of lipids in the diet of flatfish have had mixed results (Guillaume *et al.* 1991). Bush (2003) used fish oil supplements in an attempt to increase the amount of digestible energy in the diet of juvenile California halibut. While growth was not directly affected, there were some signs of protein sparing for tissue synthesis. Protein retention was increased and ammonia excretion was reduced. However, boosting dietary lipid levels also led to increases in lipid deposition in the juveniles. Similar results were found with Atlantic halibut (Aksnes *et al.* 1996), turbot (Regost *et al.* 2001), and the Senegalese sole (Dias *et al.* 2004). While there is room for further work on protein sparing and other aspects of nutrition in flatfish, fortunately effective commercial diets are already available.

3.6 Commercial trials

Feeds with the exception of the diet manufactured in the laboratory for the above nutritional studies were obtained from various commercial sources. Three commercial feeds were used for general rearing trials; Biokyowa™ for smaller juveniles, Silver Cup™ for juveniles 5–10 grams, and EWOS™ Alpha#2 and Pacific#3 for fish larger than 10 grams. Fish were fed on a sliding scale with the

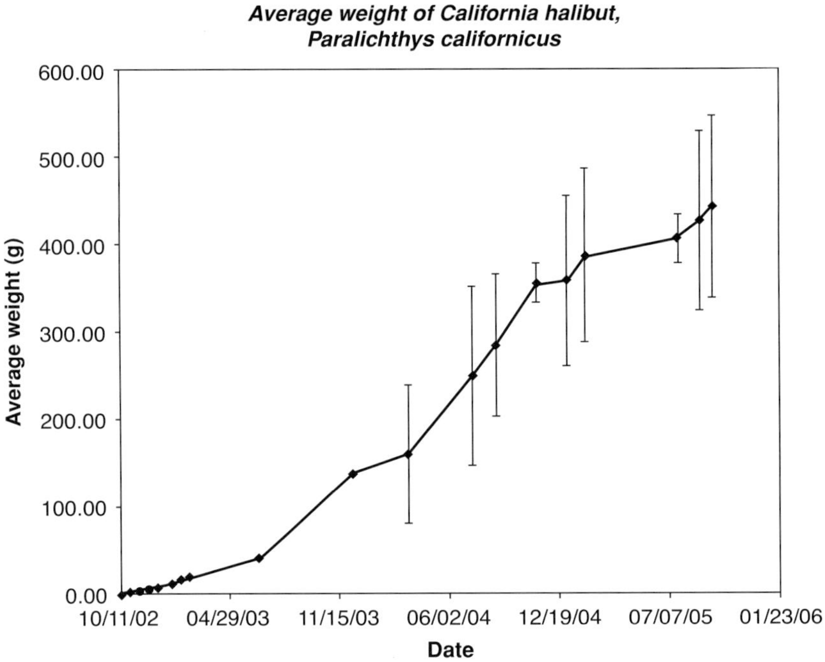

Figure 3.4 Growth in average wet weight of California halibut from metamorphosis to termination of growout in the BML commercial prototype recirculation system (unpublished data).

smaller juveniles receiving approximately 2% of body weight per day while the larger fish received 1%. Figure 3.4 shows the growth of California halibut over a 3-year period.

These fish were originally reared on the UC Davis campus and then transferred to the BML commercial prototype recirculation system. In that, growth was similar in both the rectangular and the square tanks, data from all the tanks were combined. Survival in all the tanks was above 90% for the period the fish were in the BML system with the system containing a total biomass of 215 kg of fish at the end of the experiment. As can be seen on the graph in Figure 3.4, overall growth was less than expected and the rate of growth slowed noticeably in the later part of 2004 and through most of 2005. Several reasons probably account for this slower growth. There were some problems with the system principally with failure of the heater unit causing reduced temperatures in the system until replacement could be effected. Second and likely most important, it was learned at the conclusion of the experiment in checking the sex of the fish that all (100%) were male. Sex determination was done either by stripping and noting the presence of sperm or when necessary, in a few cases, dissection followed by microscopic examination of the gonads. The male sexual basis was most likely because of the high culture temperatures used around the time of metamorphosis (Goto *et al.* 1999; Yamamoto 1999; Godwin *et al.* 2001). As with other flatfish, males tend to grow slower after becoming sexually mature at smaller sizes and at younger ages than females. In the wild, most of the California halibut males sampled were sexually mature at 1 year of age (Love and Brooks 1990).

The development of the nascent California halibut industry along the Pacific coast of the United States and Mexico will undoubtedly benefit from the successes of other commercial flatfish endeavors through the incorporation of key insights. In other developed or potential flatfish industries in which the female has a faster growth rate and reaches larger sizes, the development of all female stocks has become a priority (Yamamoto 1999; King *et al.* 2001; Luckenbach *et al.* 2002; Cal *et al.* 2006). The development of all female faster growing stocks for commercial culture of California halibut using some of the recent tools developed for other species of flatfish (Luckenbach *et al.* 2002; Morgan *et al.* 2006; Liu *et al.* 2007; Luckenbach *et al.* 2007) is probably inevitable.

Literature cited

Aksnes, A., Hjertnes, T., and Opstvedt, J. 1996. Effect of dietary protein level on growth and carcass composition in Atlantic halibut (*Hippoglossus hippoglossus* L). *Aquaculture* 145:225–233.

Allen, M.J. 1990. The biological environment of the California halibut, *Paralichthys californicus. Fish Bulletin (California Department of Fish and Game)* 174:7–29.

Alvarez-González, C.A., Cervantes-Trujano, M., Tovar-Ramirez, D., Conklin, D.E., Nolasco, H., Gisbert, E., and Piedrahita, R.H. 2005. Development of digestive enzymes in California halibut *Paralichthys californicus* larvae. *Fish Physiology and Biochemistry* 31:83–93.

Alvial, A., and Manriquez, J. 1999. Diversification of flatfish culture in Chile. *Aquaculture* 176:65–73.

Aragão, C., Conceição, L.E.C., Martins, D., Rønnestad, I., Gomes, E., and Dinis, M.T. 2004. A balanced dietary amino acid profile improves amino acid retention in post-larval Senegalese sole (*Solea Senegalensis*). *Aquaculture* 233:293–304.

Bell, J.G., McEvoy, L.A., Estevez, A., Shields, R.J., and Sargent, J.R. 2003. Optimizing lipid nutrition in first-feeding flatfish larvae. *Aquaculture* 227:211–220.

Bell, M.V., Henderson, R.J., Pirie, B.J.S., and Sargent, J.R. 1985a. Effects of dietary polyunsaturated fatty acid deficiencies on mortality growth and gill structure in the turbot *Scophthalmus maximus. Journal of Fish Biology* 26:181–192.

Bell, M.V., Henderson, R.J., and Sargent, J.R. 1985b. Changes in the fatty-acid composition of phospholipids from turbot *Scophthalmus maximus* in relation to dietary polyunsaturated fatty acid deficiencies. *Comparative Biochemistry and Physiology B* 81:193–198.

Björnsson, B. 1994. Effects of stocking density on growth rate of halibut (*Hippoglossus hippoglossus* L.) reared in large circular tanks for three years. *Aquaculture* 123:259–270.

Bolker, J.A., and Hill, C.R. 2000. Pigmentation development in hatchery-reared flatfishes. *Journal of Fish Biology* 56:1029–1052.

Bromage, N., Porter, M., and Randall, C. 2001. The environmental regulation of maturation in farmed finfish with special reference to the role of photoperiod and melatonin. *Aquaculture* 197:63–98.

Bromley, P.J. 1980. Effect of dietary protein, lipid and energy content on the growth of turbot (*Scophthalmus maximus* L.). *Aquaculture* 19:359–369.

Brown, N. 2002. Flatfish farming systems in the Atlantic region. *Reviews in Fisheries Science* 10:403–419.

Bush, D. 2003. The effect of dietary lipid content on protein utilization in California halibut, *Paralichthys californicus*. Thesis. University of California, Davis, CA.

Bush, D., Muguet, J.B., Gisbert, E., Rounds, J., Merino, G., Conklin, D.E., and Piedrahita, R.H. 2002. Effect of handling stress on egg viability of California halibut *Paralichthys californicus* [Abstract]. Aquaculture America 2002, World Aquaculture Society, San Diego, CA, pp. 43.

Caddell, S.M., Gadomski, D.M., and Abbott, L.R. 1990. Induced spawning of the California halibut, *Paralichthys californicus* (Pisces: Paralichthyidae) under artificial and natural conditions. *Fish Bulletin (California Department of Fish and Game)* 174: 175–198.

Cal, R.M., Vidal, S., Martínez, P., Álvarez-Blázquez, B., Gómez, C., and Piferrer, F. 2006. Growth and gonadal development of gynogenetic diploid *Scophthalmus maximus*. *Journal of Fish Biology* 68:401–413.

Chen, S.L., Ji, X.S., Yu, G.C., Tian, Y.S., and Sha, Z.X. 2003. Cryopreservation of sperm from turbot (*Scophthalmus maximus)* and application to large-scale fertilization. *Aquaculture* 236:547–556.

Copeman, L.A., and Parrish, C.C. 2002. Lipid composition of malpigmented and normally pigmented newly settled yellowtail flounder, *Limanda ferruginea* (Storer). *Aquaculture Research* 33:1209–1219.

Copeman, L.A., Parrish, C.C., Brown, J.A., and Harel, M. 2002. Effects of docosahexaenoic, eicosapentaenoic, and arachidonic acids on the early growth, survival, lipid composition and pigmentation of yellowtail flounder (*Limanda ferruginea*): a live food enrichment experiment. *Aquaculture* 210:285–304.

Cowey, C.B., and Sargent, J.R. 1989. Intermediary metabolism. In: Halver, J.E. (ed.) *Fish Nutrition*, 2nd edn. Academic Press, San Diego, CA, pp. 259–329.

Cripps, S.J., and Poxton, M.G. 1992. A review of the design and performance of tanks relevant to flatfish culture. *Aquacultural Engineering* 11:71–91.

Daniels, H.V., and Gallagher, M.L. 2000. Effect of dietary protein level on growth and blood parameters in summer flounder, *Paralichthys dentatus*. *Journal of Applied Aquaculture* 10:45–52.

Daniels, H.V., and Watanabe, W.O. 2002. A Practical Hatchery Manual: Production of Southern Flounder Fingerlings UNC-SG-02-08. North Carolina Sea Grant, Raleigh, NC.

Davison, W. 1997. The effects of exercise training on teleost fish, a review of recent literature. *Comparative Biochemistry and Physiology A* 117:67–75.

De Vellis, L. 2006. Performance assessment of a prototype recirculation aquaculture system. Thesis, University of California, Davis, CA.

Dias, J., Rueda-Jasso, R., Panserat, S., da Conceição, L.E.C., Gomes, E.F., and Dinis, M.T. 2004. Effect of dietary carbohydrate-to-lipid ratios on growth, lipid deposition and metabolic hepatic enzymes in juvenile Senegalese sole (*Solea senegalensis*, Kaup). *Aquaculture Research* 35:1122–1130.

Dickey-Collas, M., and Geffen, A.J. 1992. Importance of the fatty acids 20:5 ω-3 and 22:6 ω-3 in the diet of plaice *Pleuronectes platessa* larvae. *Marine Biology (Berlin)* 113:463–468.

Dosdat, A., Metailler, R., Tetu, N., Servais, F., Chartois, H., Huelvan, C., and Desbruyeres, E. 1995. Nitrogenous excretion in juvenile turbot, *Scophthalmus maximus* (L.), under controlled conditions. *Aquaculture Research* 26:639–650.

Drawbridge, M.A., and Kent, D.B. 2001. Culture of marine finfish. In: Leet, W.S., Dewees, C.M., Klingbeil, R., and Larson, E.J. *et al.* (eds) *California's Living Marine Resources: A Status Report*. California Department of Fish and Game, pp. 510–512.

FAO. 2006. FISHSTAT Plus – Universal software for fishery statistical time series [online http://www.fao.org/fi/statist/FISOFT/FISHPlus.asp or CD-Rom]. Food and Agriculture Organization of the United Nations.

Furuita, H., Tanaka, H., Yamamoto, T., Shiraishi, M., and Takeuchi, T. 2000. Effects of n-3 HUFA levels in broodstock diet on the reproductive performance and egg and larval quality of the Japanese flounder, *Paralichthys olivaceus. Aquaculture* 187:387–398.

Furuita, H., Yamamoto, T., Shima, T., Suzuki, N., and Takeuchi, T. 2003. Effect of arachidonic acid levels in broodstock diet on larval and egg quality of Japanese flounder *Paralichthys olivaceus. Aquaculture* 220:725–735.

Gadomski, D.M., and Caddell, S.M. 1991. Effects of temperature on early-life-history stages of California halibut *Paralichthys californicus. U S National Marine Fisheries Service Fishery Bulletin* 89:567–576.

Gadomski, D., Caddell, S., Abbott, L., and Caro, T. 1990. Growth and development of larval and juvenile California halibut, *Paralichthys californicus*, reared in the laboratory. In: Haugen, C. (ed.) *The California Halibut,* Paralichthys californicus, *Resource and Fisheries*. Fish Bulletin 174. State of California, The Resources Agency, Department of Fish and Game, pp. 85–98.

Gisbert, E., Conklin, D.E., and Piedrahita, R.H. 2004a. Effects of delayed first feeding on the nutritional condition and mortality of California halibut larvae. *Journal of Fish Biology* 64:116–132.

Gisbert, E., Merino, G., Muguet, J.B., Bush, D., Piedrahita, R.H., and Conklin, D.E. 2002. Morphological development and allometric growth patterns in hatchery-reared California halibut larvae. *Journal of Fish Biology* 61:1217–1229.

Gisbert, E., Piedrahita, R.H., and Conklin, D.E. 2004b. Ontogenetic development of the digestive system in California halibut (*Paralichthys californinus*) with notes on feeding practices. *Aquaculture* 232:455–470.

Godwin, J., Luckenbach, J.A., and Borski, R.J. 2001. Temperature influences on sex determination and development in flounder. *American Zoologist* 41:1456.

Goto, R., Tatsunari, M., Kawamata, K., Matsubara, T., Mizuno, S., Adachi, S., and Yamauchi, K. 1999. Effects of temperature on gonadal sex determination in barfin flounder (*Verasper moseri*). *Fisheries Science* 65:884–887.

Guillaume, J., Coustans, M.F., Métailler, R., Person-Le Ruyet, J., and Robin, J. 1991. Flatfish, turbot, sole, and plaice. In: Wilson, R.P. (ed.) *Handbook of Nutrient Requirements of Finfish*. CRC Press, Boca Raton, FL, pp. 77–82.

Hebb, C.D., Castell, J.D., Anderson, D.M., and Batt, J. 2003. Growth and feed conversion of juvenile winter flounder (*Pleuronectes americanus*) in relation to different protein-to-lipid levels in isocaloric diets. *Aquaculture* 221:1–11.

Helland, S.J., and Grisdale-Helland, B. 1998. Growth, feed utilization and body composition of juvenile Atlantic halibut (*Hippoglossus hippoglossus*) fed diets differing in the ratio between the macronutrients. *Aquaculture* 166:49–56.

Hong, W., and Zhang, Q. 2003. Review of captive bred species and fry production of marine fish in China. *Aquaculture* 227:305–318.

Irwin, S., O'Halloran, J., and FitzGerald, R.D. 1999. Stocking density, growth and growth variation in juvenile turbot, *Scophthalmus maximus* (Rafinesque). *Aquaculture* 178:77–88.

Izquierdo, M.S., Arakawa, T., Takeuchi, T., Haroun, R., and Watanabe, T. 1992. Effect of n-3 HUFA levels in *Artemia* on growth of larval Japanese flounder (*Paralichthys olivaceus*). *Aquaculture* 105:73–82.

Jeon, I., Min, K., Lee, J., Kim, K., and Son, M. 1993. Optimal stocking density for olive flounder, *Paralichthys olivaceous*, rearing in tanks. *Bulletin of National Fisheries Research and Development Agency (Korea)* 48:57–70.

Jorgensen, E.H., and Jobling, M. 1993. The effects of exercise on growth, food utilization and osmoregulatory capacity of juvenile Atlantic salmon, *Salmo salar*. *Aquaculture* 116:233–246.

Kikuchi, K. 2001. Present status of research and production of Japanese flounder *Paralichthys olivaceus* in Japan. *Journal of Applied Aquaculture* 11:165–175.

Kikuchi, K., Takeda, S., Honda, H., and Kiyono, M. 1991. Effect of feeding on nitrogen excretion of Japanese flounder *Paralichthys olivaceus*. *Nippon Suisan Gakkaishi* 57:2059–2064.

Kim, K., Wang, X., and Bai, S.C. 2003. Reevaluation of the dietary protein requirement of Japanese flounder *Paralichthys olivaceus*. *Journal of the World Aquaculture Society* 34:133–139.

King, N., Howell, W.H., and Fairchild, E. 1998. The effect of stocking density on the growth of juvenile summer flounder *Paralichthys dentatus*. In: Howell, W.H., Keller, B.J., Park, P.K., McVey, J.P., Takayanagi, K., and Uckita, Y. (eds) *Nutrition and Technical Development of Aquaculture*. Proceedings of the 26th U.S.-Japan Aquaculture Symposium. University of New Hampshire Sea Grant, Durham, NH, pp. 173–180.

King, N.J., Nardi, G.C., and Jones, C.J. 2001. Sex-linked growth divergence of summer flounder from a commercial farm: are males worth the effort? *Journal Applied Ichthyology* 11:77–88.

Koven, W. 2003. Key factors influencing juvenile quality in mariculture: a review. *Israeli Journal of Aquaculture-Bamidgeh* 55:283–297.

Kramer, S.H. 1990. Distribution and abundance of juvenile California halibut, *Paralichthys californicus*, in shallow waters of San Diego County. *Fish Bulletin (California Department of Fish and Game)* 174:99–126.

Kristiansen, T.S., Fernö, A., Holm, J.C., Privitera, L., Bakke, S., and Fosseidengen, J.E. 2004. Swimming behaviour as an indicator of low growth rate and impaired welfare in Atlantic halibut (*Hippoglossus hippoglossus* L.) reared at three stocking densities. *Aquaculture* 230:137–151.

Lazo, J.P., Varga, D., Medina, C., Zacarias, M.S., Garcia-Ortega, A., and Pedroza-Islas, R. 2004. Experimental microdiets for California halibut larvae. *Global Aquaculture Advocate* December:44–45.

Lee, S.M., Park, C.S., and Bang, I.C. 2002. Dietary protein requirement of young Japanese flounder *Paralichthys olivaceus* fed isocaloric diets. *Fisheries Science (Tokyo)* 68:158–164.

Liu, S., Zang, X., Liu, B., Zhang, X., Arunakumara, K.K.I.U., Zhang, X., and Liang, B. 2007. Effect of growth hormone transgenic *Synechocystis* on growth, feed efficiency, muscle composition, haematology and histology of turbot (*Scophthalmus maximus* L.). *Aquaculture Research* 38:1283–1292.

Love, M.S., and Brooks, A. 1990. Size and age at first maturity of the California halibut, *Paralichthys californicus*, in the southern California bight. *Fish Bulletin (California Department of Fish and Game)* 174:167–174.

Luckenbach, J.A., Godwin, J., Daniels, H.V., and Borski, R.J. 2002. Optimization of North American flounder culture: a controlled breeding scheme. *World Aquaculture* 33:40–45, 69.

Luckenbach, J.A., Murashige, R., Daniels, H.V., Godwin, J., and Borski, R.J. 2007. Temperature affects insulin-like growth factor I and growth of juvenile southern flounder, *Paralichthys lethostigma*. *Comparative Biochemistry and Physiology – Part A: Molecular & Integrative Physiology* 146:95–104.

Madon, S.P. 2002. Ecophysiology of juvenile California halibut *Paralichthys californicus* in relation to body size, water temperature and salinity. *Marine Ecology Progress Series* 243:235–249.

May, R.C. 1971. An annotated bibliography of attempts to rear the larvae of marine fishes in the laboratory. *NOAA (National Oceanic and Atmospheric Administration) Technical Report NMFS (National Marine Fisheries Service) SSRF (Special Scientific Report Fisheries)* 632:1–24.

Morgan, A.J., Murashige, R., Woolridge, C.A., Adam Luckenbach, J., Watanabe, W.O., Borski, R.J., Godwin, J., and Daniels, H.V. 2006. Effective UV dose and pressure shock for induction of meiotic gynogenesis in southern flounder (*Paralichthys lethostigma*) using black sea bass (*Centropristis striata*) sperm. *Aquaculture* 259:290–299.

Merino, G. 2004. Bioengineering requirements for the intensive culture of California halibut (*Paralichthys californicus*). Dissertation, University of California, Davis, CA.

Merino, G.E., Piedrahita, R.H., and Conklin, D.E. 2007a. The effect of fish stocking density on the growth of California halibut (*Paralichthys californicus*) juveniles. *Aquaculture* 265:176–186.

Merino, G.E., Piedrahita, R.H., and Conklin, D.E. 2007b. Effect of water velocity on the growth of California halibut (*Paralichthys californicus*) juveniles. *Aquaculture* 271:206–215.

Merino, G.E., Piedrahita, R.H., and Conklin, D.E. 2007c. Ammonia and urea excretion rates of California halibut (*Paralichthys californicus*, Ayres) under farm-like conditions. *Aquaculture* 271:227–243.

Moser, H.G., and Watson, W. 1990. Distribution and abundance of early life history stages of the California halibut, *Paralichthys californicus*, and comparison with the fantail sole, *Xystreurys liolepis*. *Fish Bulletin (California Department of Fish and Game)* 174:31–84.

Muguet, J.B., Conklin, D.E., Piedrahita, R.H., and Lazo, J.P. 2007. Evaluation of weaning performance of California halibut (*Paralichthys californicus*) larvae using growth, survival and digestive proteolytic activity. Unpublished manuscript, 25 pp.

Muguet, J.B., Bush, D.E., Conklin, D.E., Piedrahita, R.H., and Merino, G.E. 2005. Green water culture of California halibut, *Paralichthys californicus*, larvae. *Global Aquaculture Advocate* 8(2):88, 90.

Muller-Feuga, A., Cahu, R.R., Robin, C., and Divemach, P. 2003. Use of microalgae in aquaculture. In: Støttrup, J.A. (ed.) *Live Feeds in Marine Aquaculture*. Blackwell Publishing, Oxford, pp. 253–299.

Naas, K.E., Naess, T., and Harboe, T. 1992. Enhanced first feeding of halibut larvae *Hippoglossus hippoglossus* L. in green water. *Aquaculture* 105:143–156.

Ogata, H.Y., and Oku, H. 2000. Effects of water velocity on growth performance of juvenile Japanese flounder *Paralichthys olivaceus*. *Journal of the World Aquaculture Society* 31:225–231.

Øiestad, V. 1999. Shallow raceways as a compact, resource-maximizing farming procedure for marine fish species. *Aquaculture Research* 30:831–840.

Person-Le Ruyet, J., Baudin-Laurencin, F., Devauchelle, N., Metailler, R., Nicolas, J.-L., Robin, J., and Guillaume, J. 1991. Culture of turbot (*Scopthalmus maximus*). In: McVey, J.P. (ed.) Finfish Aquaculture. CRC Press, Boca Raton, FL, Vol. II, pp. 21–41.

Pinto, W., Aragao, C., Soares, F., Dinis, M.T., and Conceicao, L.E.C. 2007. Growth, stress response and free amino acid levels in Senegalese sole (*Solea senegalensis* Kaup 1858) chronically exposed to exogenous ammonia. *Aquaculture Research* 38:1198–1204.

Power, D.M., Einarsdóttir, I., Pittman, K., Sweeney, G.E., Hildahl, J., Campinho, M.A., Silva, N., Sæle, O., Galay-Burgos, M., Smáradóttir, H., and Björnsson, B.T. 2008. The molecular and endocrine basis of flatfish metamorphosis. *Reviews in Fisheries Science* 16:95–111.

Regost, C., Arzel, J., Cardinal, M., Robin, J., Laroche, M., and Kaushik, S.J.. 2001. Dietary lipid level, hepatic lipogenesis and flesh quality in turbot (*Psetta maxima*). *Aquaculture* 193:291–309.

Reig, L., Piedrahita, R.H., and Conklin, D.E. 2007. Influence of California halibut (*Paralichthys californicus*) on the vertical distribution of dissolved oxygen in a raceway and a circular tank at two depths. *Aquacultural Engineering* 36:261–271.

Reitan, K.I., Rainuzzo, J.R., Oie, G., and Olsen, Y. 1997. A review of the nutritional effects of algae in marine fish larvae. *Aquaculture* 155:207–221.

Rocha, R.J., Ribeiro, L., Costa, R., and Dinis, M.T. 2008. Does the presence of microalgae influence fish larvae prey capture? *Aquaculture Research* 39:362–369.

Ronnestad, I., Thorsen, A., and Finn, R.N. 1999. Fish larval nutrition: a review of recent advances in the roles of amino acids. *Aquaculture* 177:201–216.

Seikai, T. 1998. Japanese flounder seed production from quantity to quality. In: Howell, W.H., Keller, B.J., Park, P.K., McVey, J.P., Takayanagi, K., and Uckita. Y. (eds) *Nutrition and Technical Development Aquaculture*. Proceedings of the 26th U.S.-Japan Aquaculture Symposium. New Hampshire University Sea Grant Program, Durham, NH, pp. 5–16.

Seikai, T. 2002. Flounder culture and its challenges in Asia. *Reviews in Fisheries Science* 10:421–432.

Sjoholt, T. 2000. The World Market for Flatfish. FAO/GLOBEFISH Research Programme volume 61. FAO, Rome.

Taylor, J.C., and Miller, J.M. 2001. Physiological performance of juvenile southern flounder, *Paralichthys lethostigma* (Jordan and Gilbert, 1884), in chronic and episodic hypoxia. *Journal of Experimental Marine Biology and Ecology* 258:195–214.

Van Der Meeren, T., Mangor-Jensen, A., and Pickova, J.. 2007. The effect of green water and light intensity on survival, growth and lipid composition in Atlantic cod (*Gadus morhua*) during intensive larval rearing. *Aquaculture* 265:206–217.

Verbeeten, B.E., Carter, C.G., and Purser, G.J. 1999. The combined effect of feeding time and ration on growth performance and nitrogen metabolism of greenback flounder. *Journal of Fish Biology* 55:1328–1343.

Villalta, M., Estévez, A., and Bransden, M.P. 2005a. Arachidonic acid enriched live prey induces albinism in Senegal sole (*Solea senegalensis*) larvae. *Aquaculture* 245:193–209.

Villalta, M., Estévez, A., Bransden, M.P., and Bell, J.G. 2005b. The effect of graded concentrations of dietary DHA on growth, survival and tissue fatty acid profile of Senegal sole (*Solea senegalensis*) larvae during the *Artemia* feeding period. *Aquaculture* 249:353–365.

Watanabe, W.O., Woolridge, C.A., and Daniels, H.V. 2006. Progress toward year-round spawning of southern flounder broodstock by manipulation of photoperiod and temperature. *Journal of the World Aquaculture Society* 37:256–272.

Wilson, R.P. 1989. Amino acids and proteins. In: Halver, J.E. (ed.) *Fish Nutrition*, 2nd edn. Academic Press, San Diego, CA, pp. 111–151.

Yamamoto, E. 1999. Studies on sex-manipulation and production of cloned populations in hirame, *Paralichthys olivaceus* (Temminck et Schlegel). *Aquaculture* 173:235–246.

Zacarias-Soto, M., Muguet, J.B., and Lazo, J.P. 2006. Proteolytic activity in California halibut larvae (*Paralichthys californicus*). *Journal of the World Aquaculture Society* 37:175–185.

Chapter 4
Culture of summer flounder

David Bengtson and George Nardi

The culture of summer flounder (*Paralichthys dentatus*) began in the 1970s with research efforts at the National Marine Fisheries Service and Environmental Protection Agency laboratories in Narragansett, Rhode Island, United States (Smigielski 1975; Klein-MacPhee 1979), based on induced spawning of wild broodstock and larval rearing using natural zooplankton, rotifers, and brine shrimp, as well as efforts at the Skidaway Institute of Oceanography in Savannah, Georgia, United States (Stickney and White 1975), based on wild postlarvae captured from the plankton. After a hiatus of over 10 years, research recommenced in the 1990s due to the drastic declines in summer flounder (and other species) fishery landings and the perception that marine finfish aquaculture in the northeastern United States might help to replenish the seafood supply. The summer flounder industry began in 1995 with the development of the GreatBay Aquafarms hatchery in Portsmouth, New Hampshire, United States, and several nascent growout facilities, but very little cultured product was ever produced and the various participants soon left the industry for economic reasons. GreatBay Aquafarms, Inc. became GreatBay Aquaculture, LLC (GBA) in 2001 and has continued to produce juveniles annually. Those juveniles were first exported in 2003 to China, which now has a thriving summer flounder industry (Li 2007), and subsequently in 2006 to Mexico, which is developing an industry. Summer flounder culture has previously been reviewed by Bengtson (1999), Bengtson and Nardi (2000) and Schwarz (2003a).

4.1 Life history and biology

Summer flounder received their name due to their habitation of inshore waters of the northeastern United States during the summer months (whereas winter flounder, *Pseudopleuronectes americanus*, occupy roughly similar waters during the winter months). A review of summer flounder habitat preferences and requirements was compiled by Able and Kaiser (1994). The biology of summer flounder has recently been reviewed in the third edition of Bigelow and Schroeder's *Fishes*

Figure 4.1 Summer flounder juveniles (0.5–1.0 g size) at the GreatBay Aquaculture hatchery. (Photo by George Nardi.)

of the Gulf of Maine (Collette and Klein-MacPhee 2004), from which the information in this section is summarized, and the reader is referred to that volume for full details. Briefly, summer flounder spawn as they move offshore in the autumn to the deeper waters where they overwinter. Offshore movement appears to begin earlier in more northerly latitudes and later farther south (e.g., the Carolinas). With fecundity ranging from about 400,000 to 4 million eggs, this species is a classic marine broadcast spawner, and the fertilized eggs and larvae drift in the waters over the continental shelf for varying periods of time. Collette and Klein-MacPhee (2004) report the diameter of fertilized eggs as about 1.0 mm and the length at hatch as 2.41–2.82 mm, although Johns *et al.* (1981) report lengths at hatch of 3.02–3.05 mm. Summer flounder enter estuaries from October to April as they undergo metamorphosis. Growth during the first summer is rapid and the juveniles reach 200–300+ mm TL by the following September. Summer flounder reach maturity at median lengths of 28 cm (females) or about 25 cm (males), suggesting that fish must be a minimum of 2 years old to spawn.

Fishery landings of summer flounder, as reported by the National Marine Fisheries Service, were typically in the range of 4,000–9,000 MT during the 1950s and 1960s, but generally declined over time to about 500–1,000 MT during the 1990s (numbers recalculated from Figure 4.1 of Bengtson and Nardi 2000). In response to perceived overfishing, draconian fishing regulations were imposed on the commercial fishery in order to rebuild the stocks. While that strategy appears to be succeeding (Terceiro 2006), strict regulations are still in place. Meanwhile, the recreational fishery for summer flounder is probably responsible for more of the total catch than is the commercial fishery. For purposes of stock assessment and management, summer flounder are considered to belong to one of two populations, one found from Cape Cod to Cape Hatteras and the other from Cape Hatteras to Florida, although other competing concepts of stock structure exist. Nevertheless, a study of population structure based on mitochondrial DNA did not indicate a break in the population at Cape Hatteras.

4.2 Broodstock husbandry

Original broodstock summer flounder are normally captured from the wild with the aid of commercial fishers or research trawls, occasionally with hook and line. The shorter the duration of the trawl, the less damage done to the fish and the better the chances of their long-term survival (personal observations). Upon removal from the trawl, these adult fish should be placed in as large a container (tank, tote) as possible on deck with flowing seawater and should subsequently be transported (usually by truck) back to the hatchery in as large a container as possible with aeration. Fish entering the hatchery should be placed in quarantine for 30 days in order to ensure that they are not bringing any significant pathogens into the facility. Schwarz *et al.* (1998) described a quarantine facility for summer flounder. Fish that are transported across state borders should have a certificate of inspection from a veterinary pathologist indicating their disease-free status (specific regulations may somewhat vary from state to state).

Because broodstock fish normally lie quietly on the bottom of the tank, they can thrive in a variety of tank shapes and sizes. Nevertheless, round or oval tanks promote better circulation of water than do tanks with corners (square, rectangular, with attendant "dead spaces"). Tanks should be greater than 1 m in diameter and the larger the better for the acclimation and spawning of the fish (personal observations). Watanabe and Carroll (2001) noted that, although wild-caught broodstock can spawn in the hatchery in their first year of captivity, spawning induction is more successful if the broodstock have been kept in the hatchery longer and that is our experience as well. They also documented natural, volitional, as well as hormone-induced, spawning.

Broodstock fish are conditioned for spawning by a combination of temperature and photoperiod cues. Given that summer flounder spawn in the autumn in nature, the conditioning regime usually involves simulation of summer conditions (warm temperature, long photoperiod), followed by decreases in both temperature and photoperiod. Hormone induction of maturation and spawning has been accomplished with carp pituitary extract (CPE) (Smigielski 1975; Bengtson 1999) and gonadotropin releasing hormone analogue (GnRHa) implants or human chorionic gonadotropin (hCG) (Berlinsky *et al.* 1997, Watanabe *et al.* 1998). Berlinsky *et al.* (1997) found that females injected with CPE gave the more reliable spawning results than those with GnRHa implants and that hCG injections yielded the least reliable results. We have found that males can usually be induced to produce milt simply by photothermal manipulation, but that females spawn more reliably with hormonal induction. Development of oocytes can be monitored by placing the females on a light table and examining the ovaries (Watanabe and Carroll 2001), or more formally by ovarian biopsy

(Berlinsky *et al.* 1997; Watanabe *et al.* 1998). Commercial production at Great-Bay Aquaculture relies on the use of photothermal control and monitoring of ovarian development using a light table. Hormones may be used to synchronize the spawning of the broodstock within a specific week. Fish that exhibit oocytes ≥500 μm are good candidates for GnRHa implants (Berlinsky *et al.* 1997), while those that exhibit a stage one or one plus ovary (King 1999), with oocytes >180 μm (Berlinsky *et al.* 1997) can often be induced to spawn with a series of CPE injections as described above.

Broodstock summer flounder can be fed either frozen fish or squid (which can be a source of pathogens into the hatchery) or on pelleted feed. Bengtson and Nardi (2000) argued that research on broodstock nutrition was needed, but to our knowledge none has been accomplished. Commercial practice is to feed a vitamin fortified trash fish diet, however GBA has been successfully using a premix, either available from INVE (Breed-M) or from Skretting (Vitalis Cal) and adding additional products such as a squid hydrolysate, water, and oil to bring the finished product up to an 18% fat diet. Until recently, the only broodstock pellet diets that have been readily available have been designed for salmon, but recently Skretting has made available their marine species pellet (Vitalis Cal pellet available in sizes up to 22 mm), on which GBA is now conducting trials.

Spawning is still mostly accomplished by hand-stripping of broodstock. This generally involves removing fish from their tank(s), squeezing them to obtain small samples of eggs and milt for microscopic examination of egg morphology and sperm motility (to assess gamete quality), and dry-stripping of gametes from the adults into separate containers, followed by mixing of the gametes in seawater (milt is activated by seawater and remains motile for less than 2 minutes). After fertilization, eggs that sink to the bottom of the incubation chamber are removed and discarded because they will not hatch and become a substrate for bacteria. Incubation can occur in any of a variety of container sizes, shapes, and configurations, but commercial embryo incubation occurs in flow-through 100-liter cylindro-conical tanks with banjo filters. Watanabe *et al.* (1998) found that lower light intensity (500 lux) and higher salinity (36 g/L) during egg incubation and prolarval development yielded larger larvae than those at higher light intensity (2,000 lux) and lower salinities (26 or 31 g/L). Watanabe *et al.* (1999) examined simultaneously the effects of temperature (16, 20, or 24°C) and salinity (22, 28, or 34 g/L) on hatching rate, yolk utilization efficiency, and notochord length at first feeding; they found poor hatching rate at the two lower salinities and an optimal temperature of 16°C for all factors combined.

4.3 Larval culture

Upon hatching, summer flounder do not have functional eyes or digestive system and simply float in their culture tanks for about 3 days before the yolk-sac is exhausted, the eyes and mouth have developed, and the larvae can begin to consume live feed (Bisbal and Bengtson 1995a). Larvae must receive feed by 5–6 dph at 20°C or they will die (Bisbal and Bengtson 1995b). The stages of embryo and larval development have been described (Martinez and Bolker 2003), as have the development of the digestive tract (Bisbal and Bengtson 1995c) and the stages of metamorphic transition in later larval stages (Keefe and Able 1993). Larvae have been cultured in a variety of types and sizes of containers, from aquaria to fiberglass tanks, but commercial-scale production occurs in 3-m × 3-m tanks with rounded corners and about 1 m of depth. Schwarz (2003b) also described a recirculation system originally designed for summer flounder larvae. Commercial production takes place in conditions of constant light. Once the larvae begin to feed, water is turned on to the tanks and the flow is gradually increased as the larvae grow. Watanabe and Feeley (2004), studying growth of summer flounder larvae under light intensities of 50–2,000 lux during the first 15 days post hatch, found that growth was maximum at 50 lux and minimum at 2,000 lux. Greenwater culture (with added algae) is demonstrably preferable to clear-water culture (no algae) (Bengtson *et al.* 1999a) and is typically practiced at commercial scale. While commercial scale operations use live algae, they are increasingly using algal concentrates or pastes available from Reed Mariculture and most recently dried algae products that include probiotics, such as the ALG product from INVE. Stocking density appears to be optimal at about 20–30 larvae per liter (Klein-MacPhee 1981; Watanabe *et al.* 1999; King *et al.* 2000). Larvae in commercial culture are reared at temperatures of 19°C or less; while salinity varies during the larval stage, it is kept constant during metamorphosis.

First-feeding larvae are provided with rotifers (*Brachionus plicatilis*) beginning at 3 dph. The larvae do not need S-type rotifers and will readily consume L-type. Enrichment of the rotifers with n-3 highly unsaturated fatty acids (n-3 HUFA) improves survival and growth of the larvae (Baker *et al.* 1998). Willey *et al.* (2003) studied enrichment of rotifers with arachidonic acid (AA; 20:4n-6) at 0, 3, 6, 9, or 12% of total fatty acids in the enrichment emulsion; they found no significant differences in survival or growth of summer flounder larvae, but those enriched with 6% had significantly higher survival in a salinity tolerance test. Summer flounder larvae can begin to feed on brine shrimp when they reach about 5.0 mm TL (Koelbl 2000), which occurs about 12–18 dph, depending on the growth rate of the larvae. There is considerable inter-individual variation in growth rates of larvae, even those from the same set of parents and growth rates and/or growth variation expand around 20–22 dph; for larvae reared individually in bowls and repeatedly measured, there was no significant correlation between larval length at 8 or 9 dph (two experiments) and length at 30 dph (Katersky *et al.* 2008). Variation in growth was highly correlated with variation in food consumption for larvae reared individually in bowls (Koelbl 2000), so it is likely that differences in larval size are due to the different feeding capabilities among larvae. Average consumption rates by summer flounder larvae on rotifers and brine shrimp were first measured by Bengtson *et al.* (1999b), but were based on consumption during a 12L:12D light regime, so that feeding rates under a 24L:0D light regime are approximately double those numbers. The commercial practice is to wait until the larvae can comfortably consume enriched *Artemia* before making the transition.

When the larvae can make the transition from rotifers to brine shrimp (*Artemia* spp.), enrichment with essential fatty acids is again important. Bisbal and Bengtson (1991) found that larvae grew significantly better when brine shrimp were enriched with n-3 HUFAs. Willey *et al.* (2003) found that AA enrichment of brine shrimp did not affect survival, growth, or salinity tolerance of larvae, although those that had been fed rotifers enriched with AA at the 6% level prior to the brine shrimp experiment had given significantly better performance in all those variables at the end of the experiment than did larvae fed rotifers with 0% AA enrichment. Excellent enrichment products are commercially available from both INVE and Skretting. Since *Artemia* are recognized as key vectors for introducing *Vibrio* sp. into the culture tanks and larvae, it is critical to thoroughly rinse the ration before feeding. In addition, it may be advisable to harvest the hatch, rinse and clean, and then set up the enrichment in an effort to keep the *Vibrio* population in check.

The transition from live to formulated feed has been advanced by the development of a commercial hatchery. Initial reports indicated that survival and growth were better if larvae were weaned at 57 dph vs. 45 dph and with a gradual (7 days) vs. immediate (0 day) transition from live to formulated feed (Bengtson *et al.* 1999a), that formulated diets only began to be reasonably effective at 42 dph and weaning was not enhanced by larval treatment with thyroid hormone to hasten digestive tract development (Bengtson *et al.* 2000), and that a variety of weaning strategies did not significantly enhance weaning even after 42 dph (Musche 2003). Nevertheless, the commercial development of improved larval diets and industry practices has enabled GBA to wean their larvae from brine shrimp starting on day 25 and fully weaned by day 37.

A great deal is known about the process of metamorphosis in summer flounder, especially endocrine influences, due to the work of Dr. Jennifer Specker and her graduate students. Thyroid hormone is necessary for metamorphosis (Schreiber and Specker 1998) and immersion of larvae in thyroid hormone accelerates the development of the digestive tract (Huang *et al.* 1998; Soffientino and Specker 2001, 2003) and gill structure related to osmoregulation (Schreiber and Specker 1999a, 1999b, 2000). Given that summer flounder larvae in nature undergo metamorphosis as they enter estuaries, several studies were conducted on salinity influences on metamorphosis and survival. Larvae are able to survive and grow well at salinities as low as 8 g/L (Specker *et al.* 1999) and cortisol is necessary for their seawater tolerance (Veillette *et al.* 2006). An attempt to develop a novel approach to synchronizing metamorphosis by fluctuating salinity and immersion in thyroid hormone showed some promise (Gavlik *et al.* 2002; Gavlik and Specker 2004), but its commercial applicability is yet to be demonstrated.

Due to the variation in larval growth rates mentioned above, summer flounder metamorphose and settle to the bottom over an extended period of time, leading to considerable cannibalism if the size groups are not separated (the separation process is known as grading). Recently settled summer flounder can consume siblings up to 40% of their own length (Francis and Bengtson 1999). If individuals are removed from a tank shortly after they settle (during the extended settlement period) and placed in separate containers, their postsettlement growth rates over about 6 weeks are statistically indistinguishable, although the later-settling fish are always smaller than the earlier-settling (Simlick *et al.* 2000). Nevertheless, even within these groups, size differences begin to appear during the juvenile stage and periodic regrading is necessary. The first juveniles to settle have higher food consumption rates than do the last settlers (Getchis and Bengtson 2006), but oxygen consumption rates of the two groups are similar (Katersky *et al.* 2006), suggesting that growth rate differences are due more to increased energy (food) inputs rather than to metabolic differences.

When summer flounder larvae have been reared simultaneously in several containers, whether in aquaria, tanks, or simply glass bowls, large differences in mortality have been seen and have been attributed to disease. Burke *et al.* (1999) remarked that a viral infection was the likely cause of large mortality seen in some tanks while other tanks showed relatively little mortality during the experiment. Alves *et al.* (1999), raising eggs and larvae in replicate bowls, found survival at 10 dph ranging from 0 to 85%, but were not able to correlate the survival results with ammonia levels or bacterial levels in the water (bacterial levels in the rotifers were not analyzed). The microbial environment of the summer flounder hatchery is extremely important and has been characterized by Eddy and Jones (2002) and Gauger *et al.* (2006). Use of phytoplankton for greenwater culture and for rotifer feeding reduced the incidence of vibrios in the cultures, but enriched brine shrimp (even when rinsed) had high levels of vibrios and larval mortality increased during the brine shrimp phase of feeding (Eddy and Jones 2002). Gauger *et al.* (2006) monitored two hatcheries for bacterial pathogens and conducted experimental trials to determine pathogenicity of several bacteria by injecting summer flounder juveniles intraperitoneally; *Vibrio harveyi*, the known cause of flounder infectious necrotizing enteritis (FINE) (Soffientino *et al.* 1999), was shown to be pathogenic, whereas *Vibrio ichthyoenteri*, *Vibrio scophthalmi*, and *Photobacterium damselae* subsp. *damselae* were not. They also found that fish that suffered FINE (with 30% mortality) after being transported from a hatchery actually had a different strain of *V. harveyi* than was present at the hatchery. Eddy and Jones (2002) suggested that the use of probiotics in summer flounder hatcheries might be an effective strategy to help control the microbial environment and reduce pathogens. Both INVE (ALG product) and Eco Microbials sell probiotic products that can be used for summer flounder culture.

With regard to hatchery economics, the major operational costs are energy, labor, and feed. It is important to locate in an area that allows minimization of energy. A higher level of skilled labor and technical knowledge is required for hatchery rearing than for growout, so the average salary level at a hatchery will be higher than that at a farm. Although a much smaller amount of formulated feed is used in the hatchery than during growout, the price per kilo is large and represents along with live feed another substantial operating cost. The cost of production, depending on scale of operation, influence of location, and any subsidies, ranges from about \$0.25–1.00 for 1–2 g juveniles (Figure 4.1). Thus, if market fish were sold at \$10/kg and the average price of the fingerlings was \$0.80 each, then the cost of fingerlings represents 8% of the cost of production, compared with feed which may be around 50% or higher.

Relatively little has been done with improvement of summer flounder stocks through selective breeding. Research collaboration between GBA and

Dr. Thomas Kocher at the University of New Hampshire used microsatellite DNA to link the fastest-growing juveniles to their parents when all the juveniles were reared communally in a tank. The results showed that heritability of growth rate was large enough to warrant a selective breeding program; they also showed that most of GBA's hatchery production (after mortality and the culling of slower-growing individuals) came from a very small percentage of the females used as spawners for the production run. A formal selective breeding approach would require expanded facilities to keep separate families apart to prevent inbreeding. GBA has used these results at a small scale to develop selected F_1 and F_2 populations for faster growth in the hatchery, but has not been able to develop a full, formal program.

One interesting aspect of summer flounder biology is that females grow faster than males, leading commercial culturists to question the value of growing male fish (King *et al.* 2001) and researchers to find ways to produce all-female populations. Luckenbach *et al.* (2002) described techniques for all-female production, including temperature-dependent sex determination and meiogynogenesis via use of irradiated sperm. Attempts at hormonal induction of female populations by immersion in estradiol were studied by Specker and Chandlee (2003); uptake and clearance rates of estradiol by summer flounder larvae and juveniles were reported, but not rates of success in the production of all-female populations. GBA is currently developing the commercial protocols for the production of an all-female stock through a USDA SBIR Phase 2 award and expects the first commercial production of all-female juveniles in 2009. GBA feels that the approximately twofold advantage in growth rate for females vs. males is a more effective strategy in the short term for enhancing production than is a selective breeding program.

Chinese scientists have recently conducted experimental crosses of summer flounder and Japanese or olive flounder (*Paralichthys olivaceus*). Production of hybrids between *P. olivaceus* females and *P. dentatus* males was reported by You *et al.* (2006). Guan *et al.* (2007) identified 18°C and 25–35 g/L as optimal temperature and salinities for hatching and larval culture of such hybrids. Li *et al.* (2008) described the development of the digestive tract of these hybrids, which have been given the common name jasum. You *et al.* (2007) reported at the Aquaculture 2007 conference that the cross of *P. dentatus* females with *P. olivaceus* males produced abnormal larvae that did not survive.

The above procedures for genetic manipulation of summer flounder are intended only for purposes of commercial production for consumer markets. We note that any production of summer flounder for stock enhancement purposes should use only wild broodstock, following general recommendations for responsible marine stock enhancement (Blankenship and Leber 1996).

Experimentation with summer flounder stock enhancement has been restricted to studies by Kellison and colleagues at North Carolina State University. Kellison *et al.* (2000) conducted laboratory studies of the behavior of wild summer flounder vs. those that had been raised in a hatchery; they found that cultured fish spent significantly more time swimming and were more susceptible to predation. Antipredator conditioning of the hatchery-reared fish made them less susceptible to predation than their siblings, but still more susceptible than wild fish. When hatchery-reared summer flounder juveniles were released into nursery habitats, they were recaptured with significantly less frequency than were wild fish that were also released in those habitats; however, growth rates of hatchery-reared vs. wild summer flounder that were held in cages in the habitats were not significantly different (Kellison *et al.* 2003). Results of these studies suggest that hatchery-reared fish can obtain prey in the wild if they are protected from predators, but that they probably succumb to predation after being released. Finally, Kellison and Eggleston (2003) modeled summer flounder stock enhancement scenarios, based on their results and economic costs for hatchery production of various sizes of *P. olivaceus* and examining factors such as number released, size at release, cost at release, date of release, etc. They found that optimal results should be seen when fish of 75–80 mm TL are released in the natural nursery season in April.

4.4 Nursery culture and growout

Summer flounder juveniles are typically raised in the hatchery to a size of 10 g before being shipped to nursery or growout facilities. In the 1990s, GBA conducted a trial growout at their facility using a D-ended raceway (Figure 4.2) and recirculation and also shipped juvenile fish to two other commercial facilities that grew the fish in long, shallow, stacked raceways with recirculation. Fish that were intended for stocking into open-ocean cages were first held in flow-through tanks in nursery facilities in New York and Massachusetts. Fish stocked into cages during the first year were about 105 g at stocking and

Figure 4.2 Pilot-scale growout of summer flounder in a D-ended raceway in Portsmouth, New Hampshire, United States. (Photo by George Nardi.)

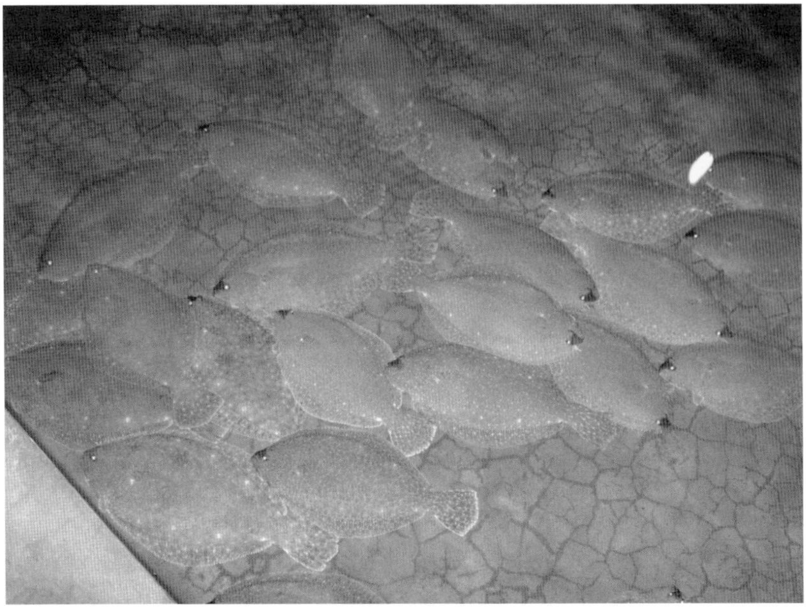

Figure 4.3 Commercial growout of summer flounder in a tank in China. (Photo by George Nardi.)

suffered high mortality in the cages because they could not tolerate the current (R. Link, personal communication). Fish stocked during the second year were larger (150 g at stocking) and survived well for about 5 months, but were unable to swim effectively when temperatures dropped below about 8°C and died due to impingement on the cage mesh (J. Gaskill, personal communication).

All of the commercial growout facilities in the United States closed because of a variety of problems, but primarily disease and economics. As part of the Open Ocean Aquaculture Demonstration Project at the University of New Hampshire, summer flounder (hatchery-reared at GBA) was the first species deployed in cages at the demonstration site in 1999, primarily because they were the only species available in sufficient numbers. Although the water was too cold for them to grow successfully, they survived and allowed the project to get underway. Sulikowski and Howell (2003) studied blood chemistry of those fish as they were transferred from a recirculation system to a flow-through system to a small coastal net pen to the open-ocean cage. Significant changes occurred as the fish were moved from recirculation to flow-through, probably due to an osmoregulatory effect (osmolarity and electrolyte changes), and in subsequent transfers, probably due to stress (cortisol and glucose changes). Another study of anesthesia on transportability of summer flounder found that tricaine methanesulfonate and metomidate were useful at reducing cortisol in summer flounder transport lasting up to 4 hours (Marcaccio and Specker 2004).

GreatBay then shipped juveniles to China (Figure 4.3), which began growing them in coastal ponds. We have found only one publication about this, reporting that fish grew from 8 to 750 g in one year (Zheng 2006). Jun Li reported at the Aquaculture 2007 conference that China produced about 76,000 tons of all flatfish (six species) in 2005, of which about half were turbot. Chinese scientists expect that summer flounder culture will thrive in colder water conditions in their country, whereas other species like southern flounder will do better in warmer water conditions.

The commercial operation in Mexico is in its infancy, as the water temperatures for flow-through operations at 14–25 degrees over much of the Baja California peninsula are conducive to both summer flounder and California halibut culture. With its location next to the large southern California market, production is expected to increase substantially. These operations will be principally land-based tank farms, both flow-through and recirculating.

Nutritional studies have shown that summer flounder juveniles require diets with relatively high protein levels, similar to other piscivorous flatfish. Daniels and Gallagher (2000) found that summer flounder fed diets with 56% protein had significantly better growth than those fed 52% protein or less. Those findings were confirmed by an unpublished study at the Universities of Rhode Island and Connecticut showing that 55% protein diets also yielded the highest growth. King (1999) found that weaning diets containing 56% or 54% protein provided significantly higher growth than did a diet containing 46% protein. Gaylord *et al.* (2003) varied lipid levels (8, 12, 16, or 20%) in a 55% protein diet and found no effect of lipid level on growth and no protein sparing effect of lipids. More recently, nutritional studies have focused on alternatives to fish meal as a protein source, for reasons of economics, availability, and environmental sustainability. Enterria (2006) found that replacement of fish meal by either soybean meal, corn gluten meal, or canola protein concentrate at levels of 20–50% did not yield equivalent growth to that of the fish meal control; however, the 40% soybean replacement reduced cost/kg of fish produced by 14%. In subsequent studies, 40% soybean replacement with added taurine and phytase provided equivalent growth as the fish meal control, although 70% replacement with those additions did not (Bengtson *et al.* 2008).

As mentioned above, FINE is a particularly problematic disease in summer flounder culture, but a number of other disease issues have been reported. Hughes and Smith (2002, 2004) reported parasites, such as *Trichodina* sp., *Amyloodinium* sp., and *Ichthyophonus hoferi*, and bacteria, such as *Vibrio anguillarum* and *Mycobacterium* sp., cause problems in summer flounder, but mentioned only that viruses have caused problems in other flounder species, not yet in summer flounder. Attempts are being made to counteract summer flounder diseases using a variety of strategies. Mowry *et al.* (2005) continuously dosed a recirculation system with hydrogen peroxide, but found that it did not improve water quality at the level tested. Gauger (2006) tried unsuccessfully to develop a vaccine against *V. harveyi* for prevention of FINE. Studies have also been conducted to determine the times for clearance of the antibiotics oxytetracycline (Chen *et al.* 2004), as well as Romet (sulfadimethoxine and ormetoprim) (Kosoff *et al.* 2007).

Very little summer flounder was ever harvested and processed in the United States, so there are no standard practices. The original market was sushi chefs, so fish were harvested and transported live to optimize quality and meet the chefs' demands. Zucker and Anderson (1998) surveyed wholesale buyers and sushi chefs to determine the factors important in their potential purchases of sushi/sashimi-grade summer flounder. They also (Zucker and Anderson 1999) developed a model to investigate the economic feasibility of summer flounder culture under a variety of production scenarios and stochastic market environments.

4.5 Summary

The culture of summer flounder is well established at the hatchery level, in terms of production techniques. Production of fish for market in the United States is currently not economically feasible at large scale, whereas production in China seems to be established, and production in Mexico is developing. Production costs need to be lowered (or market prices increased) for commercial production in the United States to be feasible. Improvements in genetics (selective breeding or all-female populations), production systems (energy efficiency), disease resistance (vaccination), and nutrition (cheaper plant-based feeds) are all ways in which production costs could be reduced. Market prices might be increased by increased demand for locally produced food, which is a major trend in the United States at the moment. Small-scale production of high-cost, high-quality summer flounder for local, specialty markets, or restaurants may enable a nascent industry to develop until production costs can be brought down. At that point, larger-scale production could begin to develop as improvements come on line.

Stock enhancement with summer flounder is really in the infant stages. Much knowledge might be applied from similar efforts with *P. olivaceus* in Japan, although it is unclear if American taxpayers would be willing to support the costs for stock enhancement for summer flounder (or any other species) in the absence of other funding mechanisms.

Literature cited

Able, K.W., and Kaiser, S.C. 1994. Synthesis of summer flounder habitat parameters. NOAA Coastal Ocean Program Decision Analysis Series No. 1. NOAA Coastal Ocean Office, Silver Spring, MD.

Alves, D., Specker, J.L., and Bengtson, D.A. 1999. Investigations into the causes of early larval mortality in cultured summer flounder (*Paralichthys dentatus* L.). *Aquaculture* 176:155–172.

Baker, E.P., Alves, D., and Bengtson, D.A. 1998. Effects of rotifer and *Artemia* fatty-acid enrichment on survival, growth and pigmentation of summer flounder *Paralichthys dentatus* larvae. *Journal of the World Aquaculture Society* 29:494–498.

Bengtson, D.A. 1999. Aquaculture of summer flounder (*Paralichthys dentatus*): status of knowledge, current research and future research priorities. *Aquaculture* 176:39–49.

Bengtson, D.A., Hossain, M.A., and Gleason, T.R.. 1999b. Consumption rates of summer flounder larvae on rotifer and *Artemia* prey during larval rearing. *North American Journal of Aquaculture* 61:243–245.

Bengtson, D.A., Lee, C.M., Slocum, M., Volson, B., Tolasa, S., and Lankin, K.F. 2008. Replacement of fish meal with plant proteins in diets for summer flounder *Paralichthys dentatus*. Abstracts of World Aquaculture 2008, Busan, Korea, May 19–23, 2008.

Bengtson, D.A., Lydon, L., and Ainley, J.D. 1999a. Green-water rearing and delayed weaning improve growth and survival of summer flounder, *Paralichthys dentatus*. *North American Journal of Aquaculture* 61:239–242.

Bengtson, D.A., and Nardi, G. 2000. Summer flounder (*Paralichthys dentatus*). In: Stickney, R.R. (ed.) *Encyclopedia of Aquaculture*. John Wiley & Sons, New York, pp. 907–913.

Bengtson, D.A., Simlick, T.L., Binette, E.W., Lovett, R.R. IV, Alves, D., Schreiber, A.M., and Specker, J.L. 2000. Survival of summer flounder (*Paralichthys dentatus*) larvae on formulated diets and failure of thyroid hormone treatment to improve performance. *Aquaculture Nutrition* 6:193–198.

Berlinsky, D.L., King, W., Hodson, R.G., and Sullivan, C.V. 1997. Hormone induced spawning of summer flounder *Paralichthys dentatus*. *Journal of the World Aquaculture Society* 28:79–86.

Bisbal, G.A., and Bengtson, D.A. 1991. Effect of dietary (n-3) HUFA enrichment on survival and growth of summer flounder, *Paralichthys dentatus*, larvae. In: Lavens, P., Sorgeloos, P., Jaspers, E., and Ollevier, F. (eds) *Larvi '91 – Fish and Crustacean Larviculture Symposium*. European Aquaculture Society, Ghent, pp. 56–57.

Bisbal, G.A., and Bengtson, D.A. 1995a. Description of the starving condition in summer flounder (*Paralicthys dentatus*) early life stages: morphometrics, histology and biochemistry. *Fishery Bulletin* 93:217–230.

Bisbal, G.A., and Bengtson, D.A. 1995b. Effects of delayed feeding on survival and growth of summer flounder, *Paralichthys dentatus*, larvae. *Marine Ecology Progress Series* 121:301–306.

Bisbal, G.A., and Bengtson, D.A. 1995c. Development of the digestive tract in larval summer flounder, *Paralichthys dentatus*. *Journal of Fish Biology* 47:277–291.

Blankenship, H.L., and Leber, K.M. 1996. A responsible approach to marine stock enhancement. In: Schramm, H.L., Jr., and Piper, R.G. (eds) *Uses and Effects of Cultured Fishes in Aquatic Ecosystems*. American Fisheries Society, Bethesda, MD, pp. 167–175.

Burke, J.S., Seikai, T., Tanaka, Y., and Tanaka, M. 1999. Experimental intensive culture of summer flounder, *Paralichthys dentatus*. *Aquaculture* 176:135–144.

Chen, C.Y., Getchell, R.G, Wooster, G.A., Craigmill, A.L., and Bowser, P.R. 2004. Oxytetracycline residues in four species of fish after 10-day oral dosing in feed. *Journal of Aquatic Animal Health* 16:208–219.

Collette, B.B., and Klein-MacPhee, G. (eds) 2004. *Bigelow and Schroeder's Fishes of the Gulf of Maine*, 3rd edn. Smithsonian Institution Press, Washington, DC.

Daniels, H.V., and Gallagher, M.L. 2000. Effect of dietary protein level on growth and blood parameters of summer flounder, *Paralichthys dentatus*. *Journal of Applied Aquaculture* 10:45–52.

Eddy, S.D., and Jones, S.H. 2002. Microbiology of summer flounder *Paralichthys dentatus* fingerling production at a marine fish hatchery. *Aquaculture* 211:9–28.

Enterria, A. 2006. Partial replacement of fish meal with plant protein sources in diets for summer flounder (*Paralichthys dentatus*). MS thesis, University of Rhode Island, Kingston.

Francis, A.W., Jr., and Bengtson, D.A. 1999. Partitioning of fish and diet selection as methods for the reduction of cannibalism in *Paralichthys dentatus* larviculture. *Journal of the World Aquaculture Society* 30:302–310.

Gauger, E., Smolowitz, R., Uhlinger, K., Casey, J., and Gomez-Chiarri, M. 2006. *Vibrio harveyi* and other bacterial pathogens in cultured summer flounder, *Paralichthys dentatus*. *Aquaculture* 260:10–20.

Gauger, E.J. 2006. Management of flounder infectious necrotizing enteritis (FINE) in cultured juvenile summer flounder. PhD diss., University of Rhode Island, Kingston.

Gavlik, S., Albino, M., and Specker, J.L. 2002. Metamorphosis in summer flounder: manipulation of thyroid status to synchronize settling behavior, growth and development. *Aquaculture* 203:359–373.

Gavlik, S., and Specker, J.L. 2004. Metamorphosis in summer flounder: manipulation of rearing salinity to synchronize settling behavior, growth and development. *Aquaculture* 240:543–559.

Gaylord, T.G., Schwarz, M.H., Davitt, G.M., Cool, R.W., Jahncke, M.L., and Craig, S.R. 2003. Dietary lipid utilization by juvenile summer flounder *Paralichthys dentatus*. *Journal of the World Aquaculture Society* 34:229–235.

Getchis, T.S. and Bengtson, D.A. 2006. Food consumption and absorption efficiency in newly settled summer flounder. *Aquaculture* 257:241–248.

Guan, J., Liu, X., Lan, C., Cai, W., Xu, Y., and Ma, S. 2007. Effects of temperature and salinity on embryo development and larva survival in crossbred F_1 of *Paralichthys olivaceus* (female) x *Paralichthys dentatus* (male). *Marine Fisheries Research/Haiyang Shuichan Yanjui* 28(3):31–37.

Huang, L., Schreiber, A.M., Soffientino, B., Bengtson, D.A., and Specker, J.L. 1998. Metamorphosis of summer flounder (*Paralichthys dentatus*): thyroid status and the timing of gastric gland formation. *Journal of Experimental Zoology* 280:413–420.

Hughes, K.P., and Smith, S.A. 2002. Clinical presentations of *Mycobacterium* sp. in summer flounder (*Paralichthys dentatus*) held in recirculating aquaculture systems. *Virginia Journal of Science* 53:58.

Hughes, K.P., and Smith, S.A. 2004. Common and emerging diseases in commercially cultured summer flounder, *Paralichthys dentatus*. *Journal of Applied Aquaculture* 14:163–178.

Johns, D.M., Howell, W.H., and Klein-MacPhee, G. 1981. Yolk utilization and growth to yolk-sac absorption in summer flounder (*Paralichthys dentatus*) larvae at constant and cyclic temperatures. *Marine Biology* 63:301–308.

Katersky, R.S., Peck, M.A., and Bengtson, D.A. 2006. Oxygen consumption of newly settled summer flounder, *Paralichthys dentatus*. *Aquaculture* 257:249–256.

Katersky, R.S., Schreiber, A.M., Specker, J.L., and Bengtson, D.A. 2008. Variance in growth and development rates in larval and metamorphosing summer flounder, *Paralichthys dentatus*. *Journal of Applied Ichthyology* 24:244–247.

Keefe, M.L., and Able, K.W. 1993. Patterns of metamorphosis in the summer flounder (*Paralichthys dentatus*). *Journal of Fish Biology* 42:713–728.

Kellison, G.T., and Eggleston, D.B. 2003. Coupling ecology and economy: modeling optimal release scenarios for summer flounder (*Paralichthys dentatus*) stock enhancement. *Fishery Bulletin* 102:78–93.

Kellison, G.T., Eggleston, D.B., and Burke, J.S. 2000. Comparative behaviour and survival of hatchery reared versus wild summer flounder (*Paralichthys dentatus*). *Canadian Journal of Fisheries and Aquatic Sciences* 57:1870–1877.

Kellison, G.T., Eggleston, D.B., Taylor, J.C., Burke, J.S., and Osborne, J.A. 2003. Pilot evaluation of summer flounder stock enhancement potential using experimental ecology. *Marine Ecology Progress Series* 250:263–278.

King, N.J. 1999. Fingerling production of summer flounder: Commercial-scale experiments studying hormonal manipulation of broodstock, larval stocking density, and weaning diet performance. MS thesis, University of New Hampshire, Durham.

King, N.J., Howell, W.H., Huber, M., and Bengtson, D.A. 2000. Effects of larval stocking density on laboratory-scale and commercial-scale production of summer flounder, *Paralichthys dentatus*. *Journal of the World Aquaculture Society* 31:436–445.

King, N.J., Nardi, G.C., and Jones, C.J. 2001. Sex-linked growth divergence of summer flounder from a commercial farm: are males worth the effort? *Journal of Applied Aquaculture* 11:77–88.

Klein-MacPhee, G. 1979. Growth, activity, and metabolism studies of summer flounder *Paralichthys dentatus* (L.) under laboratory conditions. PhD diss., University of Rhode Island, Kingston.

Klein-MacPhee, G. 1981. Effects of stocking density on survival of laboratory cultured summer flounder (*Paralichthys dentatus*) larvae. *Rapports et Procès-verbaux de la Réunion du Conseil pour l'Exploration de la Mer* 178:505–506.

Koelbl, M. 2000. Timing of feeding transition and individual consumption rates affect growth of summer flounder *Paralichthys dentatus* larvae. MS thesis, University of Rhode Island, Kingston.

Kosoff, R.E., Chen, C.Y., Wooster, G.A., Getchell, R.G., Clifford, A., Craigmill, A.L., and Bowser, P.R. 2007. Sulfadimethoxine and ormetoprim residues in three species of fish after oral dosing in feed. *Journal of Aquatic Animal Health* 19: 109–115.

Li, J., Yu, D., Xiao, Z., Liu, Q., Xu, S., and Ma, D. 2008. A rudimentary histological study on the ontogeny of the digestive tract in the hybrid flounder, jasum (*Paralichthys olivaceus* ♀ x *Paralichthys dentatus* ♂). Abstracts of World Aquaculture 2008, Busan, Korea, May 19–23, 2008.

Luckenbach, J.A., Godwin, J., Daniels, H.V., and Borski, R.J. 2002. Optimization of North American flounder culture: a controlled breeding scheme. *World Aquaculture* 33(1):40–45.

Marcaccio, N.D., and Specker, J.L. 2004. Stress in summer flounder: anesthesia mitigates transportation-induced stress response and increases post-transport performance. *Integrative and Comparative Biology* 43:928.

Martinez, G.M., and Bolker, J.A. 2003. Embryonic and larval staging of summer flounder (*Paralichthys dentatus*). *Journal of Morphology* 255:162–176.

Mowry, D.E., Schwarz, M.H., Hartman, K.H., Jahncke, M.L., and Smith, S.A. 2005. Efficacy of hydrogen peroxide in marine recirculating aquaculture systems holding summer flounder *Paralichthys dentatus*. *Journal of Applied Aquaculture* 17: 65–75.

Musche, J.M. 2003. Effects of larval rearing container size and juvenile weaning strategies on growth and survival of summer flounder, *Paralichthys dentatus*. MS thesis, University of Rhode Island, Kingston.

Schreiber, A.M., and Specker, J.L. 1998. Metamorphosis in the summer flounder (*Paralichthys dentatus*): stage-specific developmental response to altered thyroid status. *General and Comparative Endocrinology* 111:156–166.

Schreiber, A.M., and Specker, J.L. 1999a. Early larval development and metamorphosis in the summer flounder: changes in per cent whole-body water content and effects of altered thyroid status. *Journal of Fish Biology* 55:148–157.

Schreiber, A.M., and Specker, J.L. 1999b. Metamorphosis in the summer flounder *Paralichthys dentatus*: changes in gill mitochondria-rich cells. *Journal of Experimental Biology* 202:2475–2484.

Schreiber, A.M., and Specker, J.L. 2000. Metamorphosis in the summer flounder, *Paralichthys dentatus*: thyroidal status influences gill mitochondria-rich cells. *General and Comparative Endocrinology* 117:238–250.

Schwarz, M.H. 2003a. Flatfish research and production in the USA – status and perspectives. *Global Aquaculture Advocate* 6(1):73–74.

Schwarz, M.H. 2003b. A side-looped recirculation system for marine fish larval production. *Hatchery International* 4(1):27–29.

Schwarz, M.H., Jahncke, M., and Cool, R. 1998. Marine recirculating isolation/quarantine facility for summer flounder. Abstracts of the First Annual Northeast Aquaculture Conference and Exposition, p. 98.

Simlick, T.L., Katersky, R.S., Marcaccio, N., Hollis, J., and Bengtson, D.A. 2000. Post-metamorphic growth of summer flounder in laboratory culture: do early-settling larvae grow faster than late settlers? In: Flos, R., and Creswell, L. (eds) *Responsible Aquaculture in the New Millennium*. European Aquaculture Society, Oostende, Belgium, p. 651.

Smigielski, A.S. 1975. Hormone-induced spawnings of the summer flounder and rearing of the larvae in the laboratory. *Progressive Fish-Culturist* 37:3–8.

Soffientino, B., Gwaltney, T., Nelson, D.R., Specker, J.L., Mauel, M., and Gomez-Chiarri, M. 1999. Infectious necrotizing enteritis and mortality caused by *Vibrio carchariae* in summer flounder *Paralichthys dentatus* during intensive culture. *Diseases of Aquatic Organisms* 38:201–210.

Soffientino, B., and Specker, J.L. 2001. Metamorphosis of summer flounder, *Paralichthys dentatus*: cell proliferation and differentiation of the gastric mucosa and developmental effects of altered thyroid status. *Journal of Experimental Zoology* 290:31–40.

Soffientino, B., and Specker, J.L. 2003. Age-dependent changes in the response of the stomach to thyroid signaling in developmentally arrested larval summer flounder. *General and Comparative Endocrinology* 134:237–243.

Specker, J.L., and Chandlee, M.K. 2003. Methodology for estradiol treatment in marine larval and juvenile fish: uptake and clearance in summer flounder. *Aquaculture* 217:663–672.

Specker, J.L., Schreiber, A.M., McArdle, M.E., Poholek, A., Henderson, J., and Bengtson, D.A. 1999. Metamorphosis in summer flounder: effects of acclimation to low and high salinities. *Aquaculture* 176:145–154.

Stickney, R.R., and White, D.B. 1975. Ambicoloration in tank cultured flounder, *Paralichthys dentatus*. *Transactions of the American Fisheries Society* 104:158–160.

Sulikowski, J.A., and Howell, W.H. 2003. Changes in plasma cortisol, glucose, and selected blood properties in the summer flounder *Paralichthys dentatus* associated with sequential movement to three experimental conditions. *Journal of the World Aquaculture Society* 34:387–397.

Terceiro, M. 2006. Summer flounder assessment and biological reference point update for 2006. Atlantic States Marine Fisheries Commission, 64 pp. (available at http://www.asmfc.org).

Veillette, P.A., Merino, M., Marcaccio, M.D., Garcia, M.M., and Specker, J.L. 2006. Cortisol is necessary for seawater tolerance in larvae of a marine teleost, the summer flounder. *General and Comparative Endocrinology* 151:116–121.

Watanabe, W.O., and Carroll, P.M. 2001. Progress in controlled breeding of summer flounder, *Paralichthys dentatus*, and southern flounder, *P. lethostigma*. *Journal of Applied Aquaculture* 11:89–111.

Watanabe, W.O., Ellis, E.P., Ellis, S.C., and Feeley, M.W. 1998. Progress in controlled maturation and spawning of summer flounder *Paralichthys dentatus*. *Journal of the World Aquaculture Society* 29:393–404.

Watanabe, W.O., Ellis, S.C., Ellis, E.P., and Feeley, M.W. 1999. Temperature effects on eggs and yolk sac larvae of the summer flounder at different salinities. *North American Journal of Aquaculture* 61:267–277.

Watanabe, W.O., and Feeley, M.W. 2004. Light intensity effects on embryos, prolarvae, and first-feeding stage larvae of the summer flounder, *Paralichthys dentatus*. *Journal of Applied Aquaculture* 14:179–200.

because handling and anesthetizing are eliminated. This method also produces high-quality eggs.

Male flounder produce a very small volume of sperm compared to other fish. Spermiating males—especially when recently sourced from the wild—normally produce less than 0.5 mL of sperm when gentle pressure is applied to the abdomen. Hormone implants or injections or human chorionic gonadotrophin (HCG) have been ineffective in causing an increase in sperm volume or initiating spermiation (Watanabe and Carroll 2001). Proper environmental conditioning well in advance of the planned spawning date appears to be the most effective method for obtaining spermiating males among captive broodstock. Domesticated males are capable of producing 2–3 mL per individual for several consecutive days. It is best to remove the sperm from the males and place it into a small, dry container. The container should be kept on ice until the eggs are ready to be fertilized.

5.2.7 Broodstock diet and nutrition

Wild-caught broodstock, which are difficult to wean to pelleted diets, are fed thawed Atlantic silversides *Menidia menidia* or cigar minnows *Decapterus punctatus*. Vitamins are sometimes given to the fish by placing an over-the-counter multivitamin tablet into the mouth of the thawed silverside before feeding to the broodstock. Fish that have been raised in captivity should be fed a commercial pelleted feed containing a minimum of 55% protein and 12–15% fat. These commercial diets are already fortified with vitamins.

5.2.8 Biosecurity

Newly acquired broodfish should be quarantined until they are healthy and free of disease. Wild-caught broodfish often harbor fish lice (*Argulus* spp.), which can spread rapidly in a broodtank system. A recirculating aquaculture system is ideal for quarantine purposes because it allows control of temperature and photoperiod, while maintaining high water quality. The quarantine system is physically separated (room or building) from the hatchery to minimize opportunities for transfer of pathogens to hatchery stocks. Broodfish are maintained at a density of no more than 2–3/m^2 and under temperature and photoperiod conditions similar to their point of origin. After 3 days, formalin is added at 30 ppm while maintaining flow through conditions. Adult *Argulus* are picked off the fish using a pair of forceps, and then the fish are treated with $CuSO_4$ (Earth Tech stabilized copper; 5% ionic copper) at 0.3 ppm for 10 days. Fish are then treated with the antibiotic nitrofurazone (20 ppm) for 10 days. Salinity is lowered to $<$15 ppt for several days to kill monogenetic trematodes (Ellis and Watanabe 1993). When fish are healthy, usually after 30–45 days in quarantine, they are transferred to the controlled environment broodtank system.

5.3 Larviculture

Eggs hatch after a 55-hour incubation period at 17°C. After the hatch, water temperature is gradually raised to 21 or 22°C at a rate of 1°C per day (Daniels and Watanabe 2002). Recently hatched larvae do not possess fins, eyes, or mouths but develop them during the 5 days prior to first feeding. Larvae are stocked before first feeding at 10–15 per liters in 1–4 m^3 tanks supplied with 34 ppt seawater.

When southern flounder larvae were reared under controlled laboratory conditions under light intensities of 5, 50, 100, and 1,000 lx, extreme light intensity treatments (5 and 1,000 lx) appeared to exhibit osmoregulatory stress, as evidenced by a marked increase in whole body osmolality on d11ph, whereas mid-intensity treatments (50 and 100 lx) optimized growth and survival of larval southern flounder through d15 posthatching (Henne and Watanabe 2003) In another study (Moustakas *et al.* 2004), growth of southern flounder larvae increased with increasing photoperiod and was significantly greater at 24 L and 18 L than at 12 L or 6 L, with no effects on larval survival. Based on these results, larvae are reared under a photoperiod of at least 18 hours light and at a light intensity of approximately 100–300 lx at the water surface during the photophase to maximize feeding, growth, and survival (Henne and Watanabe 2003; Moustakas *et al.* 2004).

Laboratory studies comparing turbulence levels created by adjusting the rates of diffused aeration in rearing tanks, growth of southern flounder larvae was maximized under low turbulence levels, while survival was maximized at high turbulence levels. The data suggested that, in prefeeding and early feeding-stage larvae (which have weak swimming ability), higher turbulence levels improved buoyancy and prevented sinking. In feeding-stage larvae (which are relatively stronger swimmers), higher turbulence levels caused excessive swimming, osmoregulatory stress, and slower growth. Based on these results, turbulence levels are maintained relatively high during the embryo prefeeding (yolksac stage) and first-feeding stages to maintain buoyancy and survival and then decreased for mid- to late-feeding—and premetamorphic stage larvae to optimize prey encounters and feeding efficiency (Daniels and Watanabe 2002; Mangino and Watanabe 2007).

Most culturists prefer adding algae to the tank water at a density of 300,000–500,000 cells/L. The yolk is completely absorbed by first feeding (4–5 days posthatch), but the oil droplet will remain for several days. Rotifers are fed at about 20 per mL until about 15–20 days posthatching when they begin to eat *Artemia* nauplii (Table 5.1). *Artemia* are fed at 1 per mL initially, then up to 5 per mL through metamorphosis until the start of weaning at day 55 posthatch. Metamorphosis begins on around day 30 at 21°C and takes about two weeks to complete. The entire larviculture period takes place in full strength seawater (33 ppt) at temperatures between 17 and 21°C. Early stage larval southern flounder are not entirely euryhaline, showing reduced survival and markedly lower growth rates at 25 ppt compared with 34 ppt (Henne and Watanabe, 2003; Moustakas *et al.* 2004; Mangino and Watanabe 2007). Euryhalinity is not acquired until after metamorphosis as these fish transition from a pelagic to a benthic mode of existence, coincident with the recruitment to estuaries, which are the primary nursery grounds for these fish (Moustakas *et al.* 2004). Egg hatch is optimized at 17°C while larval survival and time to completion of metamorphosis are optimum at 21°C (Van Maaren and Daniels 2001).

Pond rearing of larvae also appears to be feasible (Jenkins and Smith 1999), but suitable harvest methods are still being developed.

Recently metamorphosed flounder can be weaned onto dry feeds by gradually reducing the *Artemia* ration from 5 per mL to 0 per mL over a 2–3 week period (Daniels and Hodson 1999). Weaning feeds should have particles that range from 250 to 500 microns and have an orange or reddish coloration—similar to the *Artemia*. Once they are successfully weaned (about 60 days posthatch), they are graded by size to reduce cannibalism and stocked into nursery tanks at 700 individuals/m^2. At this time, the fish are about 2.5 cm (1 inch) long and weigh about 0.25 g. Overall survival from egg through metamorphosis typically averages 40% (Daniels and Watanabe 2002).

5.4 Growout

5.4.1 Facilities

Many flatfish, such as turbot, halibut, and Japanese flounder, are cultured in outdoor land-based tanks or in indoor recirculating systems. Tank sizes and shape vary considerably but generally are 6.10–9.14 m (20–30 feet) in diameter with some manner of protection from the sun, such as black shade cloth. With water depth of only 0.45–0.91 m (1.5–3 feet) and constant water inflow, the entire tank volume is exchanged every 2–3 hours. Southern flounder culture has been done indoors in large (7–8 m diameter) tanks. If tanks cannot be housed indoors, a solid, opaque covering over the tank should be used. Flounder prefer low light intensities and can develop skin ulcerations when left in tanks exposed to direct sunlight. In cases where flounder are cultured outdoors, many managers use industrial shade cloth, similar to that used in plant nurseries, with a minimum rating of 80% (blocks out 80% of sunlight).

Indoor culture offers other advantages for recirculating systems. When nutrient-rich culture water is exposed to sunlight, it encourages the growth of algae. These algae will overgrow the bacteria in the biological filter and dramatically reduce the effectiveness of the filter. Therefore, if the flounder are being cultured outside, not only do the culture tanks need to be covered but the biological filter also must be completely shielded from sunlight. In our experience, two 40 W fluorescent lights placed above each tank provides sufficient light for the manager to see the fish but does not encourage algae growth or cause too much stress on the fish.

5.4.2 Environmental conditions: the first 3 months—25 mm fingerlings

Fingerling stocking densities should be about 700 fish /m^2 to give fish some room to move, to reduce the risk of cannibalism, and to encourage rapid

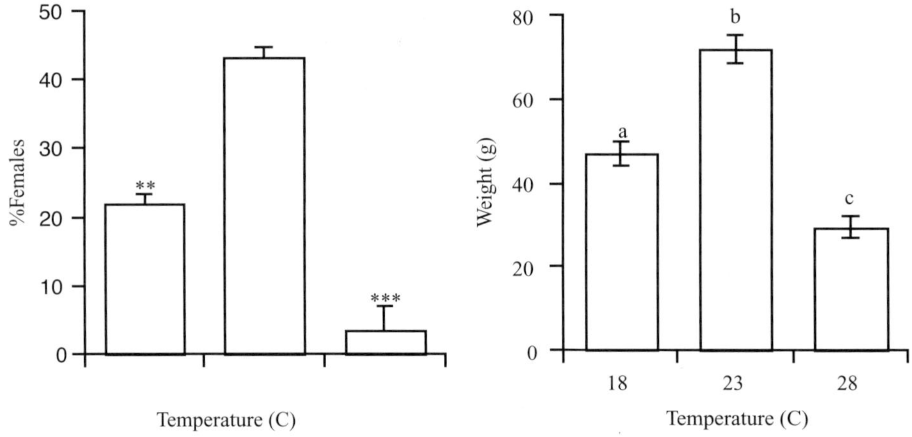

Figure 5.2 Effect of temperature on sex differentiation of southern flounder (Luckenbach *et al.* 2003).

growth (Daniels *et al.* 2006). Some cannibalism is inevitable but can be controlled through the use of lower stocking densities, grading by size, and frequent feeding. Southern flounder are cannibalistic and feed aggressively at the surface. This aggressive behavior leads to uneven growth rates that necessitate frequent grading by hand to optimize survival. Fingerlings may need to be graded 3–4 times during the few months it takes them to grow from 2 to 10 g. Suggest how to grade flounder? Larger fish do not require such frequent grading.

5.4.3 Temperature

It is imperative that the water temperature remains as close to 23°C (73°F) as possible during the first month or until the fish reach 75 mm in length. Flounder sex is determined by water temperature (Figure 5.2). Females grow approximately three times faster than males (Figure 5.3) (Luckenbach *et al.* 2003). As with other flatfish, the sex of the southern flounder is not determined until after metamorphosis (reviewed by Luckenbach *et al.* 2009). Although the precise time window has not been identified for southern flounder.

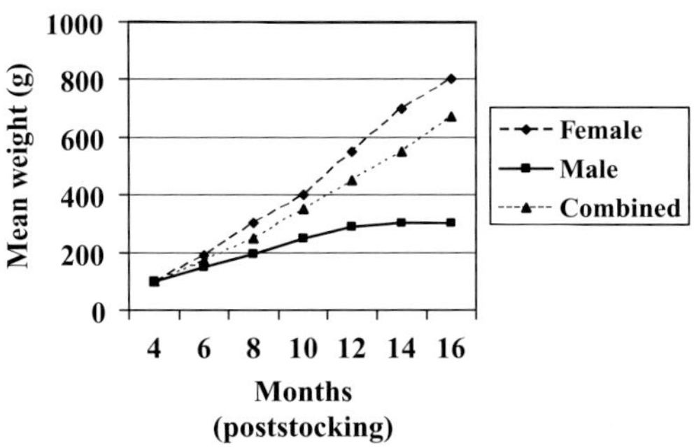

Figure 5.3 Differences in growth rate of female and male southern flounder during a production cycle. The Combined line represents the value derived from monthly sampling.

Optimum temperature to produce the highest percentage of females is approximately 23°C. Temperatures significantly higher or lower than 23°C (18°C or 28°C) likely will result in a higher percentage of males in the overall population. High stocking densities may also shift the population toward males. To achieve optimum growth and profitability of cultured flounder, hatcheries should produce all female flounder fingerlings. Fisheries managers considering the use of southern flounder for purposes of stock enhancement would likely prefer a population of fingerlings that more closely matched the sex ratios found in nature.

5.4.4 Salinity

While premetamorphic southern flounder larvae are stenohaline with impaired survival at reduced salinities of 24 g/L (Henne and Watanabe 2003; Moustakas *et al.* 2004), recently metamorphosed flounder are extremely tolerant of low salinity and can be raised in freshwater with high hardness and alkalinity (both greater than 200 ppm). Twelve-week studies have shown that juvenile southern flounder growth rates (3 g initial wt.) are not significantly different at 0 ppt versus 20 ppt up to a size of 60 g (Daniels and Borski 1999). Based on the conditions of their natural environment, it is likely that these particular flounder will tolerate water temperatures in excess of 28°C during growout. Van Maaren (1999) found the critical thermal maximum (CTM), lethal thermal tolerance (LT), and oxygen consumption rates (μg O_2/g fish/min) of juvenile southern flounder. Compared to the 14 species examined by Tsuchida (1995), the southern flounder has a relatively high thermal tolerance, similar to the sea bass (*Lateolabrax japonicus*) and the black seabream (*Acanthopagrus schlegeli*). The preferred temperature for juvenile southern flounder, as calculated by the combination of the thermal tolerance tests and the oxygen consumption rates tests, falls within the range of 25–29°C (Van Maaren 1999). Anecdotal evidence from commercial and recreational fishers indicates that southern flounder are routinely caught in shallow waters that exceed 30°C during the summer months. Tolerance to water temperatures below 4°C (38°F) can be improved by increasing salinity to full strength seawater (33 ppt).

Several studies have shown that juvenile southern flounder grow well at low salinities both in the hatchery and in natural water bodies (Lasswell *et al.* 1977; Daniels and Borski 1999). One study (Daniels *et al.* 2002, unpublished data) examined growth and survival of the same cohort of flounder grown at 0.5 ppt and 33 ppt (see Figure 5.4) and found that production was not affected by this difference in salinity.

5.4.5 Water quality

Water quality in recirculating systems is monitored daily. Table 5.2 shows the water parameters monitored for southern flounder in freshwater recirculating systems and provides optimum values for each parameter.

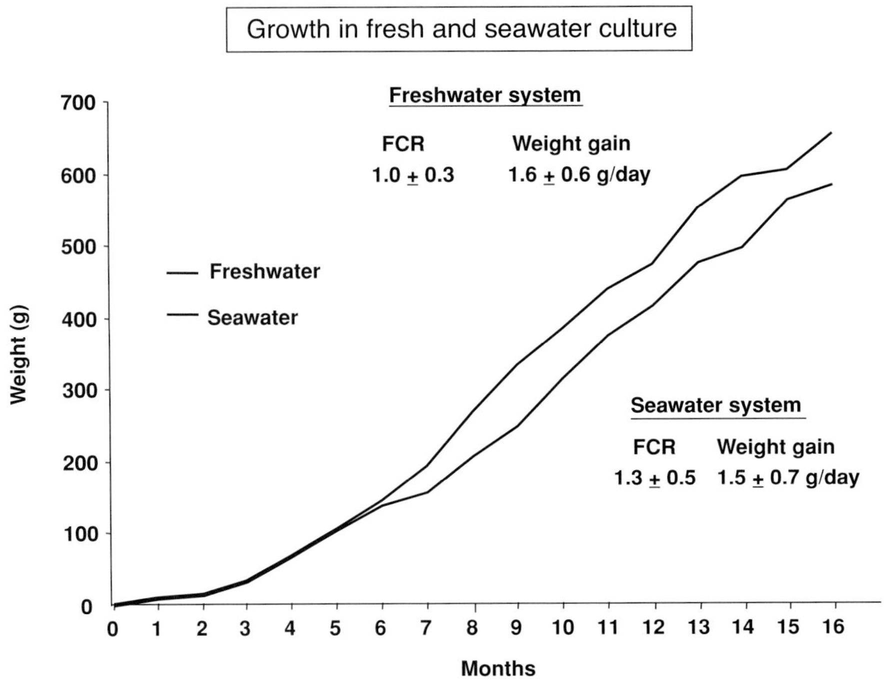

Figure 5.4 Comparison of growth rates for southern flounder grown in different salinities in water reuse systems. Fish were from the same cohort and split into the different systems at 3 months of age. Differences in growth and feed conversion were not significant at the end of the study.

5.4.6 Diets and nutrition

Juvenile flounder have a protein requirement of around 50–55%—similar to other flatfish (Gao *et al.* 2005; Gonzalez *et al.* 2005; Alam *et al.* 2009). Flounder diets typically contain 6–14% lipid and <17% ash. Alam *et al.* (2009) suggested

Table 5.2 Water quality and system maintenance throughout the growout cycle

Parameter	Optimum value
Physical	
Water depth	18–28 inches (45–70 cm)
Water exchange rate	100% every 3–4 hours (7–10 times per day)
Water temperature	Above 20°C and below 28°C best between 23 and 25°C
Chemical	
Ammonia	<1.0
Nitrite	<20
Alkalinity	200 ppm
Hardness (calcium)	200 ppm
pH	6.8–7.5
CO_2	<10 ppm
DO	>6.8 ppm (80% saturation)
Salinity	Start at 20 ppt, then gradually reduce to 0.5 ppt within first month[a]

[a]The addition of sodium bicarbonate and calcium chloride to the system water will maintain salinities at 0.5 ppt. This is the lowest the system will go and still maintain the proper balance of ions.

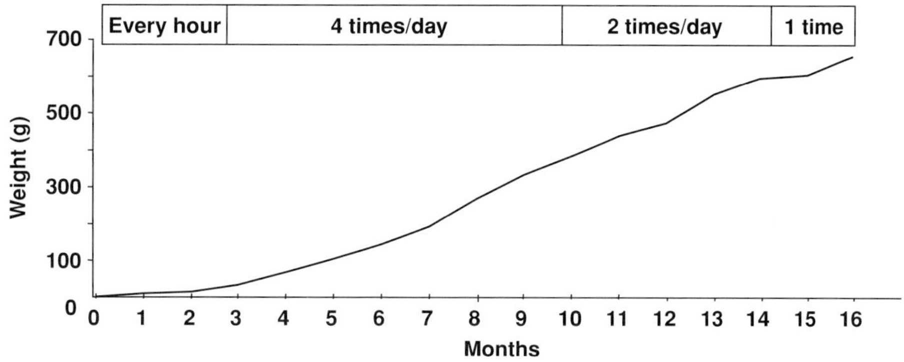

Figure 5.5 Frequency of feeding based on age of southern flounder during a production cycle.

that a combination of 50% dietary protein and 10% lipid was optimal for juvenile southern flounder fed a fishmeal-based diet reared in a recirculating system. Fingerlings are fed 3–4 times per day while larger fish are fed only 1–2 times per day (Figure 5.5).

Feeding rates should be about 6–10% of body weight during the juvenile stage (3–10 cm), 2% at 20 cm, 1.5% at sizes larger than 30 cm, and less than 0.5% during the winter when water temperatures fall below 12°C (Table 5.3).

Although growout feeds will likely be expensive because of the high price of protein, flounder are very efficient feed converters with conversion values below 1.5:1 (Gao *et al.* 2005; Alam *et al.* 2009). Therefore, the high price of the feed will be offset by the lower amount needed to produce each pound of fish.

5.4.7 Feeding

Feeding is one of the more time-consuming and satisfying responsibilities for the growout manager. This is the opportunity to observe feeding activity and get an idea of the health of the population, see differences in size of fish, and how the fish are growing.

Frequency of feeding varies depending on the age of the fish; Figure 5.4 shown previously shows how daily feeding frequency declines as fish age. It is important to feed fish to satiation (fullness) at every opportunity.

Table 5.3 Daily feeding rate for flounder of different sizes

Fish size		Daily feeding rate (%)[a]
Length (cm)	Weight (g)	Dry pellet
2.5–5	0.25–2	15
10–15	8–32	5
20–25	90–180	2–4
>25	>150	1

[a]Weight of feed/weight of fish.

Table 5.4 Size of feed used during production of southern flounder based on age poststocking

Feed pellet size (mm)	Fish age (months poststocking)
0.1–1.5 (crumble)	0–1
2 mm	0.5–2.5
3	1.5–3.5
5	2.5–5
7	4.5–6.5
9	6–9
13	8–remaining
16	11–remaining

Either extruded (floating) feed or sinking fee can be used. Pellet size depends on fish size (Table 5.4).

5.4.8 Stocking and splitting final growout: 7.5 cm to market size

Management activities during the majority of the growout stage are focused on the need to regularly split the tank populations to maintain optimum growth rates (Figure 5.6).

Survival through this first stage should exceed 85% and will be the largest single drop compared to the rest of the growout cycle. After this stage, mortality

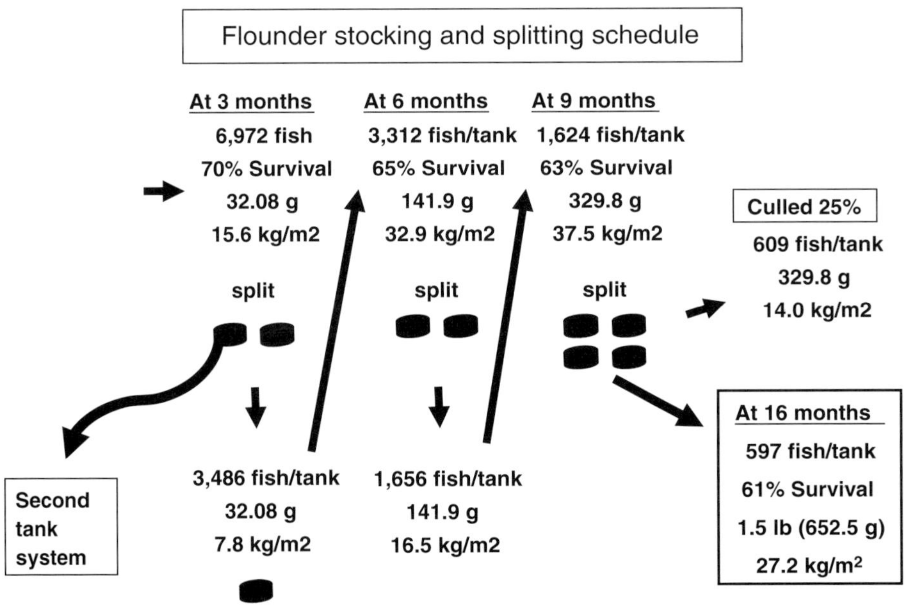

Figure 5.6 Schematic showing a typical stocking schedule and splitting of tank populations to optimize tank use and growth.

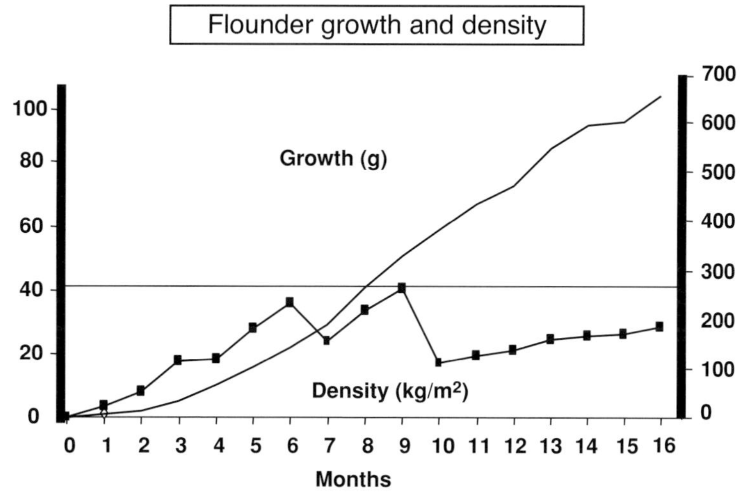

Figure 5.7 Growth of southern flounder during a production cycle in a water reuse system. Regular splitting of the tank populations was used to maintain densities below 40 kg/m^2.

should be minimal during the rest of the growout cycle until the fish reach market size. Tanks are stocked to obtain a final density of 15–25 kg/m^2 of 0.6–1.0 kg fish. A graphic depiction of the effect of splitting on tank densities is shown in Figure 5.7.

5.5 Diseases

Although no specific diseases have been reported for southern flounder, it is likely that they will be susceptible to the full range of bacterial infections (*Aeromonas*, *Staphylococcus*, Vibriosis, etc.) that infect other cultured flounders (Noga 2010, this volume). *Edwardsiella tarda* is a specific and persistent pathogen of Japanese flounder. Viruses such as epidermal hyperplasia (herpes virus) and nervous necrosis (striped jack nervous necrosis virus) also are found in cultured Japanese flounder and may cause similar problems in cultured southern flounder. Wild broodstock frequently import a host of parasites into the hatchery and should be quarantined and treated for at least several weeks prior to their introduction into the main water system. External parasites such as *Argulus* (sea lice) and leeches have been reported and intestinal worms also are common in wild broodstock. Another parasite of concern is Marine Ich. This parasite is quite aggressive and can kill a large number of fish rapidly. As the name implies, it is mostly found in saltwater with salinities greater than 3 ppt. The most cost-effective treatment for Marine Ich is to lower the salinity of the water below 3 ppt—a good management option with southern flounder.

5.6 Marketing

Flatfish have an established market worldwide. Southern flounder are nearly identical in appearance to Japanese flounder and summer flounder and therefore would be sold to the same markets. Southern flounder and summer flounder are not reported separately in fisheries statistics, so the total U.S. southern flounder catch is not known. The commercial catch of flounder in general has declined from 20,000 MT in the mid-1980s to a mandated quota of approximately 5,000 MT since 1998. Most flounder are sold processed into fillets, but a substantial amount, as much as one-third of the volume, are now marketed whole (bled), fresh killed for distribution in the Asian, principally Japanese and Korean markets. High-quality fish bled on ice are sold at \$8.00–12.00/kg for 0.5–1.0 kg fish, \$16.00/kg for 1–2 kg fish, and \$20.00–25.00/kg for >2 kg fish.

Because of the wide distribution of this market, many different market niches exist. Each niche requires a different size of fish or presentation (i.e., whole on ice, filleted, live, etc.). Because each market is unique it is not possible to declare one size of flounder as "market size". Therefore, the standard measure of "time to market" depends greatly on the targeted markets.

Likely, the best marketing plan for a growout facility would include a variety of markets that require different sizes of fish. This would take advantage of the different growth rates of the fish, allowing the producer to sell fish that are culled from the system, and smooth out the cash flow of the overall operation.

Flounder grown in fresh water will need to be purged prior to sale. Freshwater systems have the tendency to impart off-flavors to fish that are easily detected by the consumer (Drake *et al.* 2006). These flavors are not present in saltwater systems. These flavors are easily eliminated by purging the fish for 2–3 days in saltwater (15 ppt salinity or greater). After this time, the fish taste identical to those that were cultured in saltwater (Drake *et al.*, unpublished data).

5.7 Hatchery economics

Breakeven costs for southern flounder fingerlings (2.5 cm length) are highly dependent on survival and the number of production runs in the hatchery. The economics of a closed recirculating system used for southern flounder culture showed a breakeven cost of approximately \$0.34 per fish with a single annual production run and 40% survival from egg (Daniels *et al.* 2006). Increased egg availability throughout the year would lead to more frequent production cycles and could lower the breakeven cost to \$0.25 per fish. Currently, there is no established market for southern flounder fingerlings given the few growout facilities in production.

5.8 Production economics

To date, research has focused on the biological, environmental, and engineering aspects of southern flounder culture, but relatively little published information is available on the production economics and profitability of summer flounder aquaculture using a recirculating growout system. The economics of southern flounder growout using has been investigated using computer spreadsheet simulation models (Dumas 2006). The analysis incorporates biological and engineering information from recent near-commercial scale RAS field trials conducted at NCSU and UNCW Watanabe 2005, unpublished data). The model is used to evaluate the production economics and profitability of a commercial scale southern flounder recirculating aquaculture system.

As an example, we consider a simulated, scaled-up, RAS facility based on the NCSU and UNCW facilities. The simulated facility is assumed to occupy 3.5 acres (1.40 ha) of inland, rural land with a land value of $28,000 ($8,000 per acre, $20,000 per ha). It assumes that the facility owner currently owns the land and manages and operates the system with a staff of three technical assistants. The facility consists of a four, insulated, sheet metal buildings, three of which are 33,600 ft^2, with the fourth building 16,800 ft^2.

A recirculating aquaculture system (RAS) is constructed inside each building. A total of fifty-six 27 foot-diameter, steel, vinyl-lined tanks hold fish and water to a depth of 4 feet (17,333 gallons/tank @ 4 ft depth). These tanks are arrayed inside the buildings in two-tank subsystems supported by state-of-the-art RAS components, including particle trap and swirl separators, drum screen filters, trickling biological filters, UV sterilizers, heat pumps, protein skimmers, and oxygen cones. Fish wastes removed from the tank water by the biological filtration components are processed into compost material on site using a GeoTube sludge bag system (Losordo 2006). It is assumed that these capital costs are financed using a combination of the owner's personal funds (50%) and bank funding (50%). Bank funds are obtained via a 10-year construction term loan at an interest rate of prime plus 0.50%, assumed to be 8.75% (the rate circa November 2006 in eastern North Carolina, United States).

The facility purchases all female southern flounder fingerlings at an average size of 2,250 fingerlings/lb. (0.2 g/fingerling) from a separate, specialized flounder hatchery firm at a price of $1.25 per fingerling. Based on field trials at NCSU and using optimal growth rates obtainable with all-female populations, it is assumed that the fingerlings grow to a harvestable size of 1.5 lbs (682 g) in 14 months (420 days). Each production cycle begins with 90,000 fingerlings weighing a total of 39.60 lbs split between two RAS tanks. As the fish grow, the population is divided among more and more tanks to maintain a stocking density of approximately 40 kg/m^2 of tank bottom. It takes 1 year for the facility to "ramp-up" production to steady-state conditions. Based on field trials, it is assumed that 20% fingerling mortality occurs in the first month after stocking. In addition, 25% of the slowest-growing fish are culled at the end of month 9 (day 270). Fish are fed a commercial pelleted diet ($0.66/kg; $0.30/lb), and feed conversion ratio is 1:0. At the end of the 14-month growth cycle, 83,160 pounds of marketable-sized (1.5 lb/fish) fish are harvested from the first cohort. Each succeeding cohort produces an additional 83,160 pounds ready for harvest every 2 months thereafter. We assume that niche market buyers purchase the fish live at the farm gate for $5.00/lb (The buyer pays the cost of live hauling the fish by truck to market.), based on the average selling price received by a flounder aquaculture growout facility in North Carolina that grows ocean caught flounder to market size.

Farm revenue in the first year of operation is zero, as no fish are harvested. Farm revenue in the second and each subsequent year is $2,423,520, based on six harvested cohorts per year. Interest costs during construction (preproduction) are $67,117. Total costs in the first year of production are $1,365,674. Total costs in the second and each subsequent year are $1,706,233. Net returns (Total Revenue minus Total Costs) to the facility owner (before taxes) are negative in the first year of production but positive in the second and each subsequent year. We assume a project life of 10 years, based on a useful life of 10 years for the facility buildings and equipment. If we assume a financial discount rate of approximately 7% per year, the net present value (NPV) of the stream of net returns over 10 years is approximately $3,024,112 in present year (2006) dollars. The internal rate of return (IRR) of the project is 38%. The discounted payback period (DPP) is 38 months. The breakeven price per pound is $3.98 over the 10-year project. Debt service, fingerling cost, and labor account for 50.0%, 21.2%, and 9.1%, respectively, of total annual costs.

5.9 Summary: industry constraints and future expectations

5.9.1 Genetics of culture versus enhancement

Production of foodfish will require fingerling populations with a high proportion of females to optimize growout economics. Broodstock breeding programs will likely focus on using gynogenetic broodstock to produce all-female fingerlings (Morgan *et al.* 2006). These fingerlings would likely not be suitable for stock enhancement purposes where fisheries stock managers would likely require a mixed-sex population of fish that would closely match the genetic structure of the native stocks.

5.10 Conclusions

Southern flounder culture is still in its very early stages of development as an industry in the United States. Fundamental information about growout and economics has been generated from research-scale and demonstration studies but commercial-scale data are lacking. Hatchery production of southern flounder fingerlings has received the majority of attention from researchers and significant strides have been made in this field. Further research is needed to develop methods for the mass production of monosex female fingerlings since females grow three times faster than males. Production of all-female fingerlings will require highly specialized hatcheries with captive broodstock and temperature-controlled environments during the sex determination stage. In addition to sex control, studies are needed to develop improved cultivars for commercial farming through selective breeding. Considering their bottom dwelling behavior, methods to improve stocking and production densities per unit of tank volume will be important toward reducing production costs. Environmental and nutritional requirements for growout have not yet been well defined, but growout systems will likely be similar to those already employed in the culture of other flatfish such as summer flounder and Japanese flounder. Work is needed to develop diets that substitute alternative protein sources to fishmeal in southern flounder diets.

Literature cited

Alam, M.S., Watanabe, W.O., and Daniels, H.V. 2009. Effect of different dietary protein and lipid levels on growth performance and body composition of juvenile southern flounder (*Paralichthys lethostigma*) reared in a recirculating aquaculture system. *Journal of the World Aquaculture Society* 40(4):513–521.

Berlinsky, D.L., King, W.V, Smith, T.I.J., Hamilton, R.D., II, Holloway, J., Jr., and Sullivan, C.V. 1996. Induced ovulation of southern flounder *Paralichthys lethostigma* using gonadotropin releasing hormone analogue implants. *Journal of the World Aquaculture Society* 27:143–152.

Blandon, I.R., Ward, R., King, T.L., Karel, W.J., and Monaghan, J.P., Jr. 2001. Preliminary genetic population structure of southern flounder, *Paralichthys lethostigma*, along the Atlantic Coast and Gulf of Mexico. *Fisheries Bulletin* 99:671–678.

Burke, J.S., Miller, J.M., and Hoss, D.E. 1991. Immigration and settlement pattern of *Paralichthys dentatus* and *P. Lethostigma* in an estuarine nursery ground, North Carolina, USA. *Netherlands Journal of Sea Research* 27:393–405.

Daniels, H.V., and Borski, R. 1999. Effects of low salinity on growth and survival of southern flounder (*Paralichthys lethostigma*) larvae and juveniles. UJNR Technical Report. New Hampshire Sea Grant, Durham, NH.

Daniels, H.V., and Hodson, R.G. 1999. Weaning success of southern flounder juveniles: effect of changeover period and diet type on growth and survival. *North American Journal of Aquaculture* 61:47–50.

Daniels, H.V., Murashige, R., and Shewmon, L.N. 2006. Intensive culture of southern flounder in a closed recirculating system. In: *Proceedings of the 5th Conference on Recirculating Aquaculture*. Virginia Polytechnic and State University, Blacksburg, VA, pp. 88–95.

Daniels, H.V., and Watanabe, W.O. 2002. *Production of Southern Flounder Fingerlings*. North Carolina Sea Grant Publication no. UNC-SG-02–08. Raleigh, NC.

Drake, S.L., Drake, M.A., Daniels, H.V., and Yates, M.D. 2006. Sensory properties of wild and aquacultured southern flounder (*Paralichthys lethostigma*). *Journal of Sensory Studies* 21:218–227.

Dumas, C. 2006. Southern flounder growout manual. In: *Southern Flounder Aquaculture Workshop*, March 16–17, North Carolina State University, Raleigh, NC, pp. 85–109.

Ellis, E.P., and Watanabe, W.O. 1993. The effects of hyposalinity on eggs, juveniles and adults of the marine monogenean, *Neobenedenia melleni*. Treatment of ectoparasitosis in seawater cultured tilapia. *Aquaculture* 117:15–27.

Gao, Y., LV, J., and Lin, Q. 2005. Effects of protein levels on growth, feed utilization, nitrogen and energy budget in juvenile southern flounder, *Paralichthys lethostigma*. *Aquaculture Nutrition* 11:427–433.

Gonzalez, S., Craig, S.R., McLean, E., Schwarz, M.H., and Flick, G. J. 2005. Dietary protein requirement for southern flounder, *Paralichthys lethostigma*. *Journal of Applied Aquaculture* 17(3): 37–50.

Henne, J.P., and Watanabe, W.O. 2003. Effects of light intensity and salinity on growth, survival and osmoregulatory ability of southern flounder larvae *Paralichthys lethostigma*. *Journal of the World Aquaculture Society* 34:450–465.

Jenkins, W.E., and Smith, T.I.J. 1999. Pond nursery production of southern flounder (*Paralichthys lethostigma*) and weaning to commercial diets. *Aquaculture* 176:173–180.

Lasswell, J.L., Garza, G., and Bailey, W.H. 1977. Status of marine fish introductions into the freshwaters of Texas. Texas Parks and Wildlife Report PWD 3000–35, Austin, TX.

Losordo, T. 2006. Southern flounder growout manual. In: *Southern Flounder Aquaculture Workshop*, March 16–17, North Carolina State University, Raleigh, NC, pp. 85–109.

Luckenbach J.A., Godwin, J., Daniels, H.V., and Borski, R.J. 2003. Gonadal differentiation and effects of temperature on sex determination in southern flounder (*Paralichthys lethostigma*). *Aquaculture* 216:315–327.

Luckenbach, J.A., Borski, R.J., Daniels, H.V., Godwin, J. 2009. Sex determination in flatfishes: mechanisms and environmental influences. *Seminars in Cell and Developmental Biology* 20(3):256–263.

Mangino, A., and Watanabe, W.O. 2007. Combined effects of turbulence and salinity on growth, survival and whole-body osmolality of larval southern flounder. *Journal World Aquaculture Society* 37:407–418.

Morgan, A.J., Murashige, R., Woolridge, C.A., Luckenbach, J.A., Watanabe, W.O., Borski, R.J., and Daniels, H.V. 2006. Effective UV dose and pressure shock for induction of meiotic gynogenesis in southern flounder (*Paralichthys lethostigma*) using black sea bass (*Centropristis striata*) sperm. *Aquaculture* 259:290–299.

Moustakas, C. Th., Watanabe, W.O., and Copeland, K.A. 2004. Combined effects of photoperiod and salinity on growth, survival, and osmoregulatory ability of southern flounder *Paralichthys lethostigma. Aquaculture Journal* 229:159–179.

Reagan, R.E., Jr., and Wingo, W.M. 1985. Species profiles: life histories and environmental requirements of coastal fishes and invertebrates (Gulf of Mexico) – southern flounder. U.S. Fish and Wildlife Service Biological Report 82(11.30). U.S. Army Corps of Engineers TR EL-82–4.9.

Smith, T.I.J., McVey, D.C., Jenkins, W.E., Denson, M.R., Hewyard, L.D., Sullivan, C.V., and Berlinsky, D.L. 1999. Broodstock management and spawning of southern flounder, *Paralichthyus lethostigma. Aquaculture* 176:87–99.

Stokes, G.M. 1977. *Life History Studies of Southern Flounder* (*Paralichthys lethostigma*) *and Gulf Flounder (P. albigutta) in the Aransas Bay Area of Texas*. Texas Parks and Wildlife Department Technical Series 25.

Tsuchida, S. 1995. The relationship between upper temperature tolerance and final preferendum of Japanese marine fish. *Journal of Thermal Biology* 20:35–40.

Van Maaren, C.C. 1999. The effects of different temperatures on larval southern flounder *Paralichthys lethostigma* survival, growth and development and on juvenile southern flounder thermal tolerance and oxygen consumption. MSc thesis. North Carolina State University, Raleigh, NC.

Van Maaren, C.C., and Daniels, H.V. 2001. Effects of temperature on egg hatch, larval growth and metamorphosis for hatchery-reared southern flounder *Paralichthys lethostigma. Journal Applied Aquaculture* 11:21–33.

Watanabe, W.O., and Carroll, P.M. 2001. Progress in controlled breeding of summer flounder, *Paralichthys dentatus*, and southern flounder, *P. Lethostigma*. 2001. *Journal of Applied Aquaculture* 11:89–111.

Watanabe, W.O., Carroll, P.M., and Daniels, H.V. 2001. Sustained natural spawning of southern flounder *Paralichthys lethostigma* under an extended photothermal regime. *Journal of the World Aquaculture Society* 32:153–166.

Watanabe, W.O., Woolridge, C.A., and Daniels, H.V. 2006. Progress toward year-round spawning of southern flounder broodstock by manipulation of temperature and photoperiod. *Journal of the World Aquaculture Society* 37:256–272.

Chapter 6
Culture of winter flounder

Elizabeth A. Fairchild

The winter flounder, *Pseudopleuronectes americanus*, is a commercially and recreationally important right-eyed flatfish found along the northwestern Atlantic coast, ranging from Georgia, United States, to Labrador, Canada (Scott and Scott 1988; Figure 6.1). It is a long-living flatfish and can reach up to a maximum age and length of 15 years and 65 cm, respectively (Bigelow and Schroeder 1953; Fields 1988). Maximum weight can be as much as 3 kg, resulting in the thickest fillets of all native New England flatfish (Bigelow and Schroeder 1953; Lux 1973).

Winter flounder are federally managed in the United States as three separate stocks, and in Canada as three separate divisions. As with most groundfish species, catches have declined precipitously in recent years in both the countries. The goal of the current management program is to reduce fishing mortality to levels that will allow stocks to rebuild above minimum biomass thresholds, and then remain at or near target biomass levels. While it is hoped that these more stringent fisheries regulations will allow winter flounder populations to rebuild to historic levels, recovery may take decades. Culturing winter flounder for market and rearing juvenile fish for stock enhancement are viable options for reducing fishing pressure and rebuilding this species (Waters 1996; Litvak 1999; Howell and Litvak 2000), and both have been the focus of research for well over 100 years. As early as 1890, wild broodstock were captured in fyke nets near Woods Hole, MA, and spawned in government hatcheries (Bean 1890). Many of the basic techniques for culturing winter flounder have been developed since then, though on a small, experimental scale. The reader is referred to Litvak (1999) and Howell and Litvak (2000) for a more thorough review of winter flounder aquaculture techniques developed before 1999.

6.1 Life history and biology

The biology and ecology of winter flounder have been well studied (see reviews by Klein-MacPhee 1978; Able and Fahay 1998; Pereira 1999; Collette and

Figure 6.1 Adult winter flounder *Pseudopleuronectes americanus*.

Klein-MacPhee 2002). Habitat characteristics of winter flounder vary among life stages and affect their distributions. Winter flounder reside in depths from the shallow subtidal to 37 m (McCracken 1963). Winter flounder are eurythermal, euryhaline, and freeze resistant (Pearcy 1962; Duman and DeVries 1974). Growth rates are highly correlated to temperature. In the wild, fish reach maturity at 26–30 cm, depending on location (O'Brien *et al.* 1993); maturity appears to be a function of size rather than age (O'Brien *et al.* 1993). The peak spawning season for winter flounder varies by latitude, but generally lasts 2–3 months in the winter and spring (see DeCelles and Cadrin 2007). Southern populations may initiate spawning as early as December (Lobell 1939), while peak spawning in Newfoundland does not occur until May (Kennedy and Steele 1971). Except for the Georges Bank (GB) stock which spawns offshore, reproductively isolated adult populations typically undergo onshore migrations into specific estuaries or coastal embayments where spawning occurs (Lobell 1939; Perlmuter 1947; Saila 1961) and the young-of-the-year (YOY) remain for their first two years before moving offshore (Pereira *et al.* 1999).

6.2 Broodstock husbandry

Wild adult fish are readily available, and so, most researchers capture, spawn, and release their broodstock each year. Broodstock are captured easily by otter trawl (Smigielski 1975; Stoner *et al.* 1999; Butts and Litvak 2007a, 2007b Fairchild *et al.* 2007), fyke nets, fish traps (Ben Khemis *et al.* 2000; Plante *et al.* 2002; Plante *et al.* 2003; Mercier *et al.* 2004), and SCUBA (Burton and Idler 1987; Shangguan and Crim 1999). Ripe females are easily recognizable by their distended ovaries, which encompass the entire ventral side of the fish during spawning season. Spermiating males can be identified by the extrusion of milt when gentle pressure is applied to the abdomen.

Because winter flounder are divided into distinct localized populations associated with the specific bays and estuaries along the northwestern Atlantic (Klein-MacPhee 1978), broodstock origin affects zygote characteristics and development. Fish from anthropogenically contaminated areas produced eggs with a high incidence of abnormalities (Perry *et al.* 1991). In addition, the stock origin of parental fish is important since there are latitudinal differences in growth rates (Buckley *et al.* 1991a; Butts and Litvak 2007a, 2007b). For instance, larvae sired by Georges Bank males developed faster and were significantly larger during both early and late larval development than those sired by Passamaquoddy Bay, NB, males (Butts and Litvak 2007a, 2007b). It is paramount that broodstock are healthy as egg quality and survival to metamorphosis are correlated to both paternal (Butts and Litvak 2007a) and maternal health (Buckley *et al.* 1991b). Both maternal size (Chambers and Leggett 1996) and paternal sperm volume (Butts and Litvak 2007b) are positively correlated to egg size and fertilization success, respectively.

Though cultured flatfish are not immune to diseases caused and/or exacerbated by bacteria, viruses, nutrition, and environmental factors (Mulcahy 2002), few immunological studies have been conducted with winter flounder. One detected disease is nodavirus (Nervous Necrosis Virus) which was found (albeit infrequently) in wild winter flounder from Passamaquoddy Bay, NB, Canada (Barker *et al.* 2002b). Nodavirus could pose a potential threat to the successful development of a commercial winter flounder aquaculture industry since an outbreak would be pandemic and cause high mortality, especially in cultured larvae and juveniles (Gagne *et al.* 2004). Care must be taken to prevent bringing contaminated broodstock into the hatchery. In order to develop appropriate biosecurity measures, a nonlethal method for screening broodstock for nodavirus should be developed (Gagne *et al.* 2004).

Wild winter flounder are known to carry ectoparasites such as *Gyrodactylus* spp. and *Trichodina* spp. which can easily spread to other fish in a crowded, hatchery environment; these ciliates can be removed by treating fish in a formalin bath (1:4,000) for 1 hour, once/week for 3 weeks (Barker *et al.* 2002a). To prevent infections, topical antibiotics like 1% methylene blue solution (Plante *et al.* 2002, 2003) or 10% iodine solution (Fairchild, unpublished data) are applied externally to small wounds that probably occurred during capture of broodstock.

To date, no one has domesticated winter flounder broodstock. This is imperative for the success of a commercial-scale operation. Through domestication, fish can be selected for fast growth and disease resistance, among other desirable characteristics.

6.2.1 Broodstock system design and requirements

Determining optimal broodstock requirements for winter flounder is the focus of several research laboratories. Because manipulating photoperiod and temperature do not appear to strongly affect gametogenesis in adult winter flounder (Duchemin *et al.* 2004), typically adults are kept at ambient temperatures and provided with a natural photoperiod (Plante *et al.* 2002, 2003). This is the case with the broodstock at the Institut des Sciences de la Mer de Rimouski (ISMER) in Québec, where fish are kept in tanks at a density of 5–6 kg/m^2 (or the equivalent of 20 fish/m^2) in 1 m^2 rectangular tanks, supplied with 29 ppt saline water at a flow of 5–10 L/min (Plante *et al.* 2002, 2003). At the University of New Brunswick Saint John (UNBSJ), broodstock are kept in a 2,700-L closed recirculation system in which temperature and salinity are maintained at 2–4°C and 28–30 ppt, respectively (Butts and Litvak 2007a, 2007b). Winter flounder can withstand such cold water temperatures since they synthesize antifreeze proteins and secrete them into their blood during the winter (Duman and DeVries 1974; Fletcher *et al.* 1985). At UNBSJ, approximately 15% of the water is replaced daily, and photoperiod mimics natural conditions during the spawning season at a light intensity of 10 lx (Butts and Litvak 2007a, 2007b).

Plante *et al.* (2002) examined the effects of two salinities on the survival, growth, and stress of wild winter flounder broodstock. Fish were acclimated to either a brackish salinity (15 ppt) or seawater salinity (29 ppt), and reared for 5 months. There were no differences in growth, somatic condition, hepatosomatic index, and gonadosomatic index of fish reared in different salinities. However, fish in the higher salinity treatment had higher mortality and stress indicators (plasma cortisol and plasma osmolality) than fish in the brackish salinity treatment, rendering these fish more vulnerable to opportunistic infections. Adults were captured inshore. This is interesting to note since most winter flounder stocks spawn inshore in estuaries and coastal embayment where waters may be less saline (Saila 1961; McCracken 1963). Presumably, GB broodstock, which spawn offshore (Pereira 1999), would fare better in more saline waters.

In a study to determine whether captive winter flounder broodstock are chronically stressed, Plante *et al.* (2003) compared the condition and stress responses of newly captured wild broodstock with a captive, wild broodstock kept in the hatchery for 13 months. To simulate an acute stressor, fish were dipnetted, held out of water for 3 minutes, marked, and returned to the tank. Sampling to measure hemoglobin, hematocrit, and cortisol occurred 1 hour after the acute stress event. Winter flounder held in captivity for over 1 year showed higher condition indices and similar energy reserves compared to wild fish, suggesting that chronic stress was not present in the captive broodstock. In addition, they had the same cortisol stress response as the wild fish and, it did not decrease over time, indicating that their endocrine systems were not exhausted (Seyle 1973). Though the captive broodstock fed well and was not chronically stressed, Plante *et al.* (2003) still observed an unexplainable 30% mortality loss.

6.2.2 Broodstock diet and nutrition

The maintenance ration for adult winter flounder weighing 584–801 g is 7.9 kcal/g/day (Tyler and Dunn 1976). Adults are omnivorous and will feed on a variety of items including chopped clams, squid, menhaden, and silversides

(Smigielski 1975; Stoner *et al.* 1999). At ISMER, broodstock are fed daily 1.5% of their biomass with pellets consisting of 60% commercial feed (6.0 mm Corey Feed Mill, Ltd.), 25% frozen capelin or Atlantic herring, and 15% frozen amphipods (Plante *et al.* 2002). This particular formulated pellet has 38% moisture content and 14.6 kJ/g energy content (Plante *et al.* 2002, 2003).

Mature winter flounder cease feeding during the reproductive season (Smigielski 1975). Feeding resumes for female fish once spent and for males when the spawning season is over (Stoner *et al.* 1999), though some males will continue to feed throughout this period (Smigielski 1975). However, just before and following the spawning season are nutritionally critical sensitive periods to broodstock (Burton 1994). Generally, if females are deprived of food either before or immediately after spawning, they will not be gametogenic in the subsequent spawning season (Tyler and Dunn 1976; Burton and Idler 1987; Burton 1994). This can be reversed by feeding to satiation during the months following the subsequent season when females would have spawned (Burton 1991). Alternatively, if the nutritional condition of the females is high (>1.20 condition factor), they have a higher probability of withstanding starvation and being able to spawn the following year (Burton 1994).

6.2.3 Controlled spawning

Winter flounder can be induced to spawn using several hormones; the most reliable ones are freeze-dried carp pituitary extract (CPE) (Smigielski 1975) and gonadotropic-releasing hormone analog (GnRH-A; Harmin and Crim 1992). CPE mixed at doses of 0.5 or 5 mg/454 g female body weight (BW) in a solution of sodium chloride and injected intramuscularly daily (3 or 6 days depending on dosage) results in spawning (Smigielski 1975). GnRH-A either administered in saline injections or through a sustained-release cholesterol pellet, stimulates the reproductive system of both male and female fish throughout the year except for in postspawned, sexually regressed fish (Harmin *et al.* 1995a).

Harmin and Crim (1992) successfully induced spawning in female winter flounder using either 100–120 μg GnRH-A/slow-release pellet, 40 μg GnRH-A/quick-release pellet, or 20 μg/kg BW GnRH-A in saline injected 3/week. Though all three techniques were effective, the repeated handling of fish for the saline injections caused more stress, and resulted in higher mortality than the implanted pellets. In mature female fish, GnRH-A accelerated ovarian growth by increasing plasma estradiol-17β and testosterone levels (Harmin *et al.* 1995a). In prespawning female fish, the hormone stimulated ovulation through germinal vesicle migration by increasing plasma testosterone levels (Harmin *et al.* 1995a). This hormone is most effective and reliable when used as the fish approach the time of the natural spawning season (Harmin and Crim 1992). In addition, in this scenario, both egg and larval quality are higher (Harmin and Crim 1992).

Male winter flounder begin spermiating as much as 5 months before females ovulate (Shangguan and Crim 1999); however, milt production remains low until the females begin spawning. When administered in the fall, GnRH-A stimulates the growth of the testes by increasing plasma androgen levels in male winter flounder (Harmin *et al.* 1995a). Injecting males with either 20 or 200 μg/kg BW GnRH-A caused plasma levels of testosterone and 11-ketotestosterone to increase within 12 hours and remain elevated for several days (Harmin and Crim 1993). The hormone can also cause males to spermiate several months early (Harmin *et al.* 1995a) yet only small amounts (<50 μL) of milt could be collected (Harmin and Crim 1993). Rather than increasing sperm production during the spawning season, GnRH-A treated males (110 μg/kg BW GnRH-A) had a greater volume of milt but more diluted and earlier termination of the spawning season than in non-treated males (Shangguan and Crim 1999). Hormonal treatments of GnRH-A probably do not increase total spermatozoa production since spermatogenesis occurs well before the spawning season (Harmin *et al.* 1995b). Sperm quality, based on sperm motility and egg fertility, is unaffected throughout the spawning season by use of GnRH-A (Shangguan and Crim 1999).

Initial research on the cryopreservation of sperm has been tested and shown to be successful in fertilizing winter flounder eggs (Rideout *et al.* 2003). They tested three cryoprotectants (dimethyl sulphoxide, propylene glycol, and glycerol) and two diluents (sucrose based and saline based) mixed (9 parts diluent:1 part cryoprotectant) in a factorial experiment. Sperm were diluted 1:3 with each of the six extenders, and frozen and stored on liquid nitrogen. All the diluents and cryoprotectants were successful in cryopreserving winter flounder sperm, and no toxic effects were apparent on sperm. Sperm preserved with propylene glycol (regardless of diluent type) had the highest motility recovery, followed by dimethyl sulphoxide in sucrose diluent, dimethyl sulphoxide in saline diluent, and glycerol (regardless of diluent). The postthaw motility of winter flounder sperm was not affected by adding ovarian fluid to the seawater used to activate sperm swimming behavior. In addition, cryopreservation of the sperm did not adversely affect the hatching success or larval development of winter flounder. Though this study indicates that cryopreservation of winter flounder sperm is a viable method for fertilizing eggs, these techniques need to be developed further.

Despite reports that female winter flounder rarely spawn spontaneously in the laboratory (Harmin and Crim 1992), at the University of New Hampshire (UNH), fish are reared from a volitionally spawning wild broodstock (King and Howell 1997; Fairchild *et al.* 2007). Mature fish are collected during the natural spawning season and transferred to 0.9 m diameter tanks provided with flow-through, ambient, ultraviolet filtered (0.5 μm) seawater, and covered with dark screens to filter any visual disturbances in the laboratory. Typically, one female and two or more males are stocked into each tank. Usually, within 1 week, the fish will spawn during the night (Howell and Litvak 2000). When females do not volitionally spawn, luteinizing hormone-releasing hormone (LHRH), injected at a dosage of 10μg/kg, is effective (Butts and Litvak 2007a, 2007b).

There are only two published accounts of winter flounder spawning behavior in captivity. Winter flounder are nocturnal spawners; spawning events occur from sunset to dawn with the majority happening near 21:00 hours (Breder 1922; Stoner *et al.* 1999). A spawning event appears to be initiated by a male pursuing a female (Stoner *et al.* 1999). Instead of fleeing, the female stays close to the sediment, then the pair rapidly swim in an upward tight spiral as eggs and milt are released, then return to the bottom (Breder 1922; Stoner *et al.* 1999). It is not uncommon for several males to participate in spawning (Breder 1922). Stoner *et al.* (1999) noticed that in 151 spawning events, only 22.5% were paired spawnings; the remainder involved two to six males.

Female winter flounder are batch spawners; only one group of oocytes matures each year (Burton and Idler 1984). By filming spawning events in a large (121 KL) aquarium, Stoner *et al.* (1999) observed that individual females spawned multiple times over a period of a week, and they estimated that the average female spawned 40 times during the reproductive season. However, at UNH, females either release all their eggs within one night, or release a small batch followed by all the remaining eggs during the subsequent night (Howell and Litvak 2000).

6.2.4 Collection of eggs and egg incubation systems

Female winter flounder fecundity ranges from 0.4 to 3.3 million eggs (Topp 1968). Egg diameter ranges from 0.74 to 0.85 mm, with a mean of 0.80 mm (Smigielski and Arnold 1972). Once spawned, the demersal, adhesive eggs harden into masses, which can range from 33–71 g (Ben Khemis *et al.* 2000, 2003). It used to be thought that low survival of winter flounder embryos was due to the eggs clumping together (Smigielski and Arnold 1972), so diatomaceous earth was, and still is, commonly used to prevent adhesion. For example, at ISMER, females are strip spawned into plastic containers, and the eggs are fertilized with sperm from several males (Ben Khemis *et al.* 2000, 2003). A suspension of diatomaceous earth in seawater (50 g/L) is added, and the eggs and sperm are mixed by gently swirling the container. After 5 minutes, the eggs are rinsed, 100–150 mL more diatomaceous earth suspension is added, and the eggs are rocked until egg membrane hardening is completed (approx. 30 minutes after fertilization). Five hours later, the eggs are rinsed in a 500 μm sieve, treated with iodine, and incubated.

Inhibiting egg clumping is not necessary for successful fertilization and larval survival. Fairchild *et al.* (2007) simply remove spawned out adult fish and tank covers from their spawning tanks, and allow the eggs to incubate naturally in the existing tanks. Despite eggs clumping together, high fertilization and hatching rates are standard (Howell and Litvak 2000). However, because other particles may adhere to the eggs too, daily inspection of the egg masses is necessary to ensure fungus is absent. If fungal growth appears, eggs can be dipped in an iodine bath (25 ppm solution) to disinfect. Butts and Litvak (2007a, 2007b) strip spawn females, collect the eggs, and add cryopreserved sperm. The gametes are swirled together and then sterilized seawater (8°C, 28 ppt) containing antibiotics (13 mg/L each of penicillin G and streptomycin sulphate) is added. This gamete solution rests for 20 seconds, then sperm are rinsed out, sterilized seawater is added, and the eggs are incubated.

In recirculating and flow through systems, incubating eggs are subjected to a 0L:24D photoperiod, supplied with UV-sterilized, 0.5 μm filtered sea water at a rate of 1–4 L/min, and provided gentle upwelling aeration (Ben Khemis *et al.* 2000, 2003; Mercier *et al.* 2004; Fairchild *et al.* 2007). Eggs have been incubated successfully in flat-bottomed, fiberglass, flow-through tanks (Fairchild *et al.* 2007), baskets within a water table maintained at 7°C and 28 ppt (Mercier *et al.* 2004), and in 60 liters, black, cylindroconical, fiberglass tanks at 7°C and 27 ppt (Ben Khemis *et al.* 2000, 2003). In all these systems, hatching occurred 9–10 days after fertilization. Winter flounder eggs also can be incubated successfully in static systems in temperature controlled rooms using the same water quality and photothermal conditions (Chambers and Leggett 1987; Chambers *et al.* 1988; Butts and Litvak 2007a, 2007b). In these, 25–75% of the water is replaced every 1–3 days to maintain high water quality. Stocking density ranges from <1000 to >30,000 eggs/L (Howell and Litvak 2000).

6.3 Larval culture

6.3.1 Larval system design and requirements

Upon hatching, larvae are transferred into rearing systems. At ISMER, photoperiod remains 0L:24D until mouth opening, and then is switched to a different lighting regime to facilitate visual feeding of the larvae (Ben Khemis *et al.* 2000, 2003; de Montgolfier *et al.* 2005). At UNH, a 24L:0D photoperiod is used once hatching begins to bring the photophilic larvae to the surface for easier collection (Fairchild *et al.* 2007). At UNBSJ, photoperiod also is switched to 24L:0D as larval growth rates and survival to metamorphosis under continuous light are significantly higher than those reared under ambient lighting conditions (Litvak 1999). In addition, time to metamorphosis is shortened by 5 days (Litvak 1999). Light intensity, however, does not affect larval growth or survival (Litvak 1999). A variety of different rearing systems are used by researchers (Table 6.1); all contain UV-sterilized, 0.5–10 μm filtered sea water and gentle aeration, usually in the center of the tank to create an upwelling effect, thereby keeping food suspended in the water column and oxygen levels high. In nonstatic systems, water flow is very light (0.1–0.2 L/min; Ben Khemis *et al.* 2000, 2003) since newly hatched winter flounder larvae are not strong swimmers (Sullivan 1915). In static systems, 50% of the water is changed every 1–3 days depending on the size and density of the larval tanks (King and Howell 1997; Butts and Litvak 2007a, 2007b).

6.3.2 Larval food and feeding

Greenwater

The addition of microalgae to larval rearing systems (termed "greenwater") is used to enhance water quality, provide both food for larvae and rotifers, and prevent "larval wall syndrome" (Mercier *et al.* 2004). King and Howell (1997) showed that first feeding winter flounder larvae grew faster (SGR = 14.2% vs. 12.2%) and significantly longer (8.7 vs. 7.0 mm TL) if greenwater was added to

Table 6.1 Larval rearing systems used for winter flounder.

	Tank specifications						Lighting regime		
Density (larvae/L)	Size (L)	Color	Shape	Material	Temp (°C)	Salinity (ppt)	Photoperiod	Intensity	Source
10	20	gray	cylindroconical	polyethylene	8–15	30–32	24L:0D	35 watt	King and Howell 1997
10	40	black		glass	7		14L:8D	40 watt	Chambers and Leggett 1987
11	35	clear	circular	glass	7		14L:8D	40 watt	Chambers *et al.* 1988
13	38	black			8	28–30	12L:12D		Laurence 1975, 1977
30	24	black	rectangular	plastic	6–9.5	30–31	12L:12D		Klein-MacPhee 1982
50		clear	circular	glass	8	28	24L:0D	50 lx	Butts and Litvak 2007b
60	120	dark green	cylindroconical	fiberglass			12L:12D	400 lx	Ben Khemis *et al.* 2003
77		clear	circular	glass	8	28	24L:0D	1100 lx	Butts and Litvak 2007a
100	2000	black sides, green bottom	circular	fiberglass	7–15	14–31	24L:0D	30 μE/s/m2	Howell and Litvak 2000; Fairchild *et al.* 2007
250	120	dark green	cylindroconical	fiberglass			12L:12D	400 lx	Ben Khemis *et al.* 2003
250	55	black sides, white bottom	cylindroconical	polyethylene	10	28	12L:12D	7.1 μM/s/m2	Mercier *et al.* 2004

culture water for the first 5 weeks at a concentration of 1 liter *Isochrysis galbana* (200,000 cells/mL)/6 liters culture water. In addition, larvae in the greenwater treatments initiated first feeding before larvae in the clearwater treatments.

There is no standard protocol on greenwater methodology. In the past, fish culturists always used live algae, typically grown on-site. For example, at UNH a variety of species including *Nannochloropsis* sp., *Isochrysis* sp., *Dunaliella tertiolecta*, and *Tetraselmis suecica* were cultured and added (1 liter algae/100 liters culture water) twice daily (Fairchild *et al.* 2007). With the advancement in algae processing, fish culturists now have a variety of algal products to use. At ISMER, *Chlorella* sp. is added at 1 liter algae/50 liters culture water each morning along with 100 mL of an algal paste solution (1 g diluted in 450 mL seawater; Mercier *et al.* 2004). Ben Khemis *et al.* (2000) also uses *Chlorella* sp. but at a concentration of 1 liter algae/120 liters culture water/day administered in the morning. Other culturists (Ben Khemis *et al.* 2003; de Montgolfier *et al.* 2005) add only Starter Formula Algae Paste (Innovative Aquaculture Products Ltd., Skerry Bay, NB) every morning until weaning. Butts and Litvak (2007a) use Instant Algae® (*Pavlova* sp.; 40,000 cells/mL of culture water; Reed Mariculture Inc., United States). In a preliminary study, Fairchild (unpublished data) found that winter flounder larvae cultured in Instant Algae® (*Nannochloropsis* sp.) exhibited good if not better growth rates than the fish in live algae or control (no algae) treatments, plus had superior survival. One drawback was that only fish cultured in Instant Algae® had abnormal pigmentation. However, the proportion of abnormally pigmented fish was relatively low compared to the higher survival of fish in this treatment.

Live feed

Newly hatched winter flounder larvae have an endogenous yolk sac, which is absorbed, depending on the culture temperature, 4–6 days post hatch (dph; Ben Khemis *et al.* 2000; Mercier *et al.* 2004; Fairchild, unpublished data). Due to their small gape size and undifferentiated stomach, winter flounder larvae require live prey (first *Brachionus* spp., then *Artemia* spp.) once yolk sac absorption has occurred (Ben Khemis *et al.* 2000). To retain prey in the larval tanks, water flow can be ceased during the feeding (= daylight) period, and resumed at night for complete water replacement (Ben Khemis *et al.* 2000, 2003). Typically, rotifers are provided daily at a concentration of 2–5/mL (Bertram *et al.* 1993, 1997; King and Howell 1997; Litvak 1999; Ben Khemis *et al.* 2000, 2003; Mercier *et al.* 2004; Fairchild *et al.* 2007). Protocols differ between facilities. Some dispense the daily ration of rotifers in a single morning feeding (water flow is shut off for a period of time so prey are not lost from the system), while others dispense it in multiple feedings per day. Because rotifers lack some of the essential fatty acids required for optimal growth and survival for winter flounder, they typically are enriched with a combination of the following: Starter Formula Algae Paste, live algae (e.g., *Pavlova lutheri* or *Isochrysis* sp.), Culture Selco (INVE Aquaculture, Salt Lake City, UT, USA), DHA Selco (INVE Aquaculture), and Microfeast® MB-30. To receive maximum nutritional benefits, rotifer enrichment occurs $\geq$4 hours before being fed out to the winter flounder larvae at UNH

(Fairchild *et al.* 2007). Despite the water in rotifer cultures being substantially warmer than that of winter flounder cultures, there is no need to acclimate the rotifers to this colder temperature prior to feeding. Mercier *et al.* (2004) found no substantial differences in any of the growth performance and nutritional condition parameters measured between winter flounder larvae fed rotifers from a normal rotifer culture temperature (24°C) and those acclimated overnight to the fish culture tanks (10°C).

Ben Khemis *et al.* (2000) tested whether a high performance microencapsulated diet could be used as a total or partial replacement of rotifers in first-feeding winter flounder larvae. They compared the growth, survival, and nutritional condition of larvae fed either (i) enriched rotifers, (ii) enriched rotifers and microencapsulated diet, (iii) only microencapsulated diet, or (iv) not fed (starved). Fish from the latter two feeding treatments experienced high mortality and no survival beyond day 10 posthatch indicating that newly hatched winter flounder larvae are unable to digest microencapsulated diets. Larvae fed a mixed diet had slower growth than larvae fed exclusively rotifers; however, the RNA/DNA ratios between the two treatments were similar by stomach differentiation (5.5 mm TL).

As the larvae develop and undergo stomach differentiation, rotifers are gradually replaced by *Artemia* over several days. At both UNH and ISMER, newly hatched, unenriched *Artemia* are offered at first (Ben Khemis *et al.* 2003; Fairchild *et al.* 2007). Once larvae readily eat the *Artemia*, Fairchild *et al.* (2007) discontinue the rotifers and only offer *Artemia* enriched with DC DHA Selco (INVE Aquaculture) at a concentration of 3/mL/day. Alternatively, Butts and Litvak (2007a) cofeed rotifers with *Artemia* enriched with Microfeast® MB-30 and algae (*Isochrysis* sp., Tahitian strain, and *Pavlova lutheri*).

Weaning onto formulated feeds

Weaning winter flounder from live to formulated feeds usually occurs with the completion of gut development and metamorphosis. Mean age at metamorphosis (when the left eye migrates to the right side of the head and loss of pigmentation occurs on the blind side; Chambers and Leggett 1987), is temperature-dependent; it can range from as little as 26 dph at 15°C (Lee and Litvak 1996a) to as many as 80 dph at 5°C (Laurence 1975). Mean size at metamorphosis is less variable; it ranges from 6.1 to 10.1 mm TL (Chambers and Leggett 1987; Chambers *et al.* 1988; Perry *et al.* 1991; Bertram *et al.* 1993; Jearld *et al.* 1993).

Weaning is initiated by a period of cofeeding a starter diet (150–450 μm) and *Artemia* to the fish for 5–7 days, then a wean onto the commercial diet over 7–14 days (Lee and Litvak 1996a; de Montgolfier *et al.* 2005; Fairchild *et al.* 2007). Ben Khemis *et al.* (2003) found larvae as small as 5–6.6 mm TL could be weaned onto a commercial microencapsulated diet with no adverse affects on growth rate or time to metamorphosis as long as an extended transitional cofeeding period was used. They recommend feeding the larvae the commercial diet 4 times/day, followed by rotifers for at least 8 days. Once fish are ≥6.6 mm TL, they can be weaned onto a microencapsulated diet.

6.3.3 Genetics for culture versus enhancement

Winter flounder are sexually dimorphic with females growing faster and larger than males (Nash and Geffen 2005). Many flatfish species exhibit a combination of genetic and environmental sex determination in which certain exogenous parameters (e.g., temperature, pH) interact with the genetic factors that influence sex determination, thereby influencing the male to female sex ratio (Nakamura *et al.* 1998). If winter flounder sex determination is affected by these factors, it would occur when fish are <41 mm TL as this is when the gonads differentiate (Fairchild *et al.* 2007). At UNH, cultured winter flounder cohorts do not deviate from a 1:1 sex ratio, with the exception of fish produced in 2003 (Fairchild *et al.* 2007; Fairchild unpublished data). The causes of the 2003 male dominated population are unknown, but temperature effects were unlikely. Research continues at UNH to determine if winter flounder sex determination is affected by variables like temperature, stocking density, photoperiod, and other stressors. If winter flounder sex determination is affected by such variables, there would be some economic advantage to produce a faster-growing, larger, female monoculture for growout; this could be achieved by manipulating environmental cues. In contrast, if fish are produced for stock enhancement, it is necessary that cultures result in an equal mixture of males and females so as not to disrupt the natural sex ratio. In addition, rearing all females using meiogynogenetic techniques (Howell 1995; Luckenbach *et al.* 2004) warrants research.

6.4 Nursery culture and growout

6.4.1 Juvenile system design and requirements

Juvenile winter flounder can either remain in the larval tanks or be moved to different tanks for growout. de Montgolfier *et al.* (2005) found no differences in growth and condition of juveniles reared in plankton kriesels and in rectangular raceways. Litvak (1999) recommends rearing juvenile in suspended cages within tanks to isolate the small fish from bottom-accumulating debris and to facilitate cleaning the tanks without disturbing the small fish. For juvenile culture, both aeration and water flow are increased, and drains are covered by small-size (500 μm) mesh to prevent small fish from being lost from the system. At ISMER, flow rates are 1.4 L/min (de Montgolfier *et al.* 2005). At UNH, oxygen is added to each tank to ensure dissolved oxygen does not fall below 7.0 mg/L in densely stocked tanks (Fairchild, unpublished data). Juvenile growth rates are highly variable and temperature-dependent (see Litvak 1999; Howell and Litvak 2000). Juveniles can be stocked as high as 300% (measured as ventral fish area to bottom tank area ratio) with no reduction in growth (Fairchild and Howell 2001). However, winter flounder juveniles held at high densities, even for short periods of time, display elevated levels of cortisol (Sulikowski *et al.* 2006), which suggests that stocking density can act as an environmental stressor rendering the fish more vulnerable to disease.

Both fin erosion and fin rot can occur in wild winter flounder (Bodammer 2000; Ziskowski *et al.* 2008), and fin erosion has been observed in cultured winter flounder (Fairchild and Howell 2001; de Montgolfier *et al.* 2005). Possible causes of fin erosion in cultured winter flounder may include increased aggressive behaviors due to fish size hierarchies (Fairchild and Howell 2001), high stocking density (de Montgolfier *et al.* 2005), unnatural photoperiod (Sakakura and Tsukamoto 2002), unnatural bottom surfaces, and limited food (Latremouille 2003). Fin erosion in winter flounder can be exacerbated by opportunistic bacteria such as *Vibrio anguillarum* (Levin *et al.* 1972) and *Trypanosoma* (Khan 1985); additive effects of the two may cause death (Khan 1985). Treatments for fin rot are reviewed in Latremouille (2003).

Fairchild and Howell (2001) explored the possibility that high stocking densities (up to 300% fish area to tank bottom area) caused an increase in fin nipping, and thus erosion of the caudal fin area. Although there were no differences between treatments, ranging from 50 to 300% stocking density, it was clear that the smallest fish in all treatments had badly damaged caudal fins. They concluded that a size hierarchy had been established in all tanks, and that the smallest fish had suffered from the aggressive behavior (nipping) of the larger fish. One method which may reduce aggressive behavior, even at high stocking densities is manipulating photoperiod. Studies with Japanese flounder have shown that fin nipping and other signs of aggressive behavior only occur during the day and not at night (Sakakura and Tsukamoto 2002). In preliminary photoperiod manipulation trials, survival of juvenile winter flounder was on average 10% higher each week in the 12L:12D photoperiod treatments than in the 24L:0D treatments, suggesting that juvenile winter flounder exposed to a continuous light cycle may be more stressed than those under a more natural lighting regime (Fairchild, unpublished data). While a 24L:0D photoperiod may promote growth (Litvak 1999), this constant light photoperiod regime also may increase aggressive behavior resulting in stressed and fin damaged fish, which may in turn, lead to fish with lower resistance to bacterial infection.

6.4.2 Juvenile diet and nutrition

Juvenile winter flounder are omnivores (Stehlik and Meise 2000) and will readily consume a variety of items in hatcheries including frozen mysids (Stierhoff *et al.* 2006), copepods (Meise *et al.* 2003), blood worms (chironomid larvae; C. Chambers, unpublished data), amphipods, and white worms (*Enchytraeus albidus*; M. Walsh, unpublished data). To date, there is no commercially available feed specifically manufactured for winter flounder, and amino acid requirements for winter flounder have not been determined yet. Researchers either use a coldwater marine species commercial diet or manufacture diets in-house, and feed to satiation (1–7% body weight/day; Howell and Litvak 2000; Ramsay *et al.* 2000; Fairchild and Howell 2001; Plante *et al.* 2005). At UNBSJ, an inexpensive salmonid pellet (Hi-Pro™, Corey Feed Mills, Fredericton, NB, Canada) performed as well as the more costly, specialty pellets (Nippai SFI-3, CATVIS, Hertogenbosch, The Netherlands); juvenile winter flounder specific growth rates

Table 6.2 List of ingredients and their percentage by weight of diets manufactured for juvenile winter flounder by the Canadian Department of Fisheries and Oceans in Halifax, NS.

	Percentage of diet	
Ingredient	**Fredette *et al.* (2000)**	**Ramsay *et al.* (2000)**
Fish meal (hering 70% CP)[a]	49	45
Wheat middlings	12	12
Dextrin	10	10
Herring oil	8	8
Soybean meal (48% CP)	6	6
Wheat gluten	5	5
Mineral mix	2	2
Soy lecithin	2	2
Vitamin mix[b]	2	2
Sodium alginate	1.3	1.3
Choline chloride	0.7	0.7
CMC (carboxymethylcellulose)	1	1
FinnStim[c]	1	0
Crab meal	0	5

[a]CP, crude protein.
[b]Modified Bernhardt-Tomarelli salt mix.
[c]2-trimethylammonioacetate.

(wet weight) were 3.11% per day and 2.65% per day for the respective diets (Lee and Litvak 1996b). At ISMER, a mixed diet of 40% salmonid starter feed, 10% spirulina, 25% freeze-dried amphipods, and 25% tubifex worms is used for newly-weaned juvenile fish (de Montgolfier *et al.* 2005). At the Canadian Department of Fisheries and Oceans in Halifax, NS, a base diet high in protein is manufactured in-house (Fredette *et al.* 2000; Ramsay *et al.* 2000; Table 6.2). While Ramsay *et al.* (2000) add a freeze-dried *Cancer irroratus* crab meal, Fredette *et al.* (2000) found that the commercial feeding stimulant FinnStim (2-trimethylammonioacetate) supplemented at 1% to a fish meal based diet is an effective feeding stimulant for age 1 winter flounder. Fish fed the diet with stimulant had higher daily feed consumption and growth, and slightly higher feeding efficiency compared to fish fed a control diet.

A crude protein level of 50% with a digestible protein to digestible energy ratio of 26 mg/kJ is optimal for growth of juvenile (0.9–5 g) winter flounder (Hebb *et al.* 2003). Fish fed a 50:10% protein-to-lipid level diet had both a higher specific growth rate and a better feed efficiency ratio compared to fish fed a 40:20% diet, but not significantly different from fish fed a 45:15% diet (Hebb *et al.* 2003). In addition, the protein efficiency of the higher protein level fed fish tended to be higher than in the other fish. With the rising costs of fish meal protein and oil, and the push to make aquaculture more sustainable, there is an increased demand to find alternative proteins in fish feeds. Preliminary studies indicate that a 15% fish meal replacement with either soybean meal or canola protein concentrate is plausible for juvenile winter flounder. At this replacement level, the digestibility of gross energy or crude protein in 159 g, juvenile winter flounder was not different than in fish fed a control fish meal diet; gross energy and crude protein digestibilities of the diets ranged from 91

Park, I.S., Nam, Y.K., Douglas, S. E., Johnson, S. C., and Kim, D. S. 2003. Genetic characterization, morphometrics and gonad development of induced interspecific hybrids between yellowtail flounder, *Pleuronectes ferrugineus* (Storer) and winter flounder, *Pleuronectes americanus* (Walbaum). *Aquaculture Research* 34(5): 389–396.

Pearcy, W.G. 1962. Ecology of an estuarine population of winter flounder, *Pseudopleuronectes americanus* (Walbaum). Parts I-IV. *Bulletin of the Bingham Oceanographic Society* 18(1):1–18.

Pereira, J.J., Goldberg, R., Ziskowski, J.J., Berrien, P.L., Morse, W.W., and Johnson, D.L. 1999. Essential fish habitat source document: winter flounder, *Pseudopleuronectes americanus*, life history and habitat characteristics. NOAA Technical Memorandum NMFS-NE-138, 1–39.

Perry, D.M., Hughes, J.B., and Hebert, A.T. 1991. Sublethal abnormalities in embryos of winter flounder, *Pseudopleuronectes americanus*, from Long Island Sound. *Estuaries* 14(3):306–317.

Pereira, J.J., Goldberg, R., Ziskowski, J.J., Berrien, P.L., Morse, W.W., and Johnson, D.L. 1999. Essential fish habitat source document: winter flounder, *Pseudopleuronectes americanus*, life history and habitat characteristics. NOAA Technical Memorandum NMFS-NE-138, pp. 1–39.

Perlmutter, A. 1947. The blackback flounder and its fishery in New England and New York. In: *Bulletin of the Bingham Oceanographic Collection*. Yale University, New Haven, Vol. 11, p. 92.

Plante, S., Audet, C., Lambert, Yvan, and de la Noue, Joel. 2002. The effects of two rearing salinities on survival and stress of winter flounder broodstock. *Journal of Aquatic Animal Health* 14(4):281–287.

Plante, S., Audet, C., Lambert, Y., and de la Noue, J. 2003. Comparison of stress responses in wild and captive winter flounder (*Pseudopleuronectes americanus* Walbaum) broodstock. *Aquaculture Research* 34(10):803–812.

Plante, S., Audet, C., Lambert, Y., and De la Noue, J. 2005. Alternative methods for measuring energy content in winter flounder. *North American Journal of Fisheries Management* 25(1):1–6.

Ramsay, J.M., Castell, J.D., Anderson, D.M., and Hebb, C.D. 2000. Effects of fecal collection methods on estimation of digestibility of protein feedstuffs by winter flounder. *North American Journal of Aquaculture* 62(3):168–173.

Rideout, R.M., Litvak, M.K., and Trippel, E.A. *et al.* 2003. The development of a sperm cryopreservation protocol for winter flounder *Pseudopleuronectes americanus* (Walbaum): evaluation of cryoprotectants and diluents. *Aquaculture Research* 34(8):653–659.

Saila, S.B. 1961. A study of winter flounder movements. *Limnology and Oceanography* 6(3):292–298.

Sakakura, Y., and Tsukamoto, K. 2002. Onset and development of aggressive behavior in the early life stage of Japanese flounder. *Fisheries Science* 68(4):854–861.

Scott, W.B., and Scott, M.G. 1988. Atlantic fishes of Canada. *Canadian Bulletin of Fisheries and Aquatic Sciences* 219:554–557.

Seyle, H. 1973. The evolution of the stress concept. *American Scientist* 61:692–699.

Shangguan, B., and Crim, L.W. 1999. Seasonal variations in sperm production and sperm quality in male winter flounder, *Pleuronectes americanus*: the effects of hypophysectomy, pituitary replacement therapy, and GnRH-A treatment. *Marine Biology* 134(1):19–27.

Smigielski, A.S. 1975. Hormonal-induced ovulation of the winter flounder, *Pseudopleuronectes americanus*. *Fishery Bulletin* 73 (2): 431–438.

Smigielski, A.S., and Arnold, C.R. 1972. Separating and incubating winter flounder eggs. *Progressive Fish Culturist* 34(2):113.

Stehlik, L.L., and Meise, C.J. 2000. Diet of winter flounder in a New Jersey estuary: Ontogenetic change and spatial variation. *Estuaries* 23(3):381–391.

Stierhoff, K.L., Targett, T.E., and Miller, K. 2006. Ecophysiological responses of juvenile summer and winter flounder to hypoxia: experimental and modeling analyses of effects on estuarine nursery quality. *Marine Ecology-Progress Series* 325:255–266.

Stoner, A.W., Bejda, A.J., Manderson, J.P., Phelan, B.A., Stehlik, L.L., and Pessutti, J.P. 1999. Behavior of winter flounder, *Pseudopleuronectes americanus*, during the reproductive season: laboratory and field observations on spawning, feeding, and locomotion. *Fishery Bulletin* 97(4):999–1016.

Sulikowski, J.A., Fairchild, E.A., Rennels, N., Howell, W.H., and Tsang, P.C.W. 2005. The effects of tagging and transport on stress in juvenile winter flounder, *Pseudopleuronectes americanus*: Implications for successful stock enhancement. *Journal of the World Aquaculture Society* 36(1):148–156.

Sulikowski, J.A., Fairchild, E.A., Rennels, N., Howell, W.H., and Tsang, P.C.W. 2006. The effects of transport density on cortisol levels in juvenile winter flounder, *Pseudopleuronectes americanus*. *Journal of the World Aquaculture Society* 37(1):107–112.

Sullivan, W.E. 1915. A description of the young stages of the winter flounder (*Pseudopleuronectes americanus* Walbaum). *Transactions of the American Fisheries Society* 44(1):125–136.

Topp, R.W. 1968. An estimate of fecundity of the winter flounder, *Pseudopleuronectes americanus*. *Journal of the Fisheries Research Board of Canada* 25(6):1299–1302.

Tyler, A.V., and Dunn, R.S. 1976. Ration, growth, and measures of somatic and organ condition in relation to meal frequency in winter flounder, *Pseudopleuronectes americanus*, with hypothese regarding population homeostasis. *Journal of the Fisheries Research Board of Canada* 33:63–75.

Venizelos, A., and Benetti, D.D. 1999. Pigment abnormalities in flatfish. *Aquaculture* 176:181–188.

Waters, E.B. 1996. Sustainable flounder culture and fisheries. University of North Carolina Sea Grant College Program, Raleigh, NC. Publication Number UNC-SG-96–14, pp. 1–12.

Ziskowski, J., Mercaldo-Allen, R., Pereira, J.J., Kuropat, C., and Goldberg, R. 2008. The effects of fin rot disease and sampling method on blood chemistry and hematocrit measurements of winter flounder, *Pseudopleuronectes americanus* from New Haven Harbor (1987–1990). *Marine Pollution Bulletin* 56(4):740–750.

Section 2
Europe Culture

Chapter 7
Turbot culture

Jeannine Person-Le Ruyet

7.1 Life history and biology

Turbot, *Scophthalmus maximus* Rafinesque 1810: *Psetta maxima* Linnaeus, 1758, is a marine demersal carnivorous flatfish of the Scophthalmidae family naturally distributed in European waters, from the Artic Circle to Morocco (68–30°N; 23°W–42°E). It is also found in the Mediterranean Sea and in the Black Sea, where a subspecies *Psetta maxima maeotica* has been described. It is a highly esteemed fish.

Turbot grow fast in comparison with other flatfish species (in the wild growth rate of adults is around 10 cm/year. Maximum reported size is 100 cm (25 years). There is a commercial turbot fishery in European coastal waters. In France, the legal minimum size is 30 cm and the most common size is 30–40 cm (2–3 kg). Juveniles live in shallow waters on sandy, rocky, or mixed substrate bottoms and feed mainly on crustaceans. Adults feed on other bottom-living fishes and the offshore migration of large fish to depths of 100 m is associated with spawning behavior. Females can reach maturity at 3 years of age (around 46 cm) and male when 2 years old (around 30 cm). Spawning occurs in spring and early summer, between April and August. Fecundity is high, around 3–5 million pelagic eggs (0.9–1.1 mm diameter) for 3–5 kg female.

Turbot was selected for aquaculture in the early 1970s, both in the United Kingdom and France because of its commercial value as highly esteemed food fish (over 9 euro/kg in 2006) and its high-potential growth rate (above 3 kg within 3 years) under intensive culture conditions. In both countries, research efforts were initially focused on achieving adequate egg supply from captive brood stock and on larval rearing under controlled conditions (Girin 1972, 1979; Purdom *et al.* 1972; Jones *et al.* 1981). For the first 10 years, basic growout techniques were established in the United Kingdom by commercial companies using sea-caught young juveniles. Turbot farming using hatchery-reared juveniles commenced in Spain in the late 1980s and developed rapidly (16 producers were operating in the early 1990s).

The total annual wild catch has ranged from 7,000 to 9,000 MT from 1950 with marked oscillations during the 1980s. Despite recent declines in the wild harvest, the species is not considered endangered. In comparison, aquaculture production is developing in Europe, especially in Spain; it increased from 4,700 tons in 2000 to 7,633 tons in 2006 (FAO statistic 2008). Turbot farming was introduced in Chile in the late 1980s and more recently in China (1992) and is rapidly expanding in these two countries (Alvial and Manriquez 1999; Lei 2010).

7.2 Broodstock husbandry

As each farm has developed its own strategy, only general trends are provided with focusing on the main changes from initial practices (Person Le Ruyet 1989; Person Le Ruyet *et al.* 1991) to modern practices on the basis of selective breeding (Borell *et al.* 2004).

7.2.1 Acquisition of broodstock

Initially, broodstock were obtained by growing wild-caught immature fish, 0.5–2 kg, to maturity in large tanks, at low stocking densities and fed on trash fish such as small sardines and sand eel. Mature fish were selected according to age (5 age classes) and sex (sex ratio 1:1) and the oldest fish were replaced every year by 2–3 kg males and 4–5 kg females. They were fed to satiation on trash fish (fresh or frozen). In addition, 2 months before spawning, they received a supplement of vitamins (C and E) and lipids (fish oil as HUFA, Highly unsaturated Fatty acid, source) added into trash fish to improve larvae quality.

Broodstock management programs were established around 1993 in France and since 1995–1996 in Spain using both wild and farmed fish of genetically distinct lineages. Hatchery-reared individuals with high growth rate or specific characteristics (e.g., pigmentation and external appearance) were initially selected.

7.2.2 Biosecurity

The introduction of new broodstock to the farm should be in accordance with a procedure laid down by the competent authority (veterinarian service). When necessary, they are received in quarantine facilities.

Regular inspection is required for parasite and disease control. Turbot are treated against external parasites (copepods, *Trichodina*, *Uronema*, *Scyphidia*) when necessary by directly applying specific medicines such as formalin, Neguvon®, Dichlorvos® in the tank. A parasite treatment prior to and immediately after the spawning season is recommended. The assistance of a veterinarian is recommended to establish the causes of disease problems (bacterial, viral, or nutritional origin) and take remedial action.

7.2.3 System design and requirements

Natural spawning occurs exclusively in large deep tanks (effective volume 40 m^3, 1.65 m depth) with a sandy substrate on the bottom where the natural spawning behavior of fish adapted to captivity for at least 2 years may be expressed.

When stripping is used, broodstock are generally maintained in square or circular tanks ranging from 20 to 30 m^3, 1 m depth to facilitate handling procedures and at low stocking densities, 3–6 kg/m^3.

7.2.4 Photothermal conditioning

Oocyte maturation is mainly controlled by photoperiod and occurs as photoperiod increases from 8.5 to 16 hours light/day (local temperatures are used). Gametogenesis lasts for about 5 months and may be modulated by temperature but there is no egg fertilization over 16°C. The optimal temperature of spawning is 14 ± 1°C at 15–16 hours L/9–8 hours D. By photothermal conditioning, spawning period can be managed to obtain eggs all year around with an annual spawning period of 2–3 months/group of spawning broodstock.

Gonad development is monitored when fish are stripped. Female maturational status is determined by oocyte sampling (using a biopsy pincer) for comparison of oocyte diameter with a reference scale.

7.2.5 Diet and nutrition

Specific nutritional requirements of broodstock are not well known. Feeding strategies are developed to prevent any nutritional deficiency and promote egg quality (Harel *et al.* 1994). Broodstock are most often fed commercial dry pellets supplemented on a regular basis using either trash fish or moist pellets boosted with commercial or farm specific mixtures (with vitamins and HUFA).

7.2.6 Controlled spawning

Each female may spontaneously spawn several times, up to 12 spawns/season at 3–6 days intervals. To avoid egg over-ripening, females should be stripped every 4–5 days since female ovulatory cycles are 70–90 hours. In the industry, turbot gametes are obtained by hand stripping of females and male and artificial fertilization.

7.2.7 Collection of eggs and egg incubation

When natural spawning occurs, floating eggs are collected in egg collectors placed under the water outflow. Eggs are harvested by direct netting from the water surface.

After stripping, eggs and recently collected sperm (two males/female) are mixed without seawater, and clear seawater is added 5 minutes later. The use of cryopreserved sperm is becoming more and more common. Only floating eggs (at 34–35‰ salinity) are held in small incubators until hatch; they can be directly placed in larval rearing tanks. Incubation is critical; mechanical, and thermal shocks should be avoided and UV-treated running water is generally used to ensure high water quality. The optimum thermal range for incubation is 13–15°C (limits 9–17°C) and embryonic development takes 3–7 days depending on temperature.

Hatching rate averages 70–80% from the viable morula egg stage, and 30–40% from total eggs collected by stripping. Similar results may be obtained with natural spawning.

7.3 Hatchery culture

In Europe, basic techniques in turbot larvae culture have been established since the 1980s (Person Le Ruyet 1989; Shields 1991 review). The subsequent refinements have been carried out by the industry and are often confidential.

7.3.1 Larvae system design and requirements

Larvae production techniques can be classified as either extensive or intensive. In the extensive method, newly hatched larvae are stocked into ponds or large outdoor tanks previously prepared (to enhance natural blooms and kill predators) to provide a sufficient amount of suitable prey to sustain the fish until harvest that is critical for this ecosystem approach. In modern European aquaculture, turbot larvae are produced in hatcheries under more or less controlled environmental conditions. In the semi-intensive method, newly hatched larvae are held at low densities (2–5 larvae/L) in large tanks (50 per m^3) in which different algae and zooplankton blooms are produced. It is simple in design but food availability is not easily secured. In the intensive method, tank-produced live prey (e.g., rotifer and *Artemia* mainly) of high nutritive value are added daily to the tanks.

7.3.2 Larvae culture protocols

Larval rearing

In the intensive culture method, the most common method in Europe, 1 day posthatch larvae are transferred into 20–30 per m^3 tanks at an initial stocking density of 20–30 larvae/L. Temperature is increased from incubation temperature of 13–15°C to 18–20°C within a few days.

Water quality

Flow-through systems are generally used but water can be partly reused. Water flow is continuously adjusted as larvae develop, in order to limit any risk of water quality deterioration especially when compound food is introduced. Before use, heated water must be passed through a degassing column to avoid any problems of gas oversaturation, then the water is sterilized with UV light. A good distribution of larvae and food must be assured with water circulation and gentle aeration. During the first 3–5 days, when microalgae (e.g., *Isochrysis*, *Tetraselmis*, *Nannochloris*) are added to the tank to control water quality and improve larvae behavior, stagnant water conditions are used.

Food and feeding

The very small and sensitive larvae (3 mm long and 0.1–0.2 mg weight) require the production of a reliable chain of live food consisting of the rotifer *Brachionus plicatilis* and the brine shrimp *Artemia* sp. and sometimes of unicellular algae such as Isochrysis, *Tetraselmis*. Larvae are fed from the time of mouth opening, at day 3 ph (ph = posthatching), using rotifers for 10 days and then are gradually replaced by *Artemia*. Now, compound dry food of small size (100–200 μm, 60–65% protein, and 20–25% lipid contents for *Artemia* substitutes) can be used from day 15 while cofeeding with *Artemia*. Weaning onto conventional weaning diets (200–400 μm, 52–55% protein, and 12–15% lipid contents) is completed by day 30 ph.

The nauplii of calanoid copepods seem to be the most efficient first food for turbot larvae. However, because copepods cannot be easily produced at a large scale, the biochemical composition of this natural food is used as reference both for live prey and for the formulation of compound diets, as discussed in Guillaume *et al.* (2001).

Rotifers are produced at the lowest production cost using commercial specific dry food provided by different suppliers and either semicontinuous culture or batch culture. Initially, the main rotifer food is baker's yeast plus emulsified fish oil, and lactic acid. Daily harvesting of 25% of the tank volume stabilizes the rotifer population at about 200 individuals/L. To minimize the risk of bacterial contamination, the food is distributed continuously to rotifers tanks, the culture period is limited to 2–3 weeks and when harvested, the rotifers are washed in clear seawater. Prior to use, the rotifers must be nutritionally enriched to satisfy the requirements of turbot larvae for n-3 HUFA (HUFA with a carbon chain >20) using either homemade complete enrichment mixtures (lipids, proteins, vitamins, and minerals) or different commercial emulsions. Today, selected strains of rotifers (s- or l-types) can be produced at very high density (2,000 per mL) after a 4-day batch culture and concentrations as high as 5,000 per mL can be maintained for long periods in tanks continuously supplied with recirculated water and various diets. These recent developments have greatly improved the nutritional and sanitary quality of rotifers (better control of microflora) and decreased production costs.

After hatching, *Artemia* are ongrown for 48 hours at 24°C and fed various commercial diets. As for rotifers, they should be enriched before use, the enrichment diets have improved over the years from emulsion types to encapsulated complete diets (essential nutrients, plus additives such as pigments, prophylactics, or medicines). Enriched preys of high nutritional quality can be easily distributed drop by drop to rearing tanks using peristaltic pumps or similar devices. The main difficulty is to estimate daily prey amount to avoid food restriction and overfeeding.

The use of formulated feeds is increasing in turbot hatcheries in recent years after first successes obtained with sea bass and cod larvae at a laboratory scale using a patented formulation containing 12% phospholipids (Zambonino Infante and Cahu 1999). Different efficient microdiets adapted to the specific requirement of marine fish larvae are now available, but they are more difficult to use than *Artemia*; they must be supplied in excess, which requires daily cleaning of the rearing tank (designed for live prey use), a technically difficult and time consuming task.

Microbial environment

The introduction of germs by water, eggs, larvae, or maintenance apparatus must be avoided. Filtration and sterilization of inflow water (UV, ozone) is essential for flow through and recirculating systems that are sometimes supplied with bacteria free ground water. Decontamination of eggs, using iodine (iodized polyvinylpyrrolidone at 4‰, for 5 minutes) for example, is recommended. The primary means of contamination is through live food, especially rotifers. Turbot larvae are very sensitive to the microbial environment, especially *Vibrio* and *Aeromonas* that can be controlled using adequate culture and enrichment methods and additives such as probiotics. Over feeding should be avoided in larvae culture tanks when possible (should be controlled when cofeeding with compound diets).

The practice of adding antibiotics to the larval culture tanks is not a reliable method for bacterial control, so, after each hatchery run, the facilities should be stopped, dried up for several days, and disinfected.

7.3.3 Larvae harvest

Turbot larvae are moved from larval tanks to new facilities, "weaning tanks," when still pelagic, between day 20 and 30. By day 30, depending on rearing conditions and temperature, turbot look like a small adult; the flattening of the body and right eye migration are completed, and the fish are concentrated in the upper portion of the water column, where they can be easily harvested most often by nets. Behavioral metamorphosis just beginning in larger fish that are progressively changing from pelagic to benthic status with visible swimming disturbances; the population has become completely benthic by day 40–50.

7.3.4 Weaning

Except for stock enhancement programs, turbot are weaned as soon as possible to reduce production costs. Weaning success is strongly correlated with the size of the postlarvae, food quality, and feeding procedures that could be extremely diversified. The development of efficient weaning diets in the early 1980s (Person

Le Ruyet *et al.* 1991) led to major progress in weaning larvae larger than 40 mg (day 25 ph).

System design and requirements

Although weaning can be started directly in larvae tanks using special compound diets, it is more commonly done at high stocking density (2,500 fish/m^2) in shallow circular or square tanks (0.25–0.50 m depth, 5–10 m^2) to promote feeding activity and limit water pollution. Water treatment is as described for larvae. UV treatment is still required to protect fish against vibriosis, as at this age they are very susceptible to vibriosis and cannot be efficiently protected by vaccination.

Weaning protocols and results

Most formulated dry diets available today are expanded rehydratable water-stable diets developed at IFREMER for flat fish species such as sole (*Solea solea*) and turbot (Person Le Ruyet *et al.* 1991). Fish are always fed in large excess when weaning starts to increase their chances of ingesting the pellets during their slow descent to the tank bottom. Uneaten pellets and faeces are daily flushed out. Conversely to larvae weaning procedures, following 3–4 days food ration is adjusted to apparent feed intake.

Weaning can be either progressive, with dry pellets offered continuously using an automatic feeding device plus once daily feeding of *Artemia* for 3–4 days maximum, or direct, without any *Artemia* transition. Weaning of 30-day old, high-quality juveniles is fairly routine; under good conditions, 75% of postlarvae accept high-quality commercial diets in less than 2 days and a survival rate as high as 90% can be obtained. On private farms, weaning success ranges from 60 to 80%.

7.3.5 Hatchery economics

In Europe, intensive methods are the most common in larvae culture and the main problem is still the highly variable survival rate of larvae, which usually varies from less than 10% to over 30%. Peak mortality occurs at the end of the first week when endogenous yolk reserves are completely exhausted. Another mortality peak may be observed between days 12 and 15, when *Artemia* feeding begins, sometimes leading to the complete loss of larvae for a variety of reasons (apparent starvation related to inadequate nutrition or an unfavorable microbial environment). For the industry, a mean annual survival rate of 20% at day 90 ph (1–2 g) is economically acceptable.

The initial high incidence of pigmentation abnormality or incomplete right eye migration, that becomes apparent after metamorphosis, has been partly solved by nutritional improvements. Currently, the 50% decrease in the percentage of juveniles discarded before market has greatly contributed to lowered juvenile production costs.

Live food still plays an important role in successful fingerling production, but the development of the latest generation of compound diets used in cofeeding with *Artemia* will lead to a major intensification of larval rearing for turbot farming.

Turbot fry production has advanced in Europe to the point that today the availability of high-quality juveniles is not limiting industry development as it was in the 1990s. The majority of production is shared between France and Spain and varies according to world fry demand, around 10 millions fry today. During the last decade, fingerling production markedly increased in Spain (Galicia) near growout sites and today has reached a level of about 7 millions fry. In France, fry production showed a sharp decline after 2005 due to or caused by a decrease in the export market (Spain and China) and the lack of national development of turbot farming.

7.4 Nursery culture and transition to growout

Juveniles leave the hatchery at 3–4 months (1–3 g) and are cultured in a nursery up to between 5 and 20 g or exceptionally up to 80–100 g. This prefattening period in a less controlled environment than in the hatchery lasts 3–6 months.

7.4.1 System design and requirements

Nursery production is either done in a greenhouse or an industrial building normally located on the same site as the hatcheries. Traditionally, turbot juveniles were nursed in flow through systems in shallow concrete or fiberglass square or circular tanks of about 10–30 m^2 surface area and 0.5–0.7 m depth. Today, recirculating systems (RAS) using clear water or ground water and high stocking densities (500–1,000 fish/m^2) are common. RAS allows better control of key environmental factors (e.g., temperature) and biosecurity than flow through systems and reduces the cost of heating and pumping. Another alternative system combines RAS technology and shallow raceways of a large size range (0.25 m depth maximum) that can be stacked in 3–4 layers (Oiestad 1999) for better use of the nursery surface area.

Long-term effects of major environmental factors (temperature, salinity, light, oxygen, ammonia) on turbot juveniles growth have been extensively studied and reviewed (Bœuf *et al.* 1999, Person Le Ruyet *et al.* 1991; Imsland *et al.* 2000a, 2000b, 2008). Temperature control is critical for growth; the optimal range is 16–22°C for 10 g turbot and 16–19°C for 40–50 g fish, oxygen concentration should be maintained above 6 mg/L and juvenile growth may be slightly enhanced at 20‰. Turbot are relatively tolerant of ammonia: 96h-LC 50, around 56 mg/L total ammonia nitrogen (2.5 mg/L NH_3).

7.4.2 Nursery protocols

Juvenile health and growth performance are highly dependent on water quality, nutrition, and feeding. Optimum environmental conditions are as follows: temperature, 16–18°C; salinity, 20–27‰; oxygen, 100% of air saturation. Ambient photoperiod is commonly used in nurseries and excessive light intensity (over 2000 lx) may be detrimental. Juveniles are tolerant of ammonia; threshold concentrations for growth are 0.08 and 0.14 mg/L NH3-N at pH 6.8 and 7.9, respectively. Ammonia concentrations can increase rapidly in RAS systems with excessive stocking densities.

Juveniles are fed exclusively dry pellets delivered automatically and continuously during the daylight period. Basic nutritional requirements were established in the 1980s. A standard turbot diet contains 51–52% protein and 12–13% total fat. The use of fish meal and fish oil as the main protein and lipid sources minimizes the risk of HUFA deficiency (Guillaume *et al.* 1991). Since the 1990s in Europe, convenient commercial growout diets completely replaced moist pellets for feeding turbot. Currently, the industry is placing emphasis on optimizing feed formulations and improving ingredient quality for better nutrition, health, and growth performance.

7.4.3 Nursery results

Nursery production has been stable and reliable in recent years. The average survival is over 80% and growth rates are high; fingerlings reach 20–30 g at 6 months under the optimum temperature regime. Food conversion may be as low as 1.0 during this phase of production.

Prior to leaving the nursery, turbot are most often vaccinated against vibriosis and furunculosis. When requested by growout facilities, they can be vaccinated against diseases caused by *Flexibacter* and *Streptococcus.* The list of bacterial and viral diseases is continually expanding but there are a limited number of vaccines available for commercial use. For example, significant mortalities can be caused by *Edwardsellia tarda* but at the moment there is no effective treatment besides usual prophylactic methods.

Prior to leaving the nursery, turbot are graded by size using automatic machines adapted from machines designed for grading fruits (they are held out of water during grading). The most common market size is about 20 g. Fish are transported to growout farms by road (specialized international transport companies). Airline transportation is also used when necessary.

7.5 Growout

Turbot growout relies solely on intensive production methods. However, specific protocols vary greatly from country to country and from farm to farm, so the following are only generalizations. Growth performance is also extremely variable from site to site; for purposes of this discussion, market size varies from 0.5 to over 3 kg.

7.5.1 System design and environmental conditions

Turbot can be reared at high stocking densities in a variety of land-based tanks and raceways most often located in cheap industrial buildings. Individually covered outdoor tanks are rare. Tank volume is increased from 25 m^3 to 100 m^3

as fish grow, but the useful depth is 0.70 m or less even with large fish. Farms initially had flow through rearing systems but RAS are now developing rapidly (Spain, France, Great Britain, the Netherlands). RAS have many advantages compared with other rearing systems. Irrespective of their exact design, they are mechanically sophisticated and biologically extremely complex; as a result, they require continuous water quality control to be as efficient as a flow through system (Blancheton 2000). To minimize heating costs, the tendency in RAS is to limit fresh seawater supply to 5–10% of total water volume/day and to use ground water where possible. Most often, geothermal saltwater requires special treatment to satisfy basic water quality requirements (oxygen, metal ions, pH).

Flat-bottomed cages submerged in coastal areas or floating cages can also be used for growout or holding of large fish prior to marketing. Floating cages were initially tested in sea ponds, sheltered bays, or estuaries in Great Britain and France. The use of cages requires specific physical and environmental conditions (moderate currents and optimum temperature for most of the year). Cage frames are commonly made from metal with a metal or nylon bottom. Today, turbot sea cages are being tested in North West Spain for comparison with land-based systems.

Optimal environmental requirements for growth of large fish are poorly known compared to those for juveniles because of the lack of experimental data. Field observations indicate that growth of large turbot declines rapidly below 14°C and above 20°C, therefore, the optimal range is considered to be from 16 to 18°C. The lowest and highest temperature at which feeding is stopped is around 8°C and 22°C. Large fish seem more sensitive to oxygen depletion and to any environmental stress than juveniles, which often leads to a transient decrease in appetite and subsequently in growth.

The introduction of opportunistic pathogens into growout units is prevented through treatment of the water supply (UV, ozone) and the use of vaccinated juveniles. Additional vaccination is usually conducted during growout (as the duration of protection by vaccines is limited). Disease prevention and treatment protocols differ from farm to farm but they all have the objective of limiting the use of antibiotics and other therapeutic agents. Turbot are also highly susceptible to parasite infestation, mainly *Trichodina* and *Uronema*. In some sites, production managers treat their fish with formalin baths once a month to limit parasites.

7.5.2 Diet and nutrition

A wide range of diets are suitable for growout of turbot. Commercial success was first accomplished in Great Britain by farms using trash fish (mixtures of different species were commonly used). The same feeding practice (and mainly sardines) has been used in Northwest Spain for turbot fattening for several years. The use of trash fish as feed depends on availability and price and may lead to deterioration in water quality and slow growth and poor flesh quality. French farmers were the first to use extruded pellets specially developed for turbot during the early 1980s and these are now in common use for growout everywhere.

Specially formulated diets for turbot were developed as production increased. In general, the dietary energy content of turbot diets is lower than that of other farmed fish species, and substitutes to fish meal and oil sources are increasingly used (Regost *et al.* 2001). Commercial diets for growout have high protein content, 50–54% of dry matter, and low crude lipid content, about 12%. Diets with 20% lipid can be used for the fattening phase for turbot over 1.5 kg to meet the specific Spanish market demand for fish with a high flesh fat. The development of quality labels, such as "turbot label rouge" in France, certifies high quality and traceability of the diets and fish.

Feeding protocols on turbot farms are diverse and poorly defined. Turbot are most often fed by hand twice a day during summer and once a day during winter when temperature is below 12°C. They are fed at specific times according to fish preference and the best feeding time depends on environmental conditions at the farm (turbot are very sensitive to noise). Self-feeders (demand feeders) are not well adapted to feed turbot. To facilitate visual monitoring of feeding activity, floating pellets can be used.

Apparent food conversion ratios (FCR) average 1.2–1.3 during growout. The FCR is higher for larger fish compared to small fish and is highly affected by any stress condition or disease and by the onset of maturation later in the production cycle.

7.5.3 Stocking densities and splitting

As turbot can tolerate overcrowding (up to four layers of large fish stacked on top of each other), stocking density (SD) may be high (over 100 kg/m^2 of useful surface area). Irrespective of the rearing systems, the upper operational SD is difficult to define without reference to water quality (highly dependent on water exchanges) and feeding (ration and procedures). In intensive systems, SD is about 30–35 kg/m^2 of 300 g fish, 45 kg/m^2 of 750 g fish, and up to 60–80 kg/m^2 of larger fish. Regular splitting of the tank populations is used to limit excessive SD increases. SD can be increased for short periods (during handling and during the fattening period for fish over 1 kg fish).

During growout, it is advisable to maintain fish sizes homogeneous, although laboratory data has shown that regular grading does not promote growth. Fish are measured with automatic machines and sorted into several size classes generally twice during growout. Regular handling during grading has little apparent effect on the health of the fish.

7.6 Harvesting, processing, and marketing

Fish are harvested manually and stunned prior to killing using established and accepted methods with regard to animal welfare and flesh quality for human consumption (according to European regulations for any farmed species). A rapid chilling followed by bleeding is usually applied; the use of asphyxia in air or on ice is not an appropriate euthanasia protocol for animal welfare. Electrocution or percussive blow on head can be used as stunning-killing methods (Roth *et al.* 2007). Fish are transported to the processing units on ice.

Farmed turbot are usually marketed whole and fresh and are only gutted according to specific market demand. Large fish can also be sold as fillets; this market is developing in Europe. Due to an increasing demand for live turbot in Asian countries and some European cities, a limited market for live turbot has been developing in France since the late 1990s. With special preconditioning and packaging developed by private companies, turbot can survive up to 2 days without water, and survival can be over 85% after 18 hours transport without water.

Turbot can be marketed at sizes from 0.5 kg to 4 kg, with larger fish commanding higher prices. Most farmed turbot are marketed at an average weight of about 1–1.2 kg (0.8–1.5 kg). There is also much interest in large fish (3–4 kg), mainly in Spain. Recently, the demand for pan-sized fish has increased; this market requires 500–750 g fish that are sometimes sold frozen.

Because demand is higher than supply, there is not, at the moment, any major competition between farmed and wild turbot. Wild turbot are larger and command a higher price on the market.

7.7 Production economics

Although survival of turbot can be near maximum during growout, parasitic, bacterial, and viral diseases may cause significant losses. Growout units try to avoid the introduction of opportunistic pathogens through usual prevention methods and the use of vaccinated juveniles as previously discussed. Improvements in disease prevention and control have led to reduced use of antibiotics and other medicines, but practices are highly variable from site to site.

Growth is also highly variable from site to site and is mainly dependent on temperature. Turbot can reach 3 kg or more at 3 years of age. This fast growth rate was achieved several years ago on small and large farms using trash fish and heated water in Great Britain. Similar growth rates are regularly obtained on production farms using dry pellets and geothermal water when the annual temperature range is 14–19°C. In Spain, where the ambient temperature ranges from 14 to 18°C, turbot that reach 2–2.5 kg are routinely produced in less than 3 years (1 kg at 18 months). However, where the seasonal temperatures range from 9 to 19°C, a weight of 1.2–1.5 kg is obtained only after 3 years (750 g at 2 years), roughly half the maximum growth potential of turbot.

Major improvements in turbot growth may be obtained by maintaining an optimal thermal regime, by increasing oxygen to near saturation, and by feeding fish to apparent satiety (Bœuf *et al.* 1999; Person Le Ruyet 2002a, 2001b). As very efficient commercial diets are now available, they potentially have a low impact on growth performances but dietary composition is the main determinant of flesh quality (Regost *et al.* 2001). Selection of fast growth late-maturing fish or production of all-female or sterile stocks to avoid the loss in growth caused by sexual maturation are other ways to improve turbot growth. First maturity of farmed fish is at 2 years of age (less than 1 kg) for males and could be delayed

Table 7.1 Turbot production in Europe (MT).

	Spain	France	Portugal	Europe
1995	2,174	800	82	2,978
1996	2,189	850	102	2,571
1997	1,800	950	196	3,001
1998	1,969	900	188	3,087
1999	2,849	868	378	4,103
2000	3.683	908	380	4,785
2001	3,636	702	343	4,856
2002	3,847	924	386	5,267
2003	3,852	909	323	5,363
2004	4,347	949	269	6,008
2005	5,572	791	214	6,838
2006	6,419	800	185	7,633

by photoperiod manipulation. Maturation occurs one year later for females that weigh 10–20% more than males of the same age.

A classic way to improve growth over several generations is to select breeding populations for fast growth, environmental preferences, or physiological capacity to adapt to specific rearing conditions and especially to temperature. The effect of temperature on growth is functionally related to three types of hemoglobin; optimum growth occurs from 19 to 23°C due to differences in the oxygen-binding proprieties of the blood (Imsland *et al.* 2000b). Little information on genetic manipulation exists in the public domain because applied research on turbot genetics is mainly carried out by the industry. The benefits of triploidy are probably relatively low in turbot as in sea bass (Felip *et al.* 2001).

7.8 Summary: industry constraints and future expectations

Turbot production systems and protocols can be considered a mature technology in Europe. Production has been increasing from 270 MT in 1987 to 3,327 MT in 1997 and 7,633 MT in 2006 (Table 7.1). The largest producers are Spain (84% of 2006 production) and a distant second, France (18%). In Spain, production has increased steadily (70% of total production in 2000) and in France, it has stabilized at 800–900 MT since 2000. There is also a small consistent production from Portugal (185 MT in 2006). In the Netherlands, turbot production is expected to expand in near future (100 MT in 2006). The primary European market is Spain, while much smaller markets exist in France, Italy, and Germany.

In Chile, turbot farming started in 1992. Initial production of 17 MT was obtained with juveniles supplied from Europe and technical support from Great Britain. Chile produced 250 MT in 1997 and 450 MT in 2000 from juveniles supplied locally. Most of the turbot produced in Chile are exported to Asian countries and the United States.

Although turbot productivity during growout is much lower than the productivity of salmon farms, it seems likely that the industry will see market expansion in the future with the building of new rearing units in some countries (the

Netherlands, for example) and an increase in the capacity of existing farms, especially in Spain. The availability of high-quality juveniles is not limiting development, rather the high price of juveniles, 1.04–1.10 € per fish, mainly related to the low survival rate of larvae. In some countries, such as in France, turbot farming is limited by access to seawater (conflicts with tourism). There is a market demand for whole fish (about 9.2 € kg in 2006) with a high price and new markets (fillets) are developing. Profit margin seems acceptable, but as was seen for sea bass and sea bream, a decrease in market price should be expected in the future as turbot production increases.

The production efficiency of turbot farming may be improved in many ways:

- by decreasing the purchase price of juveniles which accounts for 18% of total production cost;
- by improving culture systems and automation to lower labor and feed cost (16 and 17% of total production cost);
- by better prevention and control of disease;
- by improving genetics and marketing.

Literature cited

Alvial, A., and Manriquez, J. 1999. Diversification of flatfish culture in Chile. *Aquaculture* 176:65–73.

Blancheton, J.P. 2000. Developments in recirculation systems for Mediterranean fish species. *Aquacultural Engineering* 22:17–61.

Bœuf, G., Boujard, D., and Person-Le Ruyet, J. 1999. Control of the somatic growth in turbot. *Journal of Fish Biology* 55:128–147.

Borell, Y.J., Alvarez, J., Vasquez, E., Pato, C.F., Tapia, C.M., Sanchez, J.A., and Blanco, G. 2004. Applying microsatellites to the management of farmed turbot stocks (*Scophthalmus maximus* L.) in hatcheries. *Aquaculture* 241:133–150.

Felip, A., Piferrer, F., Zanuy, S., and Carillo, M. 2001. Comparative growth performance of diploid and triploid European sea bass over the first four spawning seasons. *Journal of Fish Biology* 58:76–88.

Girin, M. 1972. Métamorphose en élevage de deux larves de turbot (*Scophthalmus maximus* L.). *Comptes Rendus De L Academie Des Sciences Paris* 275D:2933–2936.

Girin, M. 1979. *Méthodes de production des juvéniles chez trois poisons marins, le bar, la sole, et le turbot*. Rapports Scientifiques et Techniques, Publication 39, CNEXO (France), 202 pp.

Guillaume, J., Kaushik, S., Bergot, P., and Métailler, R. (eds) 2001. *Nutrition and Feeding of Fish and Crustaceans*. Springer-Praxis books in aquaculture and fisheries, Berlin, 408 pp.

Guillaume, J., Coustans, M.F., Métailler, R., Person-Le Ruyet, J., and Robin, J., 1991. Flatfish, turbot, sole and plaice. In: Wilson, R.P. (ed.) *Handbook of Nutrient Requirement of Finfish*. CRC Press, Boston, MA, pp. 77–82.

Harel, M., Tandler, A., Kissil, G.W., and Applebaum, S.A. 1994. The kinetics of nutrient incorporation into body tissues of gilthead sea bream (*Sparus aurata*) females and the subsequent effects on egg composition and egg quality. *British Journal of Nutrition* 72:45–58.

Imsland, A.K., Foss, A., Gunnarsson, S., Berntssen, M.H.G., FitzGerald, R., Bonja, S.W., Ham, E.V., Naevdal, G., and Stefansson, S.O. 2000a. The interaction of temperature and salinity on growth and food conversion in juvenile turbot (*Scophthalmus maximus*). *Aquaculture* 198:353–367.

Imsland, A.K., Foss, A., Stefansson, S.O., and Naevdal, G. 2000b. Hemoglobin genotypes of turbot (*Scophthalmus maximus*): consequences for growth and variations in optimal temperature for growth. *Fish Physiology and Biochemistry* 23:75–81.

Imsland, A.K., and Gunnarson, S. 2008. Commercial-scale validation of temperature-step rearing on growth physiology in turbot, *Scophthalmus maximus*. *Journal of the World Aquaculture Society* 39:684–691.

Jones, A., Brown, J.A.G., Douglas, M.T., Thompson, S.J., and Whitfield, R.J. 1981. Progress towards developing methods for the intensive farming of turbot (*Scophthalmus maximus* L.) in cooling water from a nuclear power station. In: Rosenthal, H., and Tiews, K. (eds) *Proceedings of the World Symposium on Aquaculture in Heated Effluents and Recirculation Systems*. May 28–30, 1980, Vol. II. Stavanger, Berlin, pp. 482–496.

Oiestad, V. 1999. Shallow raceways as a compact, resource-maximizing farming procedure for marine fish species. *Aquaculture and Research* 30:831–840.

Person Le Ruyet, J. 1989. The hatchery rearing of turbot larvae *(Scophthalmus maximus)*. In: *Cuadernos da Area de Ciencias Marinas, Publicacions do Seminario de Estudos Galegos*, Vol. 3, pp. 57–91.

Person-Le Ruyet, J. 2002a. Turbot (*Scophthalmus maximus*) grow-out in Europe: practices, results and prospects. *Turkish Journal of Fisheries and Aquatic Sciences* 2:29–39.

Person-Le Ruyet, J. 2002b. Water quality requirement for sea water fish. In: Arzul, G. (coord.) (ed.) *Aquaculture, Environment and Marine Phytoplankton*. Ifremer, Actes Colloq., Brest, May 21–23, 2001, Vol. 34, pp. 71–75.

Person Le Ruyet, J., Baudin-Laurencin, F., Devauchelle, N., Metailler, R., Nicolas, J.L., Robin, J., and Guillaume, J. 1991. Culture of turbot, *Scophthalmus maximus*. In: McVey (ed.) *Handbook of Mariculture, Finfish Aquaculture*. CRC Press, Boston, MA, Vol. 2, pp. 21–41.

Purdom, C.E., Jones, A., and Lincoln, R.F. 1972. Cultivation trials with turbot (*Scophthalmus maximus* L.). *Aquaculture* 1:213–230.

Regost, C.J., Arzel, M., Cardinal, M., Robin, J., Laroche, M., and Kaushik, S.J. 2001. Dietary lipid level, hepatic lipogenesis and flesh quality in turbot (*Psetta maxima*). *Aquaculture* 193:291–309.

Roth, B., Imsland, A., Gunnarsson, S., Foss, A., and Schelvis-Smit, R. 2007. Slaughter quality and rigor contraction in farmed turbot (*Scophthalmus maximus*): a comparison between different stunning methods. *Aquaculture* 272:754–761.

Shields, R.J. 1991. Larviculture of marine finfish in Europe. *Aquaculture* 200:55–88.

Zambonino Infante, J.L., and Cahu, C.L. 1999. High dietary lipid levels enhance digestive tract maturation and improve *Dicentrarchus labrax* larval development. *Journal of Nutrition* 129:1195–1200.

Section 3

Asia and Australia Culture

In Japanese hatcheries, which produce juveniles for restocking, broodstock maintenance, and larviculture are performed based on the policy and findings of these studies mentioned above. Particularly, most hatcheries are replacing broodstock from subcultured populations to wild-caught fish. Moreover, spawners are renewed periodically not only for disease prevention, but also for the maintenance of genetic diversity.

Literature cited

Alam, Md. S., Teshima, S.-i., Ishikawa, M., and Koshio, S. 2000. Methionine requirement of juvenile Japanese flounder *Paralichthys olivaceus*. *Journal of the World Aquaculture Society* 31:618–626.

Asahida, T., Shinotsuka, Y., Yamashita, Y., Saitoh, K., Hayashizaki, K.-i., and Ida, H. 2003. Influence of hatchery protocols on mitochondrial DNA variation in Japanese flounder juveniles. *Journal of the World Aquaculture Society* 34(2):121–132.

Deng, J., Mai, K., Ai, Q., Zhang, W., Wang, X., Xu, W., and Liufu, Z. 2006. Effects of replacing fish meal with soy protein concentrate on feed intake and growth of juvenile Japanese flounder, *Paralichthys olivaceus*. *Aquaculture* 258:503–513.

Fujinami, Y., and Kumagai, A. 2005. Influences of oxidant-disinfected seawater on survival rate and incidence of pseudoalbinism on ocular side. *Technical Report of National Center for Stock Enhancement* 5:39–42 (in Japanese).

Furuita, H., Tanaka, H., Yamamoto, T., Suzuki, N., and Takeuchi, T. 2002. Effects of high levels of n-3 HUFA in broodstock diet on egg quality and egg fatty acid composition of Japanese flounder, *Paralichthys olivaceus*. *Aquaculture* 210:323–333.

Fushimi, H. 2001. Production of juvenile marine finfish for stock enhancement in Japan. *Aquaculture* 200:33–53.

Hirano, R., and Yamamoto, E. 1992. Spawning rhythm and egg number at the individual female spawning experiment in the hirame flounder, *Paralichthys olivaceus*. *Bulletin of the Tottori Prefectural Fisheries Experimental Station* 33:18–28.

Honda, H., and Kikuchi, K. 1997. Management of a seawater recirculation fish culture system for Japanese flounder. *UJNR Technical Report* 24:165–171.

Hondo, Y., Murakami, N., Mushiake, K., and Tsuzaki, T. 2006. Broodstock management and egg collection from reared wild-broodstock of Japanese flounder *Paralichthys olivaceus* fed commercial extruded pellets. *Nippon Suisan Gakkaishi* 72(5):873–879.

Howell, B.R., and Yamashita, Y. 2005. In: Robin N. Gibson (ed.) *Flatfishes: Biology and Exploitation*, Fish and aquatic resources series 9. Blackwell publishing, Oxford, pp. 347–371.

Iwata, N., Furuta, T., Kikuchi, K., Sakaguchi, I. Okada, Y., and Kuwabara, R. 1994. Survival and growth of Japanese flounder at low salinity. Paper read at Annual Conference of the Japanese Society of Fisheries Science, October 1–4, Mie University, Tsu, Mie, Japan.

Kaji, S., and Fukunaga, T. 1999. Results of a questionnaire on the recent status of seed production and market price of recaptured Japanese flounder *Paralichthys olivaceus*. Showing abnormal coloration. *Saibai Giken* 27(2):67–101 (in Japanese).

Kikuchi, K. 1998. Blue mussels in the diet of juvenile Japanese flounder. *UJNR Technical Report* 26:269–274.

Kikuchi, K. 1999. Use of defatted soybean meal as a substitute for fish meal in diets of Japanese flounder (*Paralichthys olivaceus*). *Aquaculture* 179:3–11.

Kikuchi, K., Furuta, T., and Honda, H. 1994. Utilization of soybean meal as a protein source in the diet of juvenile Japanese flounder, *Paralichthys olivaceus*. *Suisanzoshoku* 42:601–604.

Kikuchi, K., Honda, H., and Kiyono, M. 1992. Effect of dietary protein level on growth and body composition of Japanese flounder, *Paralichthys olivaceus*. *Suisanzoshoku* 40:335–340 (in Japanese with English abstract).

Kikuchi, K., Iwata, N., and Takeda, S. 2002. Production of Japanese flounder in closed recirculating aquaculture system. *Fisheries Science* 68:851–854.

Kikuchi, K., Sato, T., and Deguchi, Y. 1998. Effect of dietary carbohydrates on the growth of juvenile Japanese flounder. *Suisanzoshoku* 46:541–546 (in Japanese with English abstract).

Kikuchi, K., Sugita, H., and Watanabe, T. 2000. Effect of dietary protein and lipid levels on growth and body composition of Japanese flounder. *Suisanzoshoku* 48:537–543.

Kikuchi, K., and Takeda, S. 2001. Present status of research and production of Japanese flounder, *Paralichthys olivaceus*. *Journal of Applied Aquaculture* 11:165–175.

Kim, K.-D., Lee, S.-M., Park, H.G., Bai, S.C., and Lee, Y.-H. 2002. Essentiality of dietary n-3 highly unsaturated fatty acids in juvenile Japanese flounder *Paralichthys olivaceus*. *Journal of the World Aquaculture Society* 33:432–440.

Kim, K., Wang, X., and Bai, S.C. 2003. Reevaluation of dietary protein requirement of Japanese flounder *Paralichthys olivaceus*. *Journal of the World Aquaculture Society* 34:133–139.

Ky, X.P. 2007. Endocrinological profile in reproduction of Japanese flounder. PhD thesis, Kitasato University, Japan.

Lee, S.-M., Cho, S.H., and Kim, K.-D. 2000. Effects of dietary protein and energy levels on growth and body composition of juvenile flounder *Paralichthys olivaceus*. *Journal of the World Aquaculture Society* 31:306–315.

Lee, S.-M., Kim, K.D., and Lall, S.P. 2003. Utilization of glucose, maltose, dextrin and cellulose by juvenile flounder (*Paralichthys olivaceus*). *Aquaculture* 221:427–438.

Lee, S.-M., Park, C.-S., and Bang, I.-C. 2002. Dietary protein requirement of young Japanese flounder *Paralichthys olivaceus* fed isocaloric diets. *Fisheries Science* 68:158–164.

Masumoto, T., Tamura, B., and Shimeno, S. 2001. Effects of phytase on bioavailability of phosphorus in soybean meal-based diets for Japanese flounder *Paralichthys olivaceus*. *Fisheries Science* 67:1075–1080.

Metropolitan Central Wholesale Market. 2008. http://www.shijou-tokei.metro.tokyo.jp/index.html (accessed on June 16, 2008) (in Japanese).

Mihelakakis, A., Yoshimatsu, T., and Kitajima, C. 1995. Change in egg size of Japanese flounder during one spawning season. *Journal of the Faculty of Agriculture, Kyushu University* 40(1/2):53–59.

Mobin, Saleh M.A., Kanai, K., and Yoshikoshi, K. 2001. Effects of feeding levels on the growth and survival of larval and juvenile Japanese flounder *Paralichthys olivaceus*. *Suisanzoshoku* 49(2): 207–218.

Mori, K., Sugaya, T., Nishioka, T., Gomez, D.K., Fujinami, Y., Oka, M., Arimoto, M., Okinaka, Y., and Nakai, T. 2005. Detection of betanodaviruses from feed fish used in marine aquaculture. Abstract of EAFP 12th International Conference, Copenhagen.

National Association for the Promotion of Productive Sea. 2008. Annual Statistics of Seed Production and Release in 2006. National Association for the Promotion of Productive Sea, Tokyo (in Japanese).

Morizane, T. 1984. Fundamental study on the culture of plaice *Paralichthys olivaceus* -II. Effect of water temperature on the growth of young fish. *Suisanzoshoku* 32:127–131 (in Japanese).

Morizane, T., and Takimoto, M. 1984. Fundamental study on the culture of plaice *Paralichthys olivaceus* -I. Growth of young fish fed different raw fish. *Suisanzoshoku* 32:121–126 (in Japanese).

Muroga, K. 2001. Voral and bacterial diseases of marine fish and shellfish in Japanese hatcheries. *Aquaculture* 201:23–44.

Mushiake, K., Nishizawa, T., Nakai, T., Furusawa, I., and Muroga, K. 1994. Control of VNN in striped jack: Selection of spawners based on the detection of SJNNV gene by polymerase chain reaction (PCR). *Fish Pathology* 29(3):177–182.

Ototake, M., and Matsusato, T. 1986. Notes on scuticociliata infection of cultured juvenile flounder *Paralichthys olivaceus. Bulletin of National Research Institute of Aquaculture* 9:65–68.

Saitoh, S., Koshio, S., Harada, H., Watanabe, K., Yoshida, T., Teshima, S.-i., and Ishikawa, M. 2003. Utilization of extruded soybean meal for Japanese flounder *Paralichthys olivaceus* juveniles. *Fisheries Science* 69:1075–1077.

Saitoh, S., Sasaki, A., Li, H.-O., Shimizu, M., and Yamada, J. 1990. Effects of low salinities on the growth and metabolism of juvenile Japanese flounder (*Paralichthys olivaceus*). *Scientific Reports of the Hokkaido Fisheries Experimental Station* 34:1–8 (in Japanese with English abstract).

Sakamoto, K., Takahashi, Y.-i., Oka, M., and Itagaki, E. 1998. Bacterial flora in still water larviculture of the Japanese flounder *Paralichthys olivaceus. Suisanzoshoku* 27(1):1–5 (in Japanese with English abstract).

Sato, T. 1998. Development of Formulated Feeds for Juvenile Japanese Flounder. Ph DThesis, Tokyo University of Fisheries, Tokyo, Japan (in Japanese).

Seikai, T. 1998. Fry production of Japanese flounder. *Farming Japan* 32(2):16–21.

Sekino, M., Saito, K., Yamada, T., Kumagai, A., Hara, M., and Yamashita, Y. 2003. Microsatellite-based pedigree tracing in a Japanese flounder *Paralichthys olivaceus* hatchery strain: implication for hatchery management related to stock enhancement program. *Aquaculture* 22:255–263.

Suzuki, S., Sekiya, S., Sugaya, T., Oetega-Villiazan Romo, M.D.M., Matsubara, T., and Taniguchi, N. 2006. Experimental study on broodstock management of barfin flounder under the concept of minimum kinship selection. Proceedings of Third International Symposium on Stock Enhancement and Sea Ranching, Seattle.

Takahashi, Y. 1994. Influence of stocking density and food at late phase of larval period on hypermelanosis on the abocular body side in Japanese flounder *Paralichthys olivaceus. Nippon Suisan Gakkaishi* 60(5):593–598 (in Japanese with English abstract).

Takahashi, Y. 1999. Reduction of the amount of food organisms supplied and simplification of rearing operations in the seedling of Japanese flounder, *Paralichthys olivaceus. Suisanzoshoku* 38(1):23–33 (in Japanese with English abstract).

Yoshimizu, M. 2006. Development of a seawater electrolyzer for disease prevention in aquaculture and food sanitation. *Nippon Suisan Gakkaishi* 72(5):831–834 (in Japanese with English abstract).

Yoshinaga, T., Kamaishi, T., Segawa, I., and Yamamoto, E. 2000. Effects of NaCl-supplemented seawater on the monogenean, *Neoheterobothrium Hirame*, infecting the Japanese flounder. *Fish Pathology* 35:97–98.

Chapter 9

Culture of olive flounder: Korean perspective

Sungchul C. Bai and Seunghyung Lee

9.1 Current status of olive flounder in Korea

Olive flounder, *Paralichthys olivaceus*, is one of the highest valued finfish in the world and has been the top product in Korea for more than 10 years. This species is an important aquaculture species due to its high growth rate, feed efficiency, tolerance to water temperature changes, resistance to diseases, and the availability of seedlings from hatcheries. Olive flounder is very popular among Koreans for its good taste and dress-out yield. It is served in several ways, with raw being the most popular of them all.

According to an FAO report (2009), there are only two countries producing olive flounder around the world, the Republic of Korea and Japan. In 2007, Korea produced 77.6% (44,245 MT) of the world olive flounder supply (57,045 MT) (Table 9.1), while Japan produced 22.4% (12,800 MT).

Total olive flounder production (culture + capture) in Korea has increased steadily during the past 25 years, from 3,881 MT in 1980 through 8,647 MT in 1995 to 44,245 MT in 2007 (Table 9.2).

While the total olive flounder production has been increasing, the capture fishery for olive flounder has leveled off. In other words, the dramatic increase in olive flounder production for the last 25 years is attributed almost solely to aquaculture. The capture fishery contributed 99.5% (3,862 MT) of total olive flounder production in 1980 while a meager 0.5% (19 MT) was from aquaculture production. In 1995, the olive flounder capture fishery dropped to 1,914 MT while aquaculture production was 6,733 MT. In 2007, an increase in capture production was recorded, but this figure (3,074 MT) was still less than the production in 1980. In the same year, aquaculture production of 41,171 MT was recorded, a 511% increase from the 6,733 MT in 1995. Out of the 140,701 MT of finfish aquaculture production in Korea in 2007, olive flounder contributed 29.3% (41,171 MT). Korean Rockfish ranked second (35,564 MT) to olive flounder, followed by sea bream (11,163 MT), Japanese eel (10,557 MT), and mullet (4,921 MT). Olive flounder and Korean rockfish made up 54.5% of the total Korean finfish aquaculture production in 2007. The dominance of

Table 9.1 World olive flounder production in 2007 (metric tons).

Country	Aquaculture	Capture	Capture and aquaculture
Republic of Korea	41,171	3,074	44,245
Japan	4,600	8,200	12,800
Total	45,771	11,274	57,045

Source: FAO (2009).

olive flounder in Korean aquaculture sector is due to the favorable government policies, geared toward production of high-value species. The value of olive flounder aquaculture production in 2007 was 472.4 million US dollars. This is a 364% increase from the 101.8 million US dollars recorded in 1995.

Most of cultured olive flounder (41,171 MT in 2007) are mainly consumed in the domestic area, and some of them are exported to Japan (3,046 MT; US$41,418,853), United States (58 MT; US$1,728,227), and Taiwan (11 MT; US$179,658). Market size of fish is near to 1 kg body weight, and usual price of 1 kg fish is about from US$10 to 15 depending on demand and supply. Korean people enjoy raw fish, and raw fish of olive flounder is mostly consumed rather than other cultured fish.

9.2 Basic biology and ecology

9.2.1 Life cycle

The olive flounder (*P. olivaceus*) is a species of large-tooth flounder native to the northwestern Pacific Ocean (Yoo 1987). Olive flounder lives in coastal areas. The optimum temperature for the fish ranges between 21 and 24°C. Olive flounder seasonally migrate from coast to offshore in order to search for spawning areas and prey availability. Olive flounder continuously grows during sexual maturation, and female fish usually grows faster and larger than male fish (NFRDI 2006).

The life cycle of olive flounder is shown in Figure 9.1.

Mature olive flounder inhabit offshore areas where the water depth is about 100 m. They spawn in shallower water (less than 70 m) from January to August depending on the water temperature (at least above 15°C) (Yoo 1987). Eggs and

Table 9.2 Olive flounder production in Korea.

	1980	1985	1990	1995	2000	2005	2006	2007
Culture (MT)	19	84	1,037	6,733	14,127	40,075	43,852	41,171
Capture (MT)	3,862	2,883	2,462	1,914	1,607	2,112	2,298	3,074
Total (MT)	3,881	2,967	3,499	8,647	15,734	42,187	46,150	44,245
Culture value (US $1,000)	11.1	69.6	32,236	101,753	168,232	346,510	481,856	472,374

Source: FAO (2009).

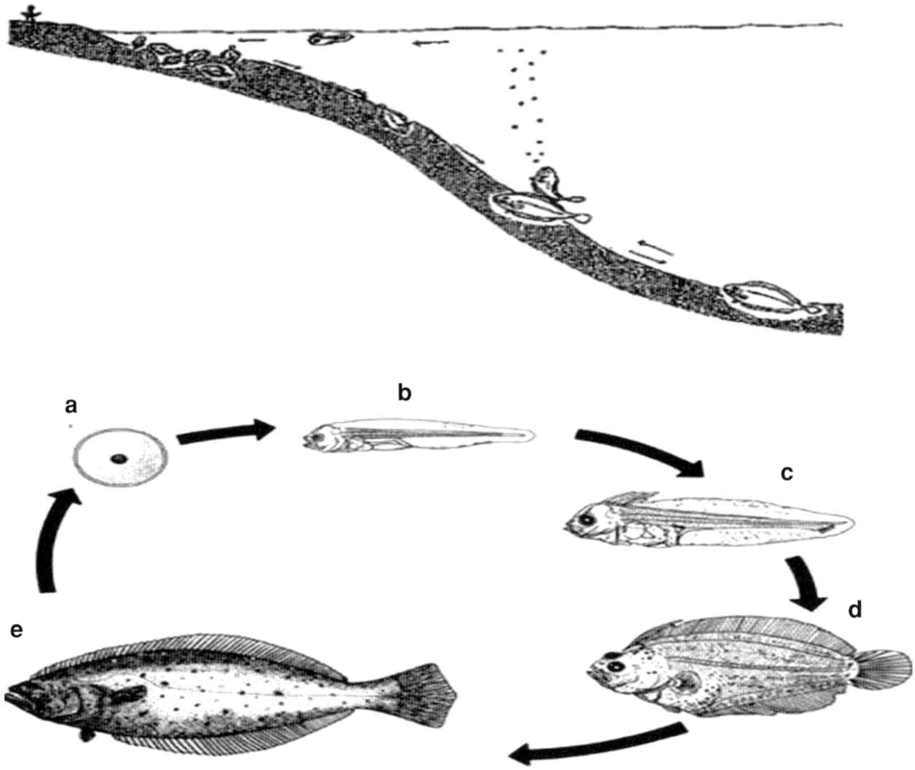

Figure 9.1 Life cycle of olive flounder. (a) fertilized egg; (b) hatched larvae; (c) before metamorphosis (7.6 mm TL); (d) after metamorphosis (14.8 mm TL); (e) adult (45 cm TL) (Bai 2007).

larvae have a planktonic life between 24 and 50 days posthatch (Yoo 1987). Metamorphosing larvae are transported to near-shore by currents and settle on sandy bottom nursery grounds less than 10 m in water depth. They spend about 2 months in their nursery ground and then migrate offshore. This migration is associated with the shift of feeding habits from mysid shrimps to fish in August when they reach 100 mm in body length. In the nursery grounds, the most important food for olive flounder is mysid crustaceans (NFRDI 2006).

9.2.2 Breeding techniques

Flounder eggs are 0.83–1.1 mm in diameter. Both eggs and yolk sac larvae are particularly difficult to rear, compared with other marine fish. At the end of the yolk sac stage, larvae start feeding and then they metamorphose, the symmetrical larvae (shaped like round fish) becoming flatfish.

Development of flounder from fertilized egg to end of metamorphosis takes 24–30 days to produce a larvae about 10–15 mm long. Survival varies from 50 to 60%, being best in batches fed rotifers, brine shrimp (*Artemia* spp.), and micropellets.

9.2.3 Broodstock and spawning

Until recently, farmers and researchers obtained eggs from either 2–3-year-old wild-caught broodstock or from 3-year-old cultured fish (Seo 2007). The natural spawning season lasts from January to May, but under controlled conditions, it also occurs from May to December (Seo 2007). For commercial purposes, control of environmental factors such as light and temperature makes it possible to have batches of fish ready to spawn throughout the year, especially in autumn. Olive flounder broodstock are held in land-based tanks rather than cages, so that they can be handled more easily for spawning. Water temperature is usually kept between 13 and 18°C (Seo 2007). Olive flounder do not release all their eggs at once. Eggs may be obtained many times from the same fish during its spawning cycle at intervals of 2–8 days. In natural populations, minimum spawning size is about 40 cm total length (TL) in 2-year-old fish. One spawning female from 45 to 60 cm TL can lay between 140,000 and 400,000 eggs several times (Seo 2007). The use of hormonal treatment and stripping is not required because ripe males and females can spawn naturally in captivity. Normally, it is expected that at least 90% of the eggs be fertilized and that over 80% of these fertilized eggs produce larvae. As more and more broodstock originate from hatchery-raised fish, this gives the opportunity for improving broodstock.

9.2.4 Egg incubation

Ripe eggs can easily be collected from spawning tanks. Once fertilized, eggs are collected into screened containers (egg collectors) and transferred to incubation tanks. Hatching rate is higher in a darkened room. Water temperature is also important; hatching rate reaches 90% at 14–16°C but drops to 60% when water temperature rises to 22°C. The incubation system varies. Volume ranges from 1 m^3 to 200 m^3, depending on the number of eggs to be incubated. In general, small incubators are used for intensive systems, juveniles being later moved to larger tanks for weaning and nursing. Larger incubators are used not only for rearing fertilized eggs but also for the production of juveniles 20–30 mm long. A continuous water flow is generally maintained through 2–20 m^3 incubators. Dead eggs and debris are siphoned daily to prevent bacterial and fungal contamination. Live eggs are immediately disinfected by using 25–30 ppm iodine solution for 15 minutes before hatching. At 20°C, hatching occurs after about 48 hours producing fragile yolk sac larvae from 2.1 to 3.0 mm long (Seo 2007). The yolk sac is relatively large but there is no functional eye or mouth yet.

9.2.5 Yolk sac larvae development

The yolk sac stage lasts for 4–5 days at 18°C water temperature, and optimum water temperature for this stage ranges between 16 and 22°C (NFRDI 2006).

During this period, larvae develop from their yolk reserves. They are sensitive to light and temperature. Larval development needs a light intensity of 400–600 lx, survival and growth being affected if light intensity is lower than 40 lx or higher than 1,000 lx (NFRDI 2006). At a water temperature of 20°C, larvae become females. To increase the proportion of males, water temperature should be maintained between 15 and 19°C. Survival generally ranges from 50 to 70%. By the end of this period, larvae are about 3.8 cm long.

9.2.6 First feeding

About 4–5 days after hatching and just before the mouth of the larvae opens, rotifers should be distributed (5 individuals/mL). In Korea, rotifer, *Brachionus niloticus* is widely used for the first 4–15 days. Then, *Artemia* nauplii are preferred.

9.2.7 Larval rearing

Larvae are fed brine shrimp (*Artemia*) nauplii hatched from dried eggs and/or copepods produced commercially. The quality of these living feeds was the subject of much research. Brine shrimp nauplii are an incomplete source of nutrients, responsible for low survival, incomplete metamorphosis, and/or abnormal pigmentation. They have to be supplemented either by copepods or by enriched brine shrimp nauplii (Jung 2007).

9.2.8 Metamorphosis

Flatfish start their life upright, like a round fish. They turn on to one side, which then becomes the belly, during metamorphosis. The eye and nostril on that side move up and over the head, joining the other eye and nostril on what now becomes the back. This extraordinary biological change usually occurs 24–50 days after hatching, depending on larval growth rate and water temperature (Yoo 1987). Not all larvae metamorphose at the same time. As bigger larvae start eating smaller ones, size grading becomes necessary. Therefore, synchronization of metamorphosis should be considered as a priority.

9.2.9 Weaning period

Weaning occurs when the diet of newly metamorphosed juveniles is changed from live food to dry feeds. The larvae are then 10–20 days old and they weigh between 20 and 100 mg. Both types of food are offered together, the supply of live food being gradually reduced. Feeding rate depends on average size of juvenile fish. Feeding frequency is 5–6 times a day. This process usually takes 30 days to be completed, by which time juvenile fish weigh about 1 g. Expected survival is about 70%. As presently practiced, weaning is a somewhat cumbersome and expensive process because juveniles require large amounts of live food until they are weaned. As far as possible, live food requirements should be reduced by helping young fish to learn to accept inert food early. This also provides a means to offer additional nutrients, which might be lacking in live food. With this technique, it is claimed that fish can be weaned at a weight of 150–200 mg with 90% survival. This is a good example of how technology for rearing marine fish larvae can be improved and made less costly in the future.

9.2.10 Nursing period

The purpose of nursing is to rear young flounders until they can be moved or sold to an on-growing system, but the size at which such transfer occurs can vary substantially. During this period, commercial feeds are fed at the rate of about 5% of body weight. To optimize growth until fish average 15 cm, water temperature should range from 18 to 23°C. Depending on location, different strategies are used in flounder hatcheries to maintain optimum water temperatures during the nursing period. In early spring or late autumn, warmer water from a power station or heated water is commonly used for this purpose. Recently, recirculation systems have become the method of choice.

9.2.11 Growout

Olive flounder is mainly produced in the flow-through systems in land-based facilities (Figure 9.2). These facilities are concentrated in the south and west

Figure 9.2 Flow-through systems for olive flounder culture at land-based facilities on the south and west coast of Korea (Bai and Okorie 2007).

Table 9.3 Optimum stocking density for each size of olive flounder in land-based tanks.

		Stocking density	
Body size (cm; a)	**Body weight (g)**	**Fish Nm/m^2 (b)**	**Total weight (kg/m^2)**
3.5	0.5	720	0.37
7.0	4.1	277	1.14
10.5	13.8	158	2.19
14.0	32.8	106	3.48
17.5	64.1	78	5.00
21.0	110.7	61	6.75
24.5	175.7	49	8.6
28.0	262.3	41	10.8
31.5	373.4	35	13.1
35.0	512.1	31	15.9
38.5	681.6	26	17.7
42.0	884.7	23	20.3

The individual data was generated by $b = 4036a^{-1.3758}$. (NFRDI 2006).

coasts of the country, mainly in Jeju-do. Each farm usually produces an average of 110 metric tons of flounder per year. Farms are located on the coast, close enough for water to be taken from the sea. Seawater is pumped directly from the open sea into the head tanks and subsequently supplied to the fish tanks after some treatment.

Olive flounder are naturally docile and not easily agitated. As a result, they subject themselves to little stress under farming conditions and, therefore, do better than more excitable species. They also like crowding together, though, as flatfish, they do not fully use the water column as do round fish such as large yellow croakers. In fact, stocking densities for flounder are usually expressed in terms of kilograms per square meter, rather than kilograms per cubic meter as they are for round fish. Optimum stocking density for flounder in each development stage is shown in Table 9.3 (NFRDI 2006).

These stocking rates do not appear to cause stress and, in this respect, they are similar to other farmed flatfish such as the European turbot and Atlantic halibut.

Olive flounder accept dry formulated feeds well and convert them efficiently, the feed conversion ratio (the weight of distributed feed per unit weight gain) being equal to 1:1 or a little more. This might be because of the naturally low metabolism and/or to a sedentary life style. If such excellent feed efficiency could be achieved in large-scale commercial systems, it would provide flounder farmers with a significant advantage from the economic point of view.

9.3 Nutrition and feeding

9.3.1 Energy requirements

Protein and energy should be kept in balance. A dietary deficiency or an excess of digestible energy can reduce growth rates of fish. A diet deficient in energy

Table 9.4 Dietary protein requirements for maximal growth of olive flounder.

Developmental stages	Size (mm, g)	Temperature (°C)	Feeding period (weeks)	Evaluation	Requirement level	References
Larvae (0–1 g)	0–41.1 mm	17–20		Weight gain	Around 60%	Bai *et al.* (2001)
Juvenile (1–50 g)	4.1–17.5 g	20–23	8	Weight gain	46.4–51.2%	Kim *et al.* (2002a)
Growing (50–300 g)	22.7–110 g	19.2	9	Weight gain	45%	Lee *et al.* (2002)

in relation to protein will mean that protein is used for energy to satisfy maintenance before growth (NRC 1993). Kim *et al.* (2004) reported that optimum dietary protein to energy ratio was 27.5 mg protein/kJ with a diet containing 45% crude protein and 16.7 kJ/g energy for juvenile olive flounder.

9.3.2 Protein requirements

The requirement for dietary protein has two components: (1) a need for indispensable amino acids that the fish cannot synthesize either at all or at a rate commensurate with its need for protein deposition or commensurate with the synthesis of a variety of other compounds with metabolic functions; and (2) a supply of either dispensable amino acids or sufficient amino nitrogen to enable the fish to synthesize them (NRC 1993). Dietary protein levels for optimal growth depend on fish size, temperature, quality of protein sources, and energy content in feeds (NRC 1993). Optimum dietary levels for each developmental stage of olive flounder are shown in Table 9.4.

Bai *et al.* (2001) reported that olive flounder larvae (0–41.1 mm) required about 60% of protein in diets for maximum growth. Kim *et al.* (2002c) showed that optimum dietary protein level in juvenile olive flounder (4.1–17.5 g) was between 46.4 and 51.2%. Lee *et al.* (2002) reported that growing olive flounder (22.7–110 g) required 45% of protein in the diet.

Several plant or animal protein sources such as dehulled soybean meal (Lim *et al.* 2004; Choi *et al.* 2004b; Kim *et al.* 2008) and fermented fisheries by-products and soybean curd residue mixture (Sun *et al.* 2007) were revealed to replace fish meal in olive flounder feed.

9.3.3 Lipids and fatty acids

Fish cannot easily synthesize either 18:2(n-6) or 18:3(n-3) fatty acids. Hence, one or both of these fatty acids must be supplied in the diet, depending on the EFA requirements of n-3 HUFAs such as EPA (eicosapentaenoic acid) and DHA (docosahexaenoic acid). To supply n-3 HUFAs in the diet, fish oil and/or squid liver oil are mainly used; however, the price of the oils is increasing now, which encourages feed companies to search for alternative lipid sources. Kim (2008) reported that olive flounder requires 1.26–1.68% of DHA and 1.10–1.59% of

Table 9.5 Vitamin requirements of olive flounder.

Developmental stages	Subject	Size (g)	Temperature (°C)	Feeding period (weeks)	Evaluation	Optimum level	References
Juvenile (1–50 g)	Vitamin C	3–16	17	12	Weight gain	93–150 mg AA/kg diet	Wang *et al.* (2002)
	Vitamin E (lipid 7, 14%)	3.9–20	22	12	Weight gain	22–60 mg TA/kg diet	Wang & Bai (2002)
	Choline	5.9–18		8	Weight gain	1,000 mg	Cui (2003)

EPA for maximum growth, and poultry oil and beef tallow could replace fish oil up to 100% with 53.9% of fish meal in the diet.

9.3.4 Carbohydrates

Research on carbohydrate nutrition in olive flounder has not been developed. Only a few studies have been done to search for better carbohydrate source in olive flounder (Lee *et al.* 2003). Additional studies on feed ingredients, which fish can utilize as an energy source, are also needed.

9.3.5 Vitamins

Results of research on vitamin requirements are shown in Table 9.5.

Optimum dietary vitamin C (L-ascorbyl-2-polyphosphate) level in juvenile olive flounder is 93 mg/kg, and symptoms of vitamin C deficiency in olive flounder are scoliosis and lordosis (Figure 9.3a and 9.3b). Optimum dietary vitamin

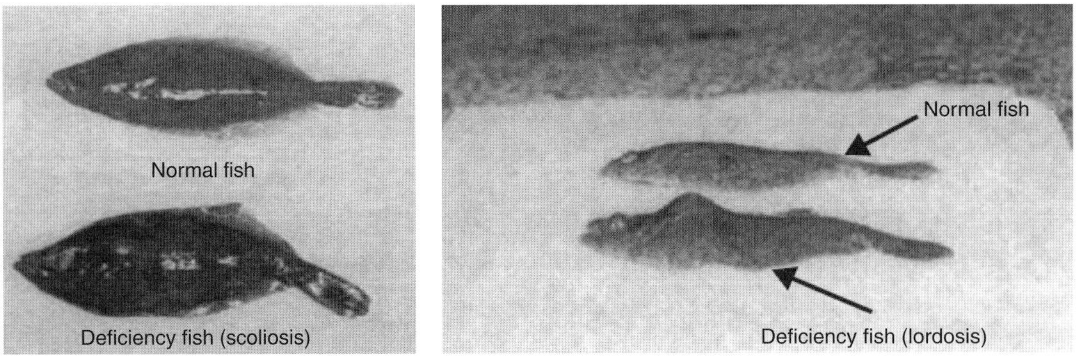

(a) (b)

Figure 9.3 (a) Vitamin C deficiency symptom of olive flounder (scoliosis; Wang *et al.* 2002); (b) Vitamin C deficiency symptom of olive flounder (lordosis; Wang *et al.* 2002).

Table 9.6 Mineral requirements of olive flounder.

Developmental stages	Subject	Size (g)	Temperature (°C)	Feeding period (weeks)	Evaluation	Optimum level	References
Juvenile (1–50 g)	P requirement	2.0–8.3	19	8	Weight gain	0.45–0.51	Wang *et al.* (2005)]
	Availability of P	4.0–12 g	22	6	Weight gain	$NaH_2PO_4 \times 2H_2O$, $Ca(H_2PO_4)$ 2 $\times H_2O$	Choi *et al.* (2005)
	Ca requirement and Ca:P ratio	1.5–7.0 g	22	6	Weight gain	Ca req.:0.6, Ca:P = 1:1	Kim (2004)

E level in olive flounder increased with increase in dietary lipid level (Wang and Bai 2002).

9.3.6 Minerals

Inorganic elements are required for the normal life processes of fish. Their main functions include the formation of skeletal structure, electron transfer, regulation of acid–base equilibrium, and osmoregulation. Minerals are important components of hormones and enzymes, and they activate enzymes (NRC 1993). Results of studies about calcium and phosphorus requirements in olive flounder are shown in Table 9.6.

Optimum dietary phosphorus level in olive flounder is suggested to be 0.45% as percentage of dry feed (unpublished data), and optimum dietary calcium to phosphorus ratio is 1:1 in olive flounder (Kim 2004). Bioavailability of $NaH_2PO_4.2H_2O$ and $Ca(H_2PO_4).H_2O$ is higher than that of other phosphorus sources in olive flounder (Choi *et al.* 2005).

Recently, the Korea government has made an effort to establish an upper limit of selenium in aquaculture feeds to prevent both cultured fish and consumers from selenium toxicity (MIFAFF 2007). Lee (2008) reported that a dietary selenium level above 7.38 mg Se/kg is likely toxic, and with a long-term feeding trial, a dietary Se level of 4.13 mg Se/kg may cause toxic effects in olive flounder.

9.3.7 Feed additives

Recently, increased attention has been paid to the use of growth and immunity stimulants, including glucans (Kim *et al.* 2006; Yoo *et al.* 2007), KH (a kind of acid hydrolysate; Choi *et al.* 2002), chlorella (Kim *et al.* 2002b), aloe (Kim *et al.* 2002a), Song-Gang stone (Choi *et al.* 2004a), and probiotics (Jeong *et al.* 2006) which are developed as feed additives to improve growth performance and/or immune responses in olive flounder (Table 9.7).

Table 9.7 Feed additives for olive flounder.

Developmental stages	Size (g)	Temperature (°C)	Feeding period (wks)	Additives	Supplementation level (%)	Evaluation	References
Juvenile (1–50 g)	18.7–28.9	11–15	2	KH	1.05	Feed efficiency	Choi *et al.* (2002)
	1.13	16–20	12	Chlorella	2	Weight gain, FE	Kim *et al.* (2002b)
	8.1	20	8	Aloe	0.5	Immune responses	Kim *et al.* (2002a)
	9.2	16	7	Glucan + Baism	0.1 + 0.9	WG, Immune responses	Yoo *et al.* (2007)
	3.2	19	6	Glucan	0.1	WG, immune responses	Kim *et al.* (2006)
	5.0	19	8	Song-Gang stone	0.5	Challenge test	Choi *et al.* (2004b)
	18.0	15	8	Probiotics	1.0×10^7 CFU	Immune response, challenge test	Jeong *et al.* (2006)

9.4 Future issues and needs for development

Over the past few decades, the central government and local governments have put a lot of money and effort into releasing juveniles and construction of artificial fishing reefs for stock enhancement of olive flounder. However, fishery production of olive flounder has not increased as a result of these efforts.

The Korean government has been pursuing a long-term aquaculture development program through the expansion of cultivated areas and the intensified development of both profitable and unexploited species. The shift toward production of high-value species has led to the emergence of olive flounder as the top aquaculture specie in Korea, since it is one of highest valued species. Improved purchasing power of consumers coupled with increased concerns about health has led to a dramatic increase in consumption of aquatic products, including olive flounder. It is expected that the industry will continue to expand, and that value-added products will be developed to meet consumers' preferences.

There has been growing concern about pollution of public waters near olive flounder culture facilities. The effluent from olive flounder farms is discharged directly into the sea. Development of highly digestible diets and treatment of effluent before discharge into the open seawater could reduce the nutrient load of the effluent.

Literature cited

Bai, S.C. 2007. Ecology and Habitat. In: Bai S.C. (ed.) *Field Manual of Olive Flounder Culture*. Vision 21 Aquaculture Forum, Busan, Korea, pp. 5–8.

Bai, S.C., Cha, Y.T., and Wang, X.J. 2001. A preliminary study on the dietary protein requirement of Japanese flounder larvae, *Paralichthys olivaceus*. *North American Journal of Aquaculture* 63:92–98.

Bai, S.C., and Okorie, O.E. 2007. Olive flounder in Korea. *Global Advocate* 10(5):74–75.

Choi, S.M., Go, S.H., Park, G.J., Lim, S.R., Yoo, G.Y., Lee, J.H., and Bai, S.C. 2004a. Utilization of Song-Gang stone as the dietary additive in juvenile olive flounder, *Paralichthys olivaceus*. *Journal of Aquaculture* 17(1):39–45.

Choi, S.M., Kim, K.W., Kang, Y.J., Wang, X.J., Kim, J.W., Yoo, G.Y., and Bai, S.C. 2005. Reevaluation of the phosphorus requirement of juvenile olive flounder *Paralichthys olivaceus* and the bioavailability of various inorganic phosphorus sources. *Journal of the World Aquaculture Society* 36(2):217–222.

Choi, S.M., Wang, X.J., Park, G.J., Lim, S.R., Kim, K.W., Bai, S.C., and Shin, I.S. 2004b. Dietary dehulled soybean meal as a replacement for fish meal in fingerling and growing olive flounder *Paralichthys olivaceus* (Temminck et Schlegel). *Aquaculture Research* 35:410–418.

Choi, Y.J., Lee, N.J., Cho, Y.J., and Bai, S.C. 2002. Identification of feeding stimulants to improve efficiency of diet for flatfish. *Journal of Korean Fishery Society* 35(2): 196–200.

Cui, H. 2003. Optimum dietary choline requirements in olive flounder, *Paralichthys olivaceus* and Korean rockfish, *Sebastes schlegeli*. MS dissertation. Pukyong National University, Busan, Korea.

Jeong, C.W., Choi, H.J., Yoo, G.Y., Lee, S.H., Kim, Y.C., Okorie, O.E., Lee, J.H., Jun, K.D., Choi, S.M., Kim, K.W., Kang, Y.J., Kang, J.C., Kong, I.S., and Bai, S.C. 2006. Effects of dietary probiotics supplementation on juvenile olive flounder *Paralichthys olivaceus*. *Journal of Korean Fishery Society* 39(6):460–465.

Jung, M.M. 2007. Feeding Organisms. In: Bai S.C. (ed.) *Field Manual of Olive Flounder Culture*. Vision 21 Aquaculture Forum, Busan, Korea, Chap. 3, pp. 22–102.

Kim, K.H., Hwang, Y.J., Kim, K.W., Bai, S.C., and Kim, D.S. 2002a. Effect of dietary aloe on chemiluminescent responses of peripheral blood phagocytes and resistance against Edwardsiella tarda Ewing and Mcwhorter 1965 in the cultured olive flounder (*Paralichthys olivaceus*). *Aquaculture Research* 33:147–150.

Kim, K.W., Bai, S.C., Koo, J.W., Wang, X.J., and Kim, S.K. 2002b. Effects of dietary chlorella ellipsoidea supplementation on growth, blood characteristics and whole-body composition in juvenile olive flounder, *Paralichthys olivaceus*. *Journal of the World Aquaculture Society* 33(4):425–431.

Kim, K.W., Wang, X.J., and Bai, S.C. 2002c. Optimum dietary protein level for maximum growth of juvenile olive flounder *Paralichthys olivaceus* (Temminck et Schlegel). *Aquaculture Research* 33:673–679.

Kim, J.H. 2004. Optimum dietary phosphorus level and Ca-P ratio in juvenile olive flounder (*Paralichthys olivaceus*). MS dissertation, Pukyong National University, Busan, Korea.

Kim, K.W., Wang, X.J., Choi, S.M., Park, G.J., and Bai, S.C. 2004. Evaluation of optimum dietary protein-to-energy ratio in juvenile olive flounder *Paralichthys olivaceus* (Temminck et Schlegel). *Aquaculture Research* 35:250–255.

Kim, Y.C. 2008. EPA and DHA requirements, and evaluation of the dietary beef tallow or poultry oil as a fish oil replacer in olive flounder, *Paralichthys olivaceus*. PhD dissertation, Pukyong National University, Busan, Korea.

Kim, Y.C., Kim, K.W., Lee, S.H., Park, G.J., Okorie, O.E. Kang, Y.J., and Bai, S.C. 2006. Effects of dietary β-1,3 glucan on growth and immune responses in juvenile olive flounder, *Paralichthys olivaceus*. *Journal of Aquaculture* 19:247–253.

Kim, Y.C., Yoo, G.Y., Wang, X.J., Lee, S.H., Shin, I.S., and Bai, S.C. 2008. Long term feeding effects of dietary dehulled soybean meal as a fish meal replacer in growing olive flounder, *Paralichthys olivaceus*. Asian-Aust. *Journal of Animal Science* 21(6):868–872.

Lee, S. 2008. Evaluation of a dietary selenium (Se) toxic level in juvenile olive flounder *Paralichthys olivaceus*. MS dissertation, Pukyong National University, Busan, Korea.

Lee, S.M., Kim, K.D., and Lall, S.P. 2003. Utilization of glucose, maltose, dextrin and cellulose by juvenile olive flounder, *Paralichthys olivaceus*. *Aquaculture* 221:427–438.

Lee, S.M., Park, C.S., and Bang, I.C. 2002. Dietary protein requirement of young Japanese flounder, *Paralichthys olivaceus* fed isocaloric diets. *Fisheries Science* 68:158–164.

Lim, S.R., Kim K.W., Choi S.M., Wang X.J., Bai S.C., and Shin I.S. 2004. Effects of dietary dehulled soybean meal as a fish meal replacer in fingerling and growing olive flounder, *Paralichthys olivaceus* (Temminck et Schlegel). *Aquaculture* 231:457–468.

Ministry for Food, Agriculture, Forestry and Fisheries (MIFAFF). 2007. Governmental Research Service Report titled "Introduction of Guidelines about Safe Upper Limit of Heavy Metals in Aquafeeds". Ministry for Food, Agriculture, Forestry and Fisheries, Seoul, Korea.

National Research Council. 1993. *Nutrient Requirements of Fish*. National Academy Press, Washinton, DC.

National Fisheries Researches and Development Institute. 2006. *Standard Manual of Olive Flounder Culture*. National Fisheries Researches and Development Institute, Busan, Korea.

Seo, J.P. 2007. Broodstock management. In: Bai S.C. (ed.) *Field Manual of Olive Flounder Culture*. Vision 21 Aquaculture Forum, Busan, Korea, pp. 9–21.

Sun, M.H., Kim, Y.C., Okorie, O.E., Devnath, S., Yoo, G.W., Lee, S.H., Jo, Y.K., and Bai, S.C. 2007. Use of fermented fisheries by-products and soybean curd residues mixture as a fish meal replacer in diets of juvenile olive flounder, *Paralichthys olivaceus*. *Journal of the World Aquaculture Society* 38(4):543–549.

Wang, X.J. and Bai, S.C. 2002. *Effects of Dietary Lipid level on Vitamin E Requirements in Olive Flounder*. World Aquaculture 2002, Beijing, China, April, pp. 22–27.

Wang, X.J., Choi, S.M., Park, S.H., Yoo, G.W., Kim, K.W., Kang, J.C., and Bai, S.C. 2005. Optimum dietary phosphorus level of juvenile Japanese flounder *Paralichthys olivaceus* reared in the recirculating system. *Fisheries Science* 71:168–173.

Wang, X.J., Kim, K.W., and Bai, S.C. 2002. Effects of different dietary levels of L-ascorbyl-2-polyphosphate on growth and tissue vitamin C concentrations in juvenile olive flounder, *Paralichthys olivaceus* (Temminck et schlegel). *Aquaculture Research* 33:261–267.

Yoo, G.Y., Lee, S.H., Kim, Y.C., Okorie, O.E., Park, G.J., Han, Y.O., Choi, S.M., Kang, J.C., Sun, M.H., and Bai, S.C. 2007. Effects of dietary β-1,3 glucan and feed stimulants in juvenile olive flounder, *Paralichthys olivaceus*. *Journal of the World Aquaculture Society* 38:138–145.

Yoo, S.K. 1987. Olive flounder culture. *Fisheries Research* 10:12–18 (in Korean).

Chapter 10

Culture of greenback flounder

Piers R. Hart

10.1 Life history and biology

The greenback flounder (*Rhombosolea tapirina*, Günther 1862) is a commercially important and highly esteemed food fish found in southern Australia and New Zealand. It is most abundant in Tasmania, Victoria, and New Zealand, but also occurs in New South Wales, South Australia, and Western Australia (Last *et al.* 1983).

R. tapirina is generally an estuarine species and is tolerant of wide variations in salinity (Last *et al.* 1983). It has been caught at depths of up to 100 m but is abundant on estuarine sandflats in shallow water of around 0–1 m depth. Juveniles occur in the highest densities from late winter to early summer in these estuaries and are daytime feeders, consuming mainly amphipods, harpacticoid copepods, and polychaete worms (Crawford 1984b).

Wild *R. tapirina* attain a length of 31–34 cm (400–500 g) at an age of III+ and can reach a length of 380 mm and a weight of 600 g, under natural conditions (Kurth 1957; Last *et al.* 1983).

First maturity occurs at a length of around 20 cm (age = I+) and the adults are serial spawners with an extended spawning season from June to October (Kurth 1957; Crawford 1984b). Descriptions of the egg and larval stages, and preliminary larval rearing trials, were first recorded by Crawford (1984a, 1986).

R. tapirina has potential for aquaculture due to its high market acceptability and demand, fluctuations in wild supply and declining catches. Increasing export values of wild caught *R. tapirina* from New Zealand suggest that there is potential in Australia to substitute farmed fish for imports and potential for both countries to produce farmed product for export to Asia.

R. tapirina is also suitable for laboratory research because of the availability of wild broodstock, ease of larval culture, and relevance to experimental research on marine fish biology, marine ecotoxicology, and system design for marine finfish aquaculture particularly larval rearing. Mondon *et al.* (2001) suggested that it might be potentially effective as an indicator species for assessing the effects of marine pollution.

10.2 Broodstock husbandry

10.2.1 Acquisition of broodstock

Broodstock can be caught with a seine net from the shores around the species range. However, this sometimes results in catches of all females, as males tend to occupy deeper water during the spawning season (Kurth 1957; Crawford 1984b). Therefore, it may be better to anticipate requirements for mature broodstock ahead of time and catch large numbers of small juveniles which can be slowly weaned to compound diets and cultured through to maturity. Mortality is generally quite high as wild-caught fish of any size are difficult to wean.

10.3 System design and requirements

Broodstock requirements for *R. tapirina* are relatively simple as they are a small species and adapt well to artificial environments after the initial difficult weaning process. Natural spawning of second-generation stocks has been recorded in tanks of 2 and 4 m^3 at the University of Tasmania over a number of years (Hart 1994; Pankhurst and Fitzgibbon 2006). Tanks are contained within a recirculation system provided with artificial lighting and temperature control.

10.4 Photothermal conditioning

Gametogenesis occurred in *R. tapirina* broodstock contained in a completely enclosed building, using artificial photoperiod and temperature control. As natural cycles were followed, it is possible that gametogenesis was controlled by endogenous rhythms. However, it is highly likely that spawning times can be controlled using photoperiod, temperature, or both, to give year-round egg production.

10.5 Monitoring gonad development

Barnett and Pankhurst (1999) found that *R. tapirina* exhibit group synchronous oocyte development, and have multiple ovulations. Reproductively mature wild fish were found from February to June. Both Kurth (1957) and Crawford (1984b) found that spawning of wild *R. tapirina* occurred between June and October and with male fish running ripe from May to October.

Catheterization to obtain oocytes samples is difficult, due to the small diameter of the genital pore but oocyte maturation can be assessed by external observation of the ripening ovaries in comparison to the stages described by Crawford (1984b). Barnett and Pankhurst (1999) modified this staging method to improve its accuracy (Table 10.1).

Endocrine changes in relation to maturation of *R. tapirina* have been described (Barnett and Pankhurst 1999; Hobby *et al.* 2000a, 2000b; Pankhurst and Riple 2000; Sun and Pankhurst 2003, 2004; Sun *et al.* 2003).

Table 10.1 Criteria for macroscopic staging of *Rhombosolea tapirina* gonads (Barnett and Pankhurst 1999).

Sex	Stage	Classification	Macroscopic appearance
Female	1	Immature	Ovary clear thread
	2	Regressed	Ovary small, semi-firm, grey-orange
	3	Vitellogenic	Ovary large, yellow-orange, vitellogenic oocytes may be visible through the epithelium
	4	Final oocyte maturation	Ovary large and plump, yellow-orange. Vitellogenic and hyaline oocytes visible through the epithelium
	5	Ovulated	Oocytes can be freely expelled from the oviduct with gentle pressure
	6	Spent	Ovary flaccid, grey, degenerating oocytes or no oocytes visible
Male	1	Immature	Testis translucent thread
	2	Spermatogenic	Testis small firm white thread
		Partially spermiated	Testis firm, white, and viscous milt expressible under pressure
		Fully spermiated	Testis plump, firm, white, and milt flows freely under gentle pressure
		Spent	Testis bloody and flaccid, no milt expressible

10.6 Diet and nutrition

Little attention has been paid to broodstock nutrition. In most studies, the broodstock were fed with fresh mussels (*Mytilus edulis*) or dry diets (38.4% protein, 16% lipid) formulated for salmonids.

Wild fish do not easily adapt to dry diets but will readily take fresh mollusc meat. Hart (1994) made a moist diet of trout pellets, fresh mollusc meat, vitamins, and dissolved gelatin powder which was frozen in blocks. This diet was successfully used as an intermediate feed to wean wild-caught adults and juveniles to dry diets over a long period of time.

10.7 Controlled spawning

Ovulation of *R. tapirina* has been induced using Ovaprim (Syndel Laboratories Inc., Vancouver, BC, Canada; [D-Arg^6, Pro^9 NEt]-sGnRH and a dopamine antagonist; domperidone) (0.5 mL/kg), administered in 2–3 intraperitoneal injections (Hart 1994), LHRHa (50 and 100 μg/kg) in a single intraperitoneal injection (Poortenaar and Pankhurst 2000), and HCG (500 IU/kg) in a single intramuscular injection (Crawford 1986) and (1,000 IU/kg) in a single intraperitoneal injection (Poortenaar and Pankhurst 2000).

Poortenaar and Pankhurst (2000) found that treatment with HCG, LHRH-a (50 μg/kg), or LHRH-a cholesterol pellet (100 μg/kg) implanted intraperitoneally, were the most effective. Of those fish that ovulated, most ovulated more than twice, and most ovulations occurred at daily intervals. Implanted

GnRH-a pellets increased milt volume of spermiating males but spermatocrit was reduced (Lim *et al.* 2004).

Natural spawning of *R. tapirina* in culture tanks has been recorded by a number of authors (Crawford 1984a; Hart 1994; Pankhurst and Fitzgibbon 2006). Spawning behavior consists of approach and courting of an ovulated female by a male, followed by synchronous swimming in mid-water prior to the release of eggs. Most spawning events occurred shortly before sunrise. Volumes of eggs produced and egg fertility was highly variable but the best batches could be reared normally. Natural spawning would decrease the handling stress associated with injection and stripping.

10.8 Collection of eggs and egg incubation

The fecundity of wild *R. tapirina* ranges from 4 to 5 million eggs/kg female body wt (Crawford 1984b). Manual fertilization of eggs can be achieved using either the "wet" or "dry" methods but the males are difficult to strip and produce very small quantities of milt. Diluting the milt about 5 × with clean seawater makes it easier to mix evenly with the eggs. The mean fertilization rate recorded by Hart (1994) was around 50%.

The number of eggs can be determined by weight or volume prior to fertilization. There are 3,500 eggs per mL or g.

Hydrated eggs of *R. tapirina* are 800 μm in diameter and contain a single oil droplet of 170 μm (Figure 10.1). Eggs described by Crawford (1986) contained multiple oil droplets possibly as a result of using HCG.

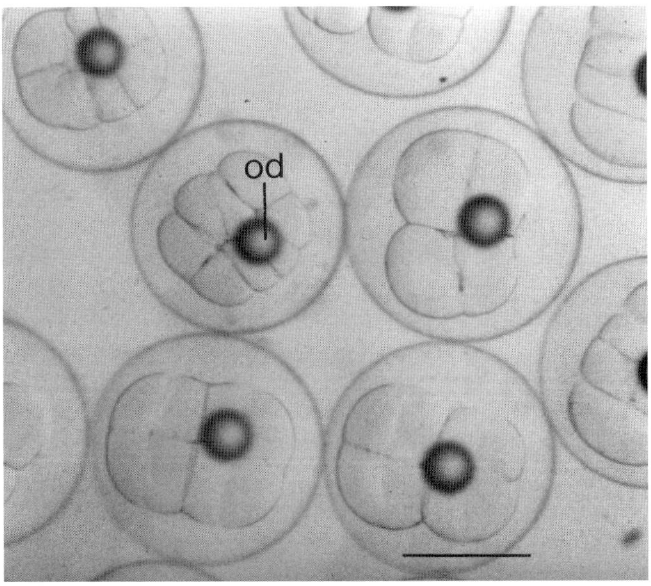

Figure 10.1 Eggs of *Rhombosolea tapirina* at the 4–8 cell stage, showing single oil droplets (od). Scale bar represents 0.5 mm.

The small egg size of *R. tapirina* has no disadvantages in commercial culture as the broodstock are small, requiring minimal holding space, and have a high fecundity. The larvae have a large mouth gape and feed readily on rotifers (*Brachionus plicatilis*), despite their small initial size.

Hatch rates of approximately 30–50% were obtained by Hart (1994). Hatch rates were improved using 1 μm filtered water for incubation. Static systems were adequate for small batches with low stocking densities (<200 per liter) but a water flow of approximately 45 L/hour was required at higher densities. Incubation in artificial seawater was highly successful in eliminating bacteria and variations between batches of natural seawater, particularly for small scale experiments.

Fertilization rates were better at salinities of 35‰ and 45‰ than at lower salinities, and eggs were buoyant at salinities above 28‰ (Hart and Purser 1995). Although eggs can be successfully incubated in a wide range of salinities (15–45‰) the salinity of the water in which the broodstock are maintained might affect the optimal salinity for fertilization. Broodstock should be maintained at the salinity in which the eggs will be fertilized. For commercial incubation, the salinity should be maintained above 28‰ as the dead eggs sink while the live eggs are buoyant. This makes the dead eggs easier to remove.

The optimal temperature for incubation of *R. tapirina* eggs is 9–12°C with a tolerance range of 9–18°C (Hart and Purser 1995). The development rate of eggs increases rapidly with temperature in a curvilinear fashion across the range of temperatures investigated. The time taken for 50% of eggs to hatch was approximately 4 days at 12°C (Figure 10.2).

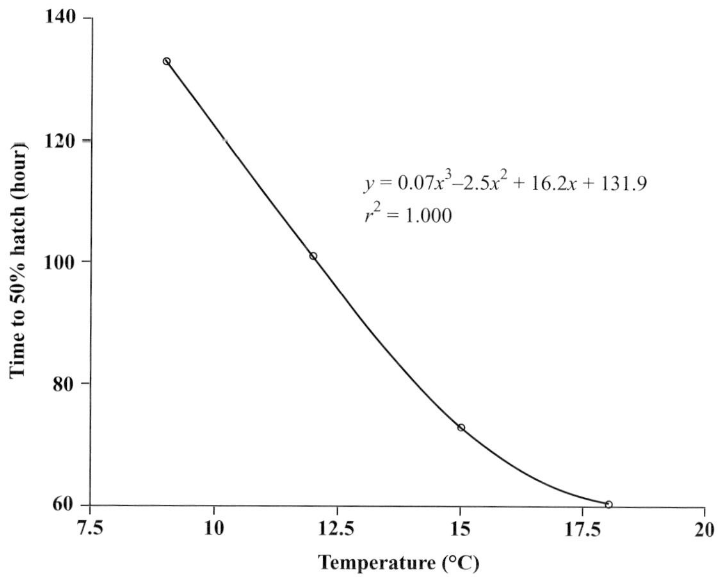

Figure 10.2 Mean times to 50% hatch at different temperatures for eggs of *Rhombosolea tapirina* incubated at 35‰ (there was no variation in hatching times between replicates at the same temperature) (Hart and Purser 1996).

Eggs can be transferred from the incubation tanks to the rearing tanks and counted by volume at the tail bud stage (around 50 hours posthatch at 12°C) without causing damage.

The optimal temperature for yolk-sac absorption in *R. tapirina* is 15°C resulting in larger larvae (2.76–3.18 mm) at complete yolk and oil absorption (5–6 days posthatch) than at higher temperatures and faster growth rates than at lower temperatures (Hart and Purser 1995).

During the yolk-sac stage, the larvae of *R. tapirina* float passively at the water surface, with the yolk-sac uppermost but by the end of this stage, they are swimming more vigorously, the eyes are black, the body is more heavily pigmented, and the digestive system is functional. Stages of development are described fully by Crawford (1986).

10.9 Larval culture

10.9.1 System design and requirements

Tank volumes from 3 liters up to 2,000 liters have been used successfully for the larval rearing of *R. tapirina*. Small tanks or systems with small water volumes have been used for experimentation as it is easier to manipulate the environment and provide large numbers of individual tanks permitting robust experimental designs with suitable replication of treatments. Most experimental systems have used recirculation or static systems with regular water exchange to allow environmental manipulation.

Black tanks with overhead lighting were used to improve the feeding efficiency of *R. tapirina* larvae. Hart *et al.* (1996) showed that long photoperiods of 18–24 hours improved growth during the early feeding stage. Light intensities of 300–1,700 lx were able to induce and maintain the feeding response.

10.10 Hatchery protocols

10.10.1 Larval rearing

A general larval rearing protocol has been developed for *R. tapirina* larvae. The pelagic eggs float at the water surface for 4 days at 12°C. Just before hatching, they are transferred to the rearing tanks and the temperature is raised to 15°C for yolk absorption. Larvae are grown in black tanks with still water for the first week and upwelling water flow for the remaining period. Continuous light is provided by overhead fluorescent lights. First feeding occurs after complete absorption of the yolk-sac (approximately day 4 posthatch at 15°C) and before the absorption of the oil droplet (approximately day 5 posthatch at 15°C). There was some evidence of feeding behavior in larvae just before complete yolk-sac absorption on day 3 posthatch, but no difference in growth rate was shown in larvae first-fed on day 3 rather than on day 4 posthatch.

The oil droplet is absorbed from around 70 hours onwards as the eyes, mouth, and gut develop. Larvae that were first fed after day 6 or 145.5 hours posthatch showed complete mortality on day 8–10 posthatch (Hart and Purser 1995).

The temperature is raised to 18°C as the larvae start feeding. Enriched rotifers are fed to the larvae at a density of 20/mL up to day 14 posthatch and instar II enriched *Artemia* are introduced from day 10 posthatch at a density of 1/mL increasing to 3 per mL by day 20. Weaning can commence after day 20 posthatch with a 10-day overlap during which the live food is reduced and dry diet increased. This protocol should result in metamorphosed and fully weaned larvae of around 11–13 mm by day 30–35 posthatch.

During the static rotifer feeding stage uneaten rotifers were flushed from the tank by turning on the water flow (180 L/hour) and using a 500 μm screen on the outlet, for 2–3 hours daily, before the addition of newly-enriched rotifers. During the *Artemia* feeding stage, a water flow rate of 45 L/hour was used and a 125 μm screen was used on the outlet to trap live *Artemia* in the tank. Uneaten *Artemia* were flushed out of the tank by changing this for a 500 μm screen for 2–3 hours daily.

Feeding of *R. tapirina* larvae can be observed initially as a "shivering" motion at the water surface rather than the s-shaped striking motion seen as the larvae increase in size. After feeding commences, the larvae go through a period up to day 15 posthatch, during which they actively feed near the water surface.

At day 15 posthatch the larvae turn black and spend more time lying on the tank bottom, coming up only occasionally to feed at the surface or in the water column. At around day 20 posthatch the larvae begin to acquire a white spotted appearance and begin to feed actively on the tank bottom. Metamorphosis is completed at approximately day 30 posthatch and a length of around 12.5 mm as the larvae take on the brownish green juvenile coloration.

The development of the digestive tract in larval *R. tapirina* was described by Hart (1994). At hatching, the gut is a simple straight tube, which differentiates into posterior intestine and anteromedian intestine, as a loop forms and feeding begins. The anteromedian intestine further differentiates into anterior intestine and median intestine, and finally the stomach increases in volume and becomes functional around day 20 posthatch (Plates 2. and 3.). A pyloric valve was identified by Hart (1994), but appears to be lost later in life as adults do not have a fully functioning stomach (Grove and Campbell 1979). This would be an interesting area for further research.

The development of the sensory organs of *R. tapirina* larvae has been described by Pankhurst and Butler (1996).

10.11 Water quality

The optimal temperature for growth of *R. tapirina* larvae is 18–20°C. Maximum survival rates were observed at 17 and 18°C (Hart *et al.* 1996).

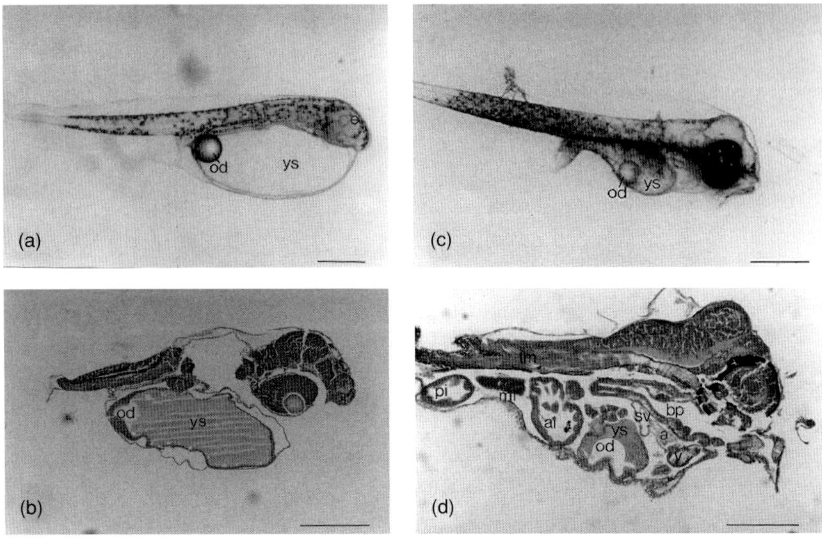

Figure 10.3 (a) External view of *Rhombosolea tapirina* larva at hatching (day 0 posthatch). ys = yolk-sac, od = oil droplet, e = eye. (b) Sagittal section of *R. tapirina* larva at day 0 posthatch. (c) External view of *R. tapirina* larva at day 3 posthatch. ys = yolk-sac, od = oil droplet. (d) Sagittal section of *R. tapirina* larva at day 3 posthatch. ys = yolk-sac, od = oil droplet, bp = buccopharyngeal cavity, ai = anterior intestine, mi = median intestine, pi = posterior intestine, sv = sinous venosus, a = atrium, v = ventricle, tm = trunk muscle. Scale bars represent 300 μm.

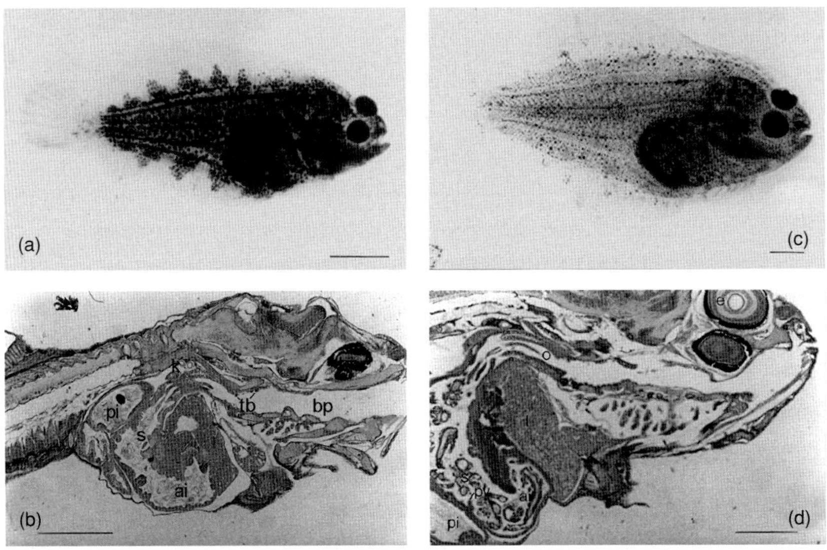

Figure 10.4 (a) External view of *Rhombosolea tapirina* larva at day 16 posthatch. Note: particularly advanced specimen showing start of eye migration, dark coloration, and completion of tail flexion. Scale bar represents 1 mm. (b) Sagittal section of *R. tapirina* larva at day 16 posthatch. bp = buccopharyngeal cavity, ai = anterior intestine, pi = posterior intestine, k = kidney, s = stomach, l = liver, tb = taste buds. Scale bar represents 500 μm. (c). External view of a fully metamorphosed *R. tapirina* juvenile at day 30 posthatch. Scale bar represents 1 mm. (d). Sagittal section of *R. tapirina* larva at day 20 posthatch showing the enlargement of the stomach. E = eye, o = oesophagus, ai = anterior intestine, pi = posterior intestine, s = stomach, l = liver, pv = pyloric valve. Scale bar represents 500 μm.

Salinity had no effect on growth rate but the survival rate was reduced at a salinity of 15‰. It appears that *R. tapirina* are not tolerant of low salinities before metamorphosis (Hart *et al.* 1996).

10.12 Food and feeding

Although *R. tapirina* are very small at first-feeding, approximately 3.0 mm, they have a large gape and are capable of ingesting rotifers. Hart (1994) used the method described by Shirota (1970) to calculate gape opening size and showed that first-feeding larvae have a gape of over 428 μm. Instar II *Artemia* nauplii are ingested when the larvae reach a length of 4.71 mm at day 9 posthatch (15.5–16°C) and a gape of 690 μm. Shaw *et al.* (2003) showed that *R. tapirina* larvae preferentially select smaller prey than rotifers up to day 13 posthatch. Rotifers were preferentially selected on day 15 and *Artemia* from day 15–18 posthatch at 14°C. Larval lengths were not given. Cox and Pankhurst (2000) concluded that *R. tapirina* larvae are primarily visual feeders.

In order to improve the levels of highly unsaturated fatty acids (HUFA), particularly ecosapentaenoic acid (EPA) and docosahexaenoic acid (DHA) in live diets, a number of enrichments have been used in studies at the University of Tasmania. The live feeds were fed to the fish twice daily with further enrichment between the first and second feeding. Better growth resulted from feeding twice per day and removal of uneaten rotifers and *Artemia* from the tanks prior to the addition of each newly enriched batch.

Initially, the most common enrichment was a mixture of two species of microalgae containing the highest levels of EPA and DHA, *Isochrysis* sp. (Tahitian clone) (0.2 and 8.3%, respectively) and *Pavlova lutheri* (19.7 and 9.4%, respectively) (Brown *et al.* 1989).

Artificial enrichment products generally resulted in improved growth. However, in some trials they also resulted in higher mortality and increased malpigmentation rates of the ocular surface when used during the rotifer feeding stage. Malpigmentation occurred on both the ocular side as complete or partial lack of pigment and on the blind side as patchy distribution of melanin. Enrichment with microalgae during the rotifer stage reduced ocular malpigmentation suggesting a deficiency in the formulation of the artificial enrichments for early larval stages. Using the microalga *Tetraselmis suecica* in aerated "green water" up to day 15 posthatch reduced malpigmentation and improved growth and survival (Hart and Saxby, unpublished data). Dark pigment on the blind side may be the result of abrasion or infection and a resultant increase in melanin. Shaw *et al.* (2006) showed that green water (*T. suecica*) improved the feeding ability of larvae at all turbidity levels from 3 to 5 Nephelometric Turbidity Units (NTU).

10.13 Formulated feeds

Weaning was initially thought to be a major issue in *R. tapirina*. Crawford (1984a) recorded 100% mortality when attempting a direct change to artificial diets. At the University of Tasmania, the early batches were weaned after metamorphosis with limited success. It is possible to wean *R. tapirina* larvae beginning at first feeding by adding both rotifers and dry diet while gradually reducing rotifers over a 10-day period. However, the growth rate is so badly compromised that many fish do not properly metamorphose (Hart and Koukouvas, unpublished data).

Hart (1994) and Hart and Purser (1996) investigated a number of factors which might affect weaning success. These authors found that premetamorphosis weaning was more successful. This may be partly because the larvae are in the water column and feeding very actively. Activity is greatly reduced after metamorphosis when the fish are benthic in habit unless starved of food when they will swim slowly around the tank and come up to the surface. This behavior was frequently observed in fish that were not accepting the manufactured diet and swam listlessly around the tank looking for *Artemia*. Once the fins of these fish become ragged and they lose condition mortality occurs, but sometimes not for a long time. If weaning is initiated on day 23 posthatch with larvae of good quality, it is possible to achieve a survival rate of 80% without greatly affecting the growth rate. A 10-day transitional stage between live and artificial diets gave the best results and resulted in fully weaned juveniles at the completion of metamorphosis around 30–40 days posthatch.

The quality of the live feeds offered to the larvae and the quality of the larvae themselves are considered to be the most important factors affecting weaning success of *R. tapirina*. Larvae fed with live feeds enriched with commercial enrichment diets grew faster and weaned more successfully than those fed on microalgae-enriched prey.

However, the use of microalgae as enrichment for rotifers and *Artemia* during the first 15 days of larval rearing had a beneficial effect on growth and survival rates which did not become apparent until after metamorphosis and weaning to an artificial diet. The reasons for this are unclear, but could be either a direct nutritional deficiency in the commercial enrichments or due to differences in gut flora, which affect digestive capacity leading to indirect nutritional deficiencies when microalgae is not present. Modern commercial enrichment products may overcome this problem.

10.14 Hatchery economics

As there has not yet been any commercial hatchery production of this species, it would be potentially misleading to try and calculate the economics of hatchery production. However, there is nothing particularly unusual that needs to be taken into account and so the hatchery economics of a standard commercial marine species should fit this one with no special adjustments.

10.15 Genetics for culture versus enhancement

Significant genetic differences exist between the Australian and New Zealand populations of *R. tapirina* based on polymorphic allozyme loci (Van Den Enden *et al.* 2000). Within Australia, *R. tapirina* from western Tasmania are genetically isolated from the population in Victoria, and northern and southeastern Tasmania, though to a much lesser extent than from the population in New Zealand. The separation of western and eastern Tasmanian populations is supported by morphometric data, and has been suggested to have occurred as a consequence of the closure of Bass Strait during the last ice age (Kurth 1957). Reductions in genetic variation (heterozygosity and alleles) were observed in two cultured cohorts when compared with wild-caught samples. No genetic variation was detected between normal and malpigmented individuals from the same culture cohort.

10.16 Nursery culture and growout

10.16.1 System design and requirements

Nursery culture has been undertaken in a number of different flat-bottomed tanks of 200 liters up to 6 m diameter. Generally, the nursery and growout stages have not been separated and the small numbers of fish that have been ongrown have remained in the same tanks through to harvest or termination of the experiment. Some shading is considered necessary to promote feeding and reduce biofouling and the water current should provide self-cleaning action, but not be excessive. Ideally, a heated recirculation system would be used to improve growth rates through the winter. It would be sensible for future commercial ventures to use similar nursery systems to those used for commercial culture of other flatfish species.

10.17 Environmental conditions

It has been shown that larvae grow better at 18°C and this possibly is also the case for juveniles and adults. A nursery phase in heated water may be beneficial in accelerating the growth rate and providing juveniles suitable for growout at an earlier stage.

Juvenile *R. tapirina* of 80–190 g are very tolerant of low salinities. Direct transfer to salinities of 3–15‰ causes a reduction in plasma osmolality over the initial 4 hours followed by stabilization to almost normal levels. Direct transfer to freshwater caused complete mortality, unless the pH was raised above 7 in which case a normal response was shown. This shows that the species would tolerate freshwater treatment baths for short periods of perhaps 2 hours without ill effects if the pH was maintained above 7 (Girling *et al.* 2003).

Early maturation of both sexes is a potential problem and further research will be required to determine the effect on growth rates. Sterilization techniques or monosex production techniques may be required to obtain profitable growth rates. Hart (1994) showed that 54% (16% female and 38% male) of *R. tapirina* reached maturity at an age of 15 months under natural temperature and light conditions. Lower numbers (32%) of mature fish were recorded in enclosed conditions and none of these were females. This phenomenon was also observed by Purser (1996) who assessed maturation in 1-year-old fish in winter and found 83.5% mature, of which 35% were females. The high fat salmonid diet caused a high degree of surplus fat storage particularly in the liver and this could have been the cue for maturation in the first year. Further research into the dietary energy requirements may help to reduce early maturation. Production of triploids is another alternative though it has not been particularly successful in reducing maturation in other flatfish. Triploidy of *R. tapirina* has been successfully induced using a 1-hour cold shock at 0°C directly after fertilization (Hart, unpublished data).

10.18 Diet and nutrition

The nutritional requirements of *R. tapirina* are poorly understood. In most cases, salmonid diets have been used with proximate composition of 50% protein and 15–18% lipid. Nutritional deficiencies have been responsible for reductions in larval growth rates, increased deformities, frequency of malpigmentation, and cataracts of the eye. Vitamin C deficiency during weaning appears to be the cause of shortened opercula and lordosis in juvenile flounder. The high fat content of trout pellets appears to be the cause of fatty accumulation in the liver and may result in reduced growth rates and early maturation.

Bransden and Carter (1999) found that the lack of acid digestive capability of large juvenile flounder did not prevent digestion of acid soluble forms of phosphorous. They also showed high apparent digestibility of nitrogen and minerals from soybean meals suggesting that high fish meal diets may not be required.

Purser (1996) and Verbeeten *et al.* (1999) found that the highest growth efficiency of small juvenile flounder was obtained by feeding 2% body weight/day. Feeding either once per day at 1 or 2% body weight or twice per day at 2% body weight has no effect on growth (Carter and Bransden 2001), but a single daily feed during the dark phase resulted in reduced growth as larger flounder are almost exclusively daytime feeders (Chen *et al.* 1999; Chen and Purser 2001). Purser and Chen (2001) showed that small flounder became accustomed to feeding at regular times and places and this could be used to optimize feed intake.

Reduced food rations result in greater size variation among small juveniles due to variations in individual food consumption (Carter *et al.* 1996; Shelverton and Carter 1998). Agonistic behavior was uncommon at all feed levels suggesting that strong dominance hierarchies were not involved. However, at low stocking density Thomas *et al.* (1998) recorded elevated cortisol levels in individual fish suggesting that dominance hierarchies can develop.

10.19 Health issues

Both adults and juveniles are susceptible to outbreaks of *Flexibacter maritimus*, though this is always associated with stress or physical damage. The adults after stripping and juveniles during weaning are particularly susceptible. Adults can be treated with acriflavine applied directly to the affected area.

Both adults and juveniles have been affected by heavy infestations of *Trichodina* sp., possibly causing low level mortality. However, wild fish have also been shown to carry large populations of *Trichodina* sp. and it may therefore not represent a major threat. Freshwater baths were unsuccessful in controlling the disease but formalin was effective.

The most potentially threatening disease is *Aeromonas salmonicida,* which was isolated from a small percentage of specimens during 1993. The strain was different from that involved in diseases of commercial importance in other countries. However, the source of the disease is unknown. Salmonids reared in the same water were unaffected.

10.20 Stocking and splitting

10.20.1 Density effects

Although a number of trials have been conducted to establish optimal stocking density for *R. tapirina* no upper limit has been demonstrated (Purser 1996; Thomas *et al.* 1998). In a trial described by Purser (1996) conducted over 249 days using fish of 36–38 g initial weight, the maximum stocking density achieved was 15.9 kg/m^2 or 61.1 kg/m^3 in tanks of 270 liters and surface area of 1 m^2. There was no significant reduction in growth at the highest density. Surface area of the tank base was considered more significant than water volume in deciding maximum stocking density for the species. A relationship between individual fish weight and floor coverage at the highest density showed that there were 1.32 layers of fish on the tank floor. These figures were used to develop a method to calculate the achievable stocking rate in different sized tanks and with fish of different ages. In theory, a stocking density of 28.8 kg/m^2 of 500 g fish would be possible in a tank of 10 m diameter.

Significant individual size variation develops among *R. tapirina* during the ongrowing stages. Initially, this appears to be because of differential weaning success between individuals and low feed intake or poor assimilation of manufactured diets by a proportion of fish. Later, the onset of maturation among some very small individuals results in further differences in growth rates. Low stocking densities also appear to result in increased size variation. Regular grading is required to reduce these effects (Purser 1996).

10.21 Marketing

The wholesale market price for *R. tapirina* is AUS$6–10 per kg. Larger fish are the most valuable, but are rarely caught. A market assessment and sensory testing of cultured flounder has been carried out and described by Purser (1996). The most acceptable market size for *R. tapirina* is 500–800 g with a 30% flesh to bone ratio. Recovery rate gilled and gutted is 95%. Fresh chilled fish command higher prices than frozen particularly in the Japanese market. The main producer of wild fish in the region is New Zealand. Initial sensory testing produced favorable results with farmed product tasting as good as frozen imports from New Zealand and better than wild-caught Australian product. Visual acceptability was impaired by a greenish tinge to the flesh.

10.22 Production economics

As with hatchery economics, it would be potentially misleading to discuss this without any actual commercial production figures to use. However, production would almost certainly require shore-based tank systems and the associated set up and running costs of either continuous pumping or recirculation.

10.23 Summary: industry constraints and future expectations

There is presently no commercial culture of *R. tapirina*. Although production of juveniles is relatively simple, the relatively small market size and low market value of the species together with concerns by the established salmon producers over *A. salmonicida* have constrained the development of an industry in Tasmania.

Improved nutrition during larval rearing, weaning and ongrowing as well as heated recirculation systems are probably required if the optimal growth rate of 500 g in 2 years is to be achieved. This goal has only been achieved in the laboratory and recirculation may prove to be too expensive commercially. There may be potential for restocking sheltered areas with hatchery-reared juveniles though it seems unlikely that this would be economically viable. However, there is still some industry interest and considerable domestic demand for the species particularly in Australia. If prices increase or catch rates fall then an industry may develop.

Literature cited

Barnett, C.W., and Pankhurst, N.W. 1999. Reproductive biology and endocrinology of greenback flounder *Rhombosolea tapirina* (Günther 1862). *Marine and Freshwater Research* 50:35–42.

Bransden, M.P., and Carter, C.G. 1999. Effect of processing soybean meal on the apparent digestibility of practical diets for the greenback flounder *Rhombosolea tapirina* (Günther). *Aquacultural Research* 30:719–723.

Brown, M.R., Jeffrey, S.W., and Garland, C.D. 1989. Nutritional aspects of microalgae used in mariculture; a literature review. CSIRO Marine Laboratories, Report No. 205, 30 pp.

Carter, C.G., and Bransden, M.P. 2001. Relationships between protein-nitrogen flux and feeding regime in greenback flounder, *Rhombosolea tapirina* (Günther). *Comparative Biochemistry and Physiology – Part A: molecular and Integrative Physiology* 130:799–807.

Carter, C.G., Purser, G.J., Houlihan, D.F., and Thomas, P. 1996. The effect of decreased ration on feeding hierarchies in groups of greenback flounder (*Rhombosolea tapirina*: Teleostei). *Journal of the Marine biological Association of the United Kingdom* 76:505–516.

Chen, W.M., and Purser, G.J. 2001. The effect of feeding regime on growth, locomotor activity pattern and the development of food anticipatory activity in greenback flounder. *Journal of Fish Biology* 58:177–187.

Chen, W.M., Purser, J., and Blyth, P. 1999. Diel feeding rhythms of greenback flounder *Rhombosolea tapirina* (Günther 1862): the role of light-dark cycles and food deprivation. *Aquacultural Research* 30:529–537.

Cox, E.S., and Pankhurst, P.M. 2000. Feeding behaviour of greenback flounder larvae, *Rhombosolea tapirina* (Günther) with differing exposure histories to live prey. *Aquaculture* 183:285–297.

Crawford, C.M. 1984a. Preliminary results of experiments on the rearing of Tasmanian flounders, *Rhombosolea tapirina* and *Ammotretis rostratus*. *Aquaculture* 42:75–81.

Crawford, C.M. 1984b. An ecological study of Tasmanian flounder. PhD thesis, University of Tasmania, Hobart, Tasmania, Australia, 181 pp.

Crawford, C.M. 1986. development of eggs and larvae of the flounders, *Rhombosolea tapirina* and *Ammotretis rostratus*. (Pisces: Pleuronectidae). *Journal of Fish Biology* 29:325–334.

Girling, P., Purser, J., and Nowak, B. 2003. Effects of acute salinity and water quality changes on juvenile greenback flounder, *Rhombosolea tapirina* (Günther, 1862). *Acta Ichthyologica et Piscatoria* 33(1):1–16.

Grove, D.J., and Campbell, G. 1979. The role of extrinsic and intrinsic nerves in the co-ordination of gut motility in the stomachless flatfish *Rhombosolea tapirina* and *Ammotretis rostrata* Günther. *Comparative Biochemistry and Physiology* 63C:143–159.

Günther, A. 1862. *A Catalogue of Fishes*. British Museum, London, Vol. 4, pp. 458–460.

Hart, P.R. 1994. Factors affecting the early life history stages of hatchery-reared greenback flounder (*Rhombosolea tapirina* Günther, 1862). PhD thesis, University of Tasmania, 231 pp.

Hart, P.R., Hutchinson, W.G., and Purser, G.J. 1996. Effects of photoperiod, temperature and salinity on hatchery-reared larvae of the greenback flounder (*Rhombosolea tapirina* Günther, 1862). *Aquaculture* 144:303–311.

Hart, P.R., and Purser, G.J. 1995. Effects of salinity and temperature on eggs and yolk sac larvae of the greenback flounder (*Rhombosolea tapirina* Günther, 1862). *Aquaculture* 136:221–230.

Hart, P.R., and Purser, G.J. 1996. Weaning of hatchery-reared greenback flounder (*Rhombosolea tapirina* Günther) from live to artificial diets: Effects of age and duration of the changeover period. *Aquaculture* 145:171–181.

Hobby, A.C., Geraghty, D.P., Pankhurst, N.W. 2000a. Differences in binding characteristics of sex steroid binding protein in reproductive and non-reproductive female rainbow trout (*Oncorhynchus mykiss*), black bream (*Acanthopagrus butcheri*), and greenback flounder (*Rhombosolea tapirina*). *General and Comparative Endocrinology* 120:249–259.

Hobby, A.C., Pankhurst, N.W., and Geraghty, D.P. 2000b. A comparison of sex steroid binding protein (SBP) in four species of teleost fish. *Fish Physiology and Biochemistry* 23:245–256.

Kurth, D. 1957. An investigation of the greenback flounder, *Rhombosolea tapirina* Gunther. PhD thesis, University of Tasmania, 106 pp.

Last, P.R., Scott, E.O.G., and Talbot, F.H. 1983. *Fishes of Tasmania*. Tasmanian Fisheries Development Authority, Hobart, Tasmania, 563 pp.

Lim, H.K., Pankhurst, N.W., and Fitzgibbon, Q.P. 2004. Effects of slow release gonadotropin releasing hormone analog on milt characteristics and plasma levels of gonadal steroids in greenback flounder, *Rhombosolea tapirina. Aquaculture* 240:505–516.

Mondon, J.A., Duda, S., and Nowak, B.F. 2001. Histological, growth and 7-ethoxyresorufin O-deethylase (EROD) activity responses of greenback flounder *Rhombosolea tapirina* to contaminated marine sediment and diet. *Aquatic Toxicology* 54:231–247.

Pankhurst, N.W., and Fitzgibbon, Q.P. 2006. Characteristics of spawning behaviour in cultured greenback flounder *Rhombosolea tapirina. Aquaculture* 253:279–289.

Pankhurst, N.W., and Riple, G. 2000. Characterization of parameters for in vitro culture of isolated ovarian follicles of greenback flounder *Rhombosolea tapirina. Comparative Biochemistry and Physiology – Part A: molecular and Integrative Physiology* 127:177–189.

Pankhurst, P.M., and Butler, P. 1996. Development of the sensory organs in the greenback flounder, *Rhombosolea tapirina. Marine and Freshwater Behaviour and Physiology* 28:55–73.

Poortenaar, C.W., and Pankhurst, N.W. 2000. Effect of luteinising hormone-releasing hormone analogue and human chorionic gonadotrophin on ovulation, plasma and ovarian levels of gonadal steroids in greenback flounder *Rhombosolea tapirina. Journal of the World Aquaculture Society* 31:175–185.

Purser, G.J. 1996. The culture performance of the greenback flounder under growout conditions. Fisheries Research and Development Corporation. Project T93/234, 66 pp.

Purser, G.J., and Chen, W.M. 2001. The effect of meal size and meal duration on food anticipatory activity in greenback flounder. *Journal of Fish Biology* 58:188–200.

Shaw, G.W., Pankhurst, P.M., and Battaglene, S.C. 2006. Effect of turbidity, prey density and culture history on prey consumption by greenback flounder *Rhombosolea tapirina* larvae. *Aquaculture* 253:447–460.

Shaw, G.W., Pankhurst, P.M., and Purser, G.J. 2003. Prey selection by greenback flounder *Rhombosolea tapirina* (Gunther) larvae. *Aquaculture* 228:249–265.

Shelverton, P.A., and Carter, C.G. 1998. The effect of ration on behaviour, food consumption and growth in juvenile greenback flounder (*Rhombosolea tapirina*: Teleostei). *Journal of the Marine Biological Association of the United Kingdom* 78(4):1307–1320.

Shirota, A. 1970. Studies on the mouth size of larval fish. *Bulletin of the Japanese Society of Scientific Fisheries* 36(4):353–367.

Sun, B., and Pankhurst, N.W. 2003. Correlation between oocyte development and plasma concentrations of steroids and vitellogenin in greenback flounder *Rhombosolea tapirina. Fish Physiology and Biochemistry* 28:367–368.

Sun, B., and Pankhurst, N.W. 2004. Patterns of oocyte growth, vitellogenin and gonadal steroid concentrations in greenback flounder. *Journal of Fish Biology* 64:1399–1412.

Sun, B., Pankhurst, N.W., and Watts, M. 2003. Development of an enzyme-linked immunosorbent assay (ELISA) for vitellogenin measurement in greenback flounder *Rhombosolea tapirina. Fish Physiology and Biochemistry* 29:13–21.

Thomas, C., Purser, J., and Hart P. 1998. Effect of stocking density on growth, feeding, and cortisol levels of greenback flounder. *Austasia Aquaculture* 12(2):17–18.

Van Den Enden, T., White, R.W.G., and Elliott, N.G. 2000. Genetic variation in the greenback flounder *Rhombosolea tapirina* Günther (Teleostei, Pleuronectidae) and the implications for aquaculture. *Marine and Freshwater Research* 51:23–33.

Verbeeten, B.E., Carter, C.G., and Purser, G.J. 1999. The combined effect of feeding time and ration on growth performance and nitrogen metabolism of greenback flounder. *Journal of Fish Biology* 55:1328–1343.

Chapter 11

Culture of turbot: Chinese perspective

Ji-Lin Lei and Xin-Fu Liu

11.1 Introduction

Turbot were first introduced in China by Yellow Sea Fisheries Research Institute, Chinese Academy of Fishery Sciences, to diversify domestic flatfish culture along the Northern coast of China (Lei and Liu 1995; Men 2002). The first batch of 200 turbot juveniles (5–6 cm in total length) were imported from the United Kingdom in 1992 for a pilot rearing study, and reached sexual maturity during the summer of 1995. Large-scale larviculture using eggs spawned from these broodstock was achieved in 1997. By the beginning of 1999, commercial-scale juvenile production began in the Shandong province along the northern coast. The first full-scale turbot growout systems were built along the coast of Bohai Bay and the Yellow Sea in greenhouses using deep saline well water (Figure 11.1).

The earlier batches of turbot were marketed as live fish in large cities along the southeast coast, including Guangzhou, Shenzhen, and Shanghai. These fish were sold for 80 USD/kg (1 USD is equal to 6.84 RMB on February 2009). In less than 10 years, the turbot farming industry has developed into one of the main mariculture industries in China, with yearly production of over 50 thousand tonnes along the Northern coast of China (Figures 11.2 and 11.3).

11.2 Broodstock husbandry

11.2.1 Acquisition of broodstock

During the early stage of turbot culture industry, most of the broodstocks were adult fish selected from the growout of hatchery-produced juveniles. The first few generations of fish originated from United Kingdom, while subsequent batches were selected from their progeny. Currently, hatchery produced juveniles intended for broodstock are imported from France, Denmark, Iceland, and Norway to avoid inbreeding.

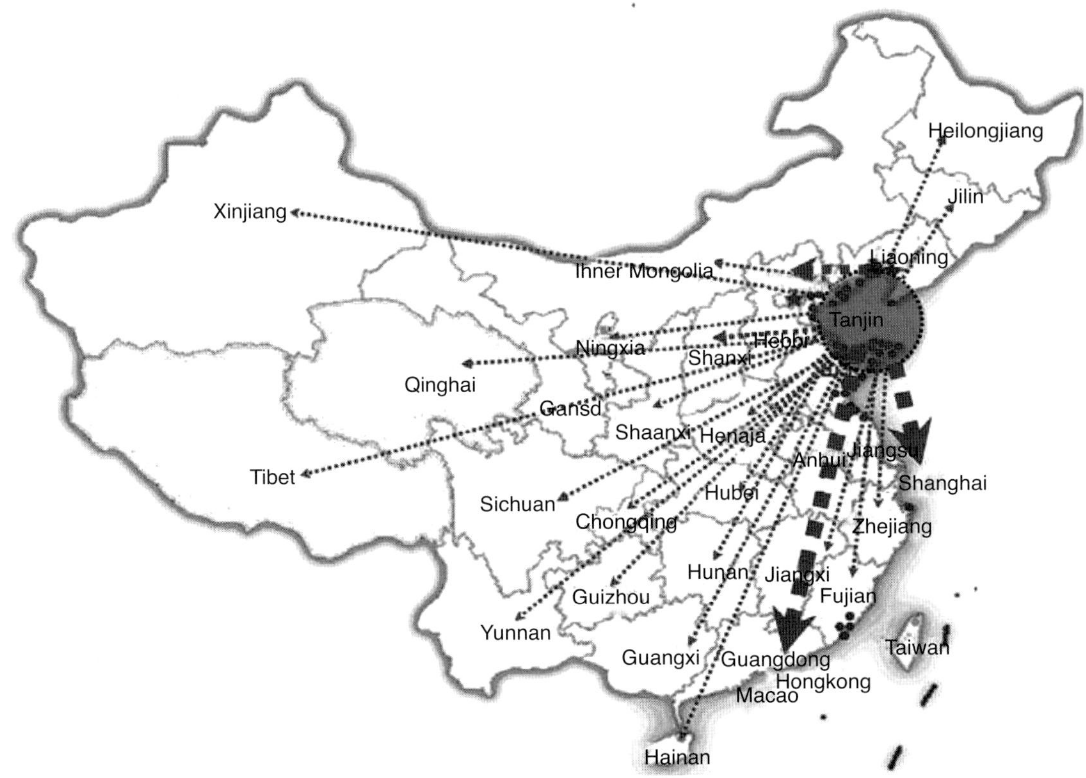

Figure 11.1 Distribution of farms and markets of turbot in China. Note. Black points in the map represent the farms and the arrows represent the distribution of markets with thick arrow as major markets.

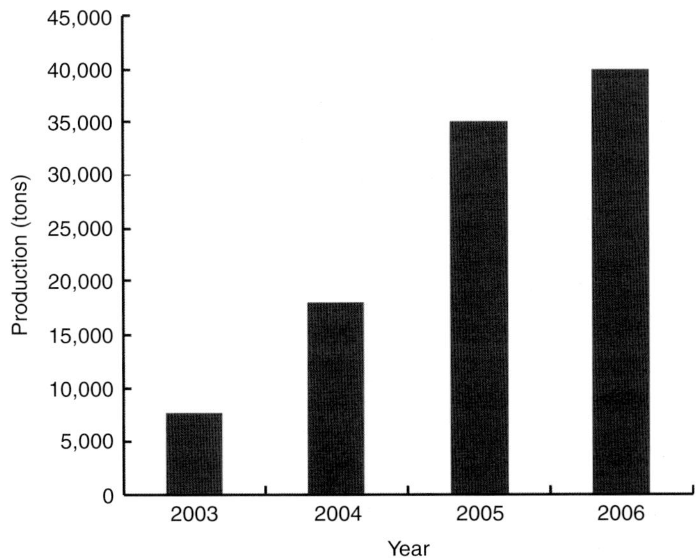

Figure 11.2 Annual production of turbot in Shandong province during 2003–2006. Note: Data from Administration of Ocean and Fisheries of Shandong Province.

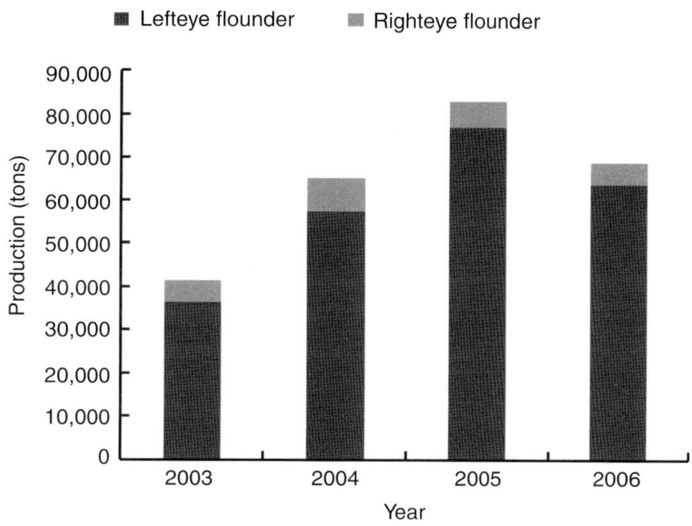

Figure 11.3 Flatfish culture production from 2003 to 2006 in China (FAO, Fishstat Plus).

Males aged 3–6 years and females from 4 to 8 years are used. Most hatcheries choose potential broodstock from populations of 2-year-old adults, so there is at least 1 year to prepare them for reproduction before the maturation protocol is started. The selection of these brooders is a subjective process totally depending on the expertise of the hatchery technicians; however, there are still several criteria to follow, e.g., normal body shape and color, vigorous swimming and feeding behavior, no history of diseases during growout, and representative of as many different populations as possible.

11.2.2 Biosecurity

The majority of turbot are farmed in land-based tanks with deep well seawater in greenhouse-type structures along the northern coast (Lei and Zhang 2001; Lei *et al.* 2002, 2005b). Several farms have cage culture operations on the southern coast where the water temperature is suitable for seasonal production (Lei and Zhang 2001; Zeng *et al.* 2006). The annual temperature fluctuations of coast waters vary from 0 to 27°C in the north and 8–36°C in the south. These temperatures well exceed the tolerance range for turbot. Therefore, it is unlikely that escaped farmed turbot pose a threat to the local ecological system.

11.2.3 System design and requirements

Most hatcheries use indoor flow-through seawater systems to hold broodstock. The tanks are constructed of concrete and range from 10 to 100 m^3 in volume, and 0.6–1.2 m in depth. They can be circular, square, or octagonal in shape. With the inlet water pipes on the top of the tank wall and drainpipe on the center of tank bottom, a circular water flow in the tank is formed easily to flush out feces and uneaten feed. Water level in the tank is controlled with an outside standpipe. Black plastic sheeting is used as a curtain to block outside light for the photothermal control of maturation.

Seawater is normally sand filtered before being pumped into the tank; some deep well seawater is clear enough to bypass sand filtration. Aeration with a regenerative blower or pure oxygen injection is employed to increase the dissolved oxygen in the water. This is especially important for well seawater that is normally deficient in dissolved oxygen. Some hatcheries also treat the water with ozone or ultraviolet light to prevent the spread of diseases.

Boilers and chillers are used to maintain temperature within the optimum range for the broodstock. Measures such as mixing deep well water and natural seawater, shifting the timing of the breeding season, or using a partial water reuse system are also adopted by some hatcheries to reduce energy cost.

Broodstock are held at 1–2 individuals/m^2 at a male to female sex ratio of between 1:1 and 2:3. Water renewal is maintained at a rate of six exchanges per day. Broodstock are kept in a quiet area insulated from the growout fish to minimize noise and the spread of infectious diseases.

11.2.4 Photothermal conditioning

Rearing water temperature and photoperiod are manipulated to control the maturation and ovulation of the broodstock. Water temperature is increased gradually from 8 to 18°C and then returned to 8°C in a normal yearly cycle. Four 60 W tungsten bulbs located about 1.0–1.5 m above the water surface provide artificial light. Light intensity at the water surface is 200–600 lx and 0 lx during the light and dark phase, respectively. Photoperiod is regulated artificially or using digital electronic timers with no dawn/dusk dimming ability, which is changed gradually from 8 hours light:16 hours dark (8 L:16 D) to 16 L:8 D, and then decreased to 8 L:16 D with the changing of temperature. Generally, broodstock will mature about 2 months after the temperature and photoperiod reach 12°C and 14 hours simultaneously. Many hatcheries will keep 5–6 batches of broodstock under different photothermal protocols to obtain fertilized eggs year round.

11.2.5 Monitoring gonad development

Maturation and ovulation of the broodstock are monitored largely by observation of morphological changes of the gonad and the behavior of the broodstock instead of through cannulation and direct oocyte measurement. Abrupt bulging of the gonad and restless swimming of the broodstock signal the impending ovulation of the females. The appearance of milt-running male by hand stripping is also a strong indication that the breeding season has begun.

11.2.6 Diet and nutrition

There are no commercial dry pellets for turbot broodstock in China (Lei *et al.* 2005a). Either chopped trash fish or moist pellets are fed to broodstock. The fish species used as broodstock feed include sand eel *Ammodytes personatus*, horse mackerel *Trachurus japonicus*, redlip croaker *Larimichthys polyactis*, white croaker *Argyrosomus argentatus*, and Japanese Spanish mackerel *Scomberomorus niphonius*. Whichever species is used, it must be of prime quality, i.e., frozen after catching as soon as possible, and stored at −30°C until use. Additives such as vitamins and lecithin are adhered to the chopped fish with boiled starch whenever it is necessary. Moist pellets are manufactured on-site by mixing a powdered commercial premix with the trash fish before extrusion through feed processor. However, it is suspected that the inclusion of fresh fish in broodstock diets is a possible vector for pathogens. More research efforts are now being directed to study the nutritional requirements of the broodstock and to develop formulated dry pellet (Ma *et al.* 2005a).

11.2.7 Controlled spawning

Artificial fertilization is currently the only method to obtain viable turbot eggs in China. Females are induced to mature and ovulate through the photothermal manipulation. With the decreasing of the price of juveniles, many hatcheries choose to buy fertilized eggs from hatcheries that are specialized to produce them to lower production costs. In order to supply fertilized eggs according to the timetable required by the customers, these hatcheries now tend to induce synchrony ovulation by hormones (LHRHa) treatment rather than natural ovulation.

11.2.8 Collection of eggs and egg incubation

Eggs are obtained by manual stripping and fertilized with pooled sperm from 2 to 3 running males. Although the timing of ovulation differs for each female, most hatcheries choose to check and strip all the ovulated females every 2 or 3 days to keep it simple. Eggs are incubated in a hatching net (0.8 × 0.6 × 0.6 m) suspended in a large tank (10–20 m^3) with flow-though water.

Semi-recirculation systems are employed by several hatcheries for egg incubation to have a better control of the hatching conditions. The stocking density of eggs is 100–200 eggs/L, and water temperature and salinity are kept at 13–15°C and 28–33 ppt, respectively.

Incubation lasts about 116 hours at 13 ± 0.2°C. Dead eggs accumulated on the bottom of the net are siphoned out once daily. One day before hatching, buoyant eggs are disinfected with effective iodine concentration of 20–30 mg/L for 10 minutes (Lei *et al.* 2005a) or glutaraldehyde concentration of 400 mg/L for 2.5 minutes (Salvesen *et al.* 1997), then transferred into the rearing tank after volumetric enumeration.

11.3 Larval culture

Turbot larviculture methods are well established among the hatcheries along the northern coast of China (Men *et al.* 2004). Annual total production of juveniles has been increasing gradually since 1999 and exceeded 60 million in 2004. It is estimated that the total juvenile production from Shandong province, where most turbot hatcheries are located, was 120 million in 2005. The abrupt increase in juvenile production caused a dramatic drop in price.

Survival rate from newly hatched larva to juvenile 2 cm in total length varies from 0 to 40% among hatcheries, and can reach an average of 10–20% in large hatcheries with good facilities and expertise. With the wide use of commercial live feed enrichments such as DHA Protein Selco® (INVE) and AlgaMac-3050® (Aquafauna, Bio-marine), the incidence of albinism and color abnormalities on the blind side of the juvenile is less than 20% for most hatcheries, and below 5% for the ones with good expertise.

11.3.1 System design and requirements

Greenwater production systems are used for the larviculture of turbot in China. Indoor tank size varies among hatcheries depending on the preference of operators, but is commonly 7–25 m^3 in volume (Lei 2005; Lei *et al.* 2005a) (Figure 11.4).

Most larviculture tanks are circular, square, or octagonal in shape, and constructed from concrete. Sand-filtered seawater is adjusted to optimum

Figure 11.4 View of larviculture system for turbot.

temperature (18–19°C) by boiler, and pumped into head tank before distribution to the larviculture system. Wherever possible, saline well water from deep aquifers is preferred. Aeration is supplied by regenerative blowers to maintain dissolved oxygen, and assist in water circulation. Some hatcheries now treat the water with ozone or ultraviolet light to prevent the horizontal transmission of diseases.

Additional rooms are also required for the culture of microalgae and rotifers, hatching of *Artemia* nauplii, and enrichment of live feed including rotifer and *Artemia* nauplii. As a result of the sharp decline in juvenile prices lately, many hatcheries have chosen to produce juveniles in spring and autumn when cheaper pond-cultured rotifers and *Artemia* are available.

11.3.2 Hatchery protocols

Larval rearing

Fertilized eggs are stocked before hatching at densities of 15–30 eggs/L in the rearing tank. When eggs are stocked, the water temperature is the same as the incubation temperature, which is commonly 13–15°C. After hatching, water temperature is increased gradually to 18–19°C over 3 days, then maintained hereafter for the remainder of larviculture.

Rearing tanks are illuminated by natural sunlight or artificial light for 14–16 hours per day, and are left dark during the night. Light intensity at the water surface is 500–4,000 lx depending on the type of light employed (natural or artificial).

Water is not exchanged until 5 days post hatching (dph). Exchange rates are increased gradually from 10%/day to 200–300%/day by 35 dph. Aeration is provided to mix gently the larvae and prey, and is adjusted according to the swimming ability of the larvae. From 10 dph, the tank bottom is cleaned every 2–3 days by siphoning after stopping the aeration for half an hour. Whenever oil film forms on the water surface, it is collected by forcing air into a floating pipe frame and removed with pieces of soft filter cloth.

Water quality

As the salinity affects the buoyancy both of fertilized eggs and yolk-sac larvae, natural seawater with a salinity of above 28 ppt is preferred for the larval rearing.

Other water quality in the larval rearing tanks, particularly the dissolved oxygen (DO), pH, and ammonia nitrogen are monitored regularly twice a day to determine if it is necessary to increase the water exchange to maintain the suitable abiotic environmental condition for the larvae. Though the semilethal concentration of DO for the larvae is as low as 3 mg/L (Lei 2005), the aeration or water exchange is increased when it falls below 6 mg/L (Zhang *et al.* 2006). Pure oxygen and air stones generating microbubble are avoided in larval tank to prevent the occurrence of gas bubble disease, especially for first feeding larvae.

In practice, optimal water quality of pH range of 7.6–8.2, and ammonia below 0.02 ppm could be maintained by a little adjustment of above water exchange protocol (Zhang *et al.* 2006).

11.3.3 Food and feeding

The feeding regime for the larviculture is enriched rotifer (*Brachionus plicatilis*), enriched *Artemia* nauplii, and micropellets. The larvae are fed with rotifer as the initial live feed at a density of 5–10 individuals/mL as soon as their digestive tract is functional, which normally happens 3 days after hatching. The feeding of rotifers lasts for 15–20 days. During this time, microalgae (*Chlorella*, *Isochrysis*, or *Nannochloropsis*) are added to the rearing water at a concentration of 5×10^5 cells/mL. *Artemia* nauplii are fed from 9 or 10 dph to 40 dph with an initial density of 0.1 to 0.2 individuals/mL, then gradually increased to a final density of 0.5–1.0 individuals/mL. The greenwater rearing environment helps to maintain the nutrition level of the rotifer and *Artemia* nauplii. Micropellets are cofed with *Artemia* from 20 dph to 40 dph; thereafter, juveniles are weaned to accept micropellets solely.

11.3.4 Formulated feeds

At present, formulated feeds are either imported from Denmark (Biomar and Aller), Japan (Hayashikane), or manufactured by domestic companies (Shengsuo and Haili). The particle size of formulated feed is increased as the growth of larvae with size of 250–400 μm fed to juvenile 20 dph, 400–600 μm to juvenile 25 dph, and 630–800 μm to juveniles over 30 dph, respectively. For the first several days of weaning, a small amount of micropellet is fed up to 10–12 times per day to familiarize the juvenile with the diet and reduce the wasting. As weaning to dry feeds progresses, the frequency of feeding is gradually reduced to 5–6 times per day.

Time for successful weaning depends on the size of juveniles at the initiation of weaning. Generally, a period of 30 days is required for larvae 6 mm TL (15 dph), 20 days for 10 mm TL (20 dph), and 10 days for 15 mm TL (30 dph). It is normal to wean the juveniles 12 mm TL (25 dph) with formulated feed size of 400–600 μm to reduce the cleaning work without any harmful effect on the survival rate of juvenile.

11.3.5 Microbial environment

Flow through water is used by hatcheries in China. Natural seawater or saline well water is pretreated by mechanical filtration, aeration, UV disinfection, and temperature adjustment, before it is used in the rearing tank. There is no attempt to reuse the outflow water from the rearing tanks. General tests of the microbial flora are made only when larvae show signs of disease.

11.3.6 Harvest

Juveniles can be harvested and transferred to nursery tanks when they reach an average of 3 cm TL. To harvest the fish, the water level of the rearing tanks is lowered and the juveniles are carefully collected with a small hand net, and transferred to the nursery tanks. Size grading and removal of deformed or discolored juveniles is carried out during the transfer. Normal individuals are grouped into large, medium, and small body sizes, and then transferred to separate nursery tanks.

11.3.7 Hatchery economics

Production costs and farm gating price for hatchery-reared turbot juveniles around 5 cm TL is currently USD 0.07–0.25 and USD 0.22–0.37 per individual, respectively. At present, there are more than 80 turbot hatcheries in China (Zhu *et al.* 2004). The annual production for large hatcheries is 4–5 million juveniles, while small hatcheries produce around 100,000–200,000 juveniles per year. Most hatcheries produce 2–3 batches of juveniles per year. As production increases, the sale price for juveniles continues to fall. This reduction in juvenile price encourages the further development of the farming industry and the continuous expansion of the market.

11.4 Nursery culture and growout

11.4.1 System design and requirement

Greenhouse-type structures are used to hold land-based turbot culture tanks. These buildings are preferred for their low cost of construction and ease of temperature control in the winter. Standard green house dimension is 90 × 15 m in surface area, 1.0–2.0 m in wall height, 4.0–4.2 m height in the top of roof (see Figures 11.5 and 11.6).

Figure 11.5 View of greenhouse for nursery and on-growing system of turbot.

Figure 11.6 View of tanks inside greenhouse for nursery and on-growing of turbot.

Greenhouses are constructed with brick walls and steel girders with fiberglass tile or a thick plastic film roof cover to resist the strong coastal winds. Two rows of 13–15 concrete tanks (either 5–6 m circular, or 5–6 m square with rounded corners) are built along both sides of a vertical wall. The sidewall of the tanks is about 1 m in height with the water inlet placed along the inside of the wall. Outlet water is flushed out through the center drain and an external standpipe that is used to control water level in tank. Wastewater is collected by a 1-m wide covered drain placed between two rows of tanks.

Two rooms 15–25 m^2 in area are attached to one end of the green house for the storage and preparation of feeds, record keeping, and cloth changing for workers. A separate area of 60–100 m^2 should be reserved between the workroom and the rearing tanks for tool storage, quarantine, disinfection, and other related operations.

Water temperature is the most important factor to be considered for site selection for turbot farming in China. Therefore, turbot farms are located only where saline well water is available. Generally, saline wells in the coastal zone can be divided into four types (Lei 2005):

1. Wells located in offshore fractured rock zones: 80–120 m in depth, clean water, and free of particles, close to natural seawater in chemical composition, annual water temperature range 11–15°C. This kind of well water is normally also suitable for larviculture.
2. Fine sand shore wells (such as sand shores along Laizhou): 18–22 m in depth, clean water, and generally free of fine particles, annual temperature ranges from 14 to 18°C.

Table 11.1 Feeding method of fresh diet for turbot.

Fish total body length (cm)	Feed amount (g)/fish wet weight (g) (%)	Feeding times per day
4	9–10	6–4
5	9–8	6–4
6	8–7	6–4
7	7–5	6–4
10–15	5–4	4–3
15–20	4–3	3
20–25	3.5–3	2
25–30	3	1
30–35	3–2.5	1

Note: When water temperature falls to 12–13°C, the feed amount is about 1% of fish body weight. When the water temperature falls less than 11°C, the amount is about 0.4–0.2% of body weight, and the feeding frequency is once every 2 days.

3. Coarse sand beach wells (such as in Haiyang): 10 m in depth, pumping capacity changing with tide, occasional occurrences of fine sand, annual temperature fluctuating from 8 to 20°C with the season, similar salinity to natural seawater.
4. Wells of silt in saline areas (such as in Tangshan, Hebei, and Changyi): over 100 m in depth and 40 ppt in salinity, annual temperature ranges from 14 to 16°C. Desired salinity is normally obtained by mixing well saline water with fresh water drilled in the same area in aquifers closer to the surface.

Dissolved oxygen in the well water is usually too low to support the survival and growth of turbot, and should be increased by aeration or adding pure oxygen. The dissolved oxygen in the inlet water should be maintained above 6 mg/L and can be raised up to 5–7 mg/L after aeration with a blower, and over 10 mg/L after the addition of pure oxygen. Generally, one airstone is used per every 3–4 m^2 of tank bottom area. There is a trend for more and more farms to change from air blowers to pure oxygen.

11.4.2 Nursery culture

Juveniles of 3 cm in total body length should be kept in the nursery tanks for an additional 1.0–1.5 months before they can be sold to farmers. When the juveniles reach 5–6 cm TL, they are selected and graded once more by body color, shape, activity, and size. Juveniles with abnormal body color, deformities, weak activity, lesion, and symptom of diseases are rejected. Some hatcheries will volunteer to get the quarantine certificate issued by the state organization to guarantee the health condition of their juveniles, though it is not a regulatory requirement by the relative administrative regulation to market the juveniles.

The main management activities of a turbot nursery include the following:

1. Water quality management: Water exchange rate is maintained at 5–10 tank volumes per day. The water salinity, pH, temperature, and oxygen

concentration are measured on a daily basis in the nursery system, and necessary adjustment of water exchange and aeration is made based on these data. Uneaten feed and feces are flushed out of the tank twice daily by removing the external standpipe to lower the water level to 10–15 cm rapidly.

2. Diseases prevention:
    - Measures such as stringent disinfection of facilities before the stocking of juveniles, and use of disinfectant footbaths in the entrance and exit of tank room, and separate and daily disinfection of utensils for the same batch of juveniles are enforced during the whole process of culture.
    - Though moist pellet and minced fish can also be used as feed for the juveniles, most hatchery only used dry pellet for its more stable quality and ease of handling and storing.
    - Behavior and feeding activity of juveniles is monitored closely. Fish with the initial signs of a disease should be quarantined immediately, and antibiotic food should be prepared and used. Bath with 20–30 ppm oxytetracycline hydrochloride plus 167 ppm formalin for one and half hour is also applied as a prophylactic measure whenever the juveniles are moved, graded, or suspected infection or disease. Commercial feed additives including herbs such as ginseng and licorice are also adopted to protect liver and improve digestion and immunity of the farmed fish.

11.4.3 Diet and nutrition

Commercial dry pellets are recommended for the farmers. The use of wet industrial fish or moist pellets is discouraged to prevent the spread of diseases.

11.5 Growout

The growout facilities for subadult and adult fish are very similar to those of juvenile nurseries, the main difference being that the tanks for growout are larger. By using the saline well water and greenhouse structures, turbot can be farmed at a reasonable cost along the coast of China where the natural seawater temperature is otherwise not suitable for the survival and growth of turbot. The investment in a standard greenhouse system with a tank area of 1,000 m^2 is less than USD 19,000 (Lei *et al.* 2002). The water temperature in the tanks is maintained from 11 to 18°C year round. Juveniles about 10 g in body weight can grow to market size of 500 g in 7–9 months in these greenhouse systems (Lei and Zhang 2001; Su *et al.* 2003).

Turbot are also cultured seasonally in cages on the southern coast when the temperature declines to the point where it is suitable for turbot in winter (Lei 2005; Zeng *et al.* 2006). Fish farmers in Fujian and Guangdong buy young turbot 100–150 g that have been grown inside greenhouses in the northern provinces, such as Shandong and Dalian, and stocked into cages at 10 kg/m^2 of cage bottom area by the end of October when coastal water temperatures drop below 21°C. Fish in cages are fed either fresh fish or dry pellets. These turbot can grow to a market size of 500–700 g by the May of the following year (approximately 6 months). Cages on the coast of Fujian and Guangzhou are more than 1,000 miles closer to the main turbot markets than the Northern production region, so the cost of transporting young fish is much less than adult fish. Cage culture also has the advantage of lower cost of pumping water and facility construction, and faster growth rate than tank-cultured. The practice of relaying turbot culture from the tanks in north to cages in the south has been demonstrated as a good way to reduce the cost of turbot culture in China. However, further research work is necessary to develop special flatfish culture cages that can resist the strong wind and currents in the southern coastal water.

11.5.1 Environmental conditions

The optimal temperature range for turbot culture is between 10 and 20°C, and a range of 14–19°C is optimal for rapid growth. Turbot can tolerate a water temperature as low as 3°C and as high as 23°C for a short time. However, when the temperature is lesser than 10°C and more than 20°C, the feeding and swimming activity of turbot begin to decline. Signs of temperature-induced stress, such as feed suspending, mucus secretion on gill and body surface, darkening of body color will appear when turbot exposed to water temperature over 23°C for an extended period. Severe skin ulcers or parasite infestations (such as *Trichodina* disease) may also follow.

In order to prevent the occurrence of diseases at high temperature, water temperature must be controlled. In addition, other measures such as water disinfection with ozone, increase of water exchange rate, and addition of vitamins in feed are also employed temporarily.

Turbot oxygen consumption is 96 mg/kg/hour at 12°C. In comparison, rainbow trout oxygen consumption is as high as 200–250 mg/kg/hour at the same temperature. Turbot farmers need to maintain DO concentrations in the culture tanks above 5 mg/L.

Turbot can survive and grow in a wide range of salinity, from 12 to 40 ppt. The optimal salinity range for turbot is 20–32 ppt and 25–30 ppt. The daily water exchange rate for the flow-through water is commonly over 600% per day. However, farmers increase exchange rates if the biomass in the tank is high.

11.5.2 Diet and nutrition

Until the late 1990s, there were few studies on nutritional requirements of flatfish in China. For turbot formulated diet, a protein level of 56, 50, and 48%, and a lipid level of 7, 10–13%, and 14% for the juvenile, young fish and on-growing adult, respectively, is recommended (Jiang *et al.* 2005; Lei 2005). Lei (2005) also suggested an optimal energy–protein ration of 65–76% and 80% in the feed for on-growing adult and growout fish 1 month before marketing, respectively. Recent research on the nutrition of turbot includes: immunity stimulating effects of dietary vitamins, lipid, mineral mixtures, and alginate oligosaccharides (Ma *et al.* 2004; Chang *et al.* 2005; Liang *et al.* 2005; Wang *et al.* 2006), activity of digestive enzymes (Chen *et al.* 2005; Fu *et al.* 2005), and food consumption and evacuation rate (Ma *et al.* 2003). These studies are very helpful for the domestic feed companies to design and produce dry pelletted feed for flatfish in China.

During the early stage of the turbot culture industry, minced wet industrial fish was fed directly, a technique which had been employed in Japanese flounder culture. Because of the difficulty of preparation and inconvenience for handling, and variation in quality, minced industrial fish was later replaced by dry pellet imported from Japan and Europe, which were designed originally for red sea bream or salmon. High price of imported diets and great market potential for farmed turbot encouraged diet trials in China. Moist pellet were first developed for turbot with the effort of research institutes and commercial companies. The new diet turned out to be very effective, and was widely accepted by farmers because it was cheaper and more convenient to process and storage. However, the rapid dissipation of the moist pellet resulted in water quality deterioration and increased incidence of diseases. Currently, imported commercial dry pellets such as Biomar® and Aller®, which are formulated especially for flatfish, are used more widely in the big company even though the high cost of these diets makes them a reluctant choice for the small farmer. Now, several institutes and companies, such as Yellow Sea Fishery Institute and Shandong Shengsuo Fishery Feeds Research Center, are endeavoring to develop high quality domestic diets for turbot and other flatfish.

Currently, the feed conversion rate (FCR) of growout turbot expressed as weight of feed to wet weight fish gained varies with the types of the feed. It is normally about 6 for wet industrial fish (Zeng *et al.* 2006), 1.83–3.3 for moist pellet (Sun and Yan 2003; Ma *et al.* 2005b; Zeng *et al.* 2006), 0.95–1.5 for domestic dry pellet (Ma *et al.* 2005b), and 0.81–1.0(Lei *et al.* 2005a) for import dry pellet such as Biomar® pellet.

Feed rate for weaning larvae is commonly about 20% of their body weight per day, and then falls to 10–15% of body weight per day when the total length of fish is about 2.0 cm, and needs to be adjusted according to feeding activity to avoid wastage. It is reduced to 1.5–2% of body weight for adult fish. When the water temperature is lower than 12°C or higher than 22°C, daily feed rate and feed frequency are reduced. Feeding methods of different types of diet are recommended in Tables 11.1–11.3. At present, hand feeding is the main feeding method, and automatic machine feeding is employed by a few big farms.

The feeding method of wet pellet diets for turbot is shown in Table 11.2, which needs to be adjusted to adapt to different rearing condition and growth stages.

Table 11.2 Feeding methods of wet pellet diet for turbot.

Weight (g)	Daily feed rate (dry pellet/wet fish weight) (%)	Feed times per day
Juvenile	15–10	4
100	10–7	3
300	5–4	2
500	3–2	1

Table 11.3 Feeding methods of dry pellet diet for turbot.

Fish weight (g)	Daily feeding rate at different water temperatures (%)						
	11–12°C	13–14°C (Spring)	13–14°C (Fall)	14–15°C (Spring)	14–15°C (Fall)	17–18°C	>19°C
10	1	1.07	1.39	1.32	1.77	1.78	1.95
50	0.64	0.71	0.83	0.85	1.04	1.06	1.09
100	0.52	0.59	0.67	0.71	0.83	0.85	0.85
200	0.43	0.49	0.53	0.59	0.66	0.68	0.66
500	0.33	0.39	0.4	0.46	0.49	0.51	0.48
1,000	0.27	0.32	0.32	0.38	0.39	0.41	0.37
5,000	0.25	0.29	0.28	0.34	0.34	0.36	0.32

Feeding methods of dry pellet diets for turbot is recommended in Table 11.3. For instance, for fish with weight of 200 g, when the water temperature falls to 17–18°C, the amount fed is about 0.68% of the fish body weight.

11.5.3 Stocking and splitting

Stocking densities of juveniles depend on water quality and water exchange rate. For juveniles of 15 g in body weight, culture density is about 2 kg/m^2. As fish grows, culture density increases to 7 kg/m^2 and 10–20 kg/m^2 for fish in body weight of 50–100 g and 600–800 g, respectively.

11.5.4 Harvesting and transportation

Most turbot are harvested between 500 and 750 g. Feeding is stopped one day before harvesting. Hand net is used to harvest the fish after water level is reduced to 15–20 cm in tank. Polyethylene bags of 40 L volume are used for transportation with 8–16 L filtered seawater in it. Pure oxygen is sealed within the bag with fish and water. Water temperature is maintained between 7 and 8°C during the transportation. During winter, 18–20 fish are put into one bag, and fewer fish per bag are used during summer. Depending on the shipping distance, either live fish vehicle or airline is employed to transport the live on-grown turbot.

11.5.5 Marketing

In the past, the majority of the farmed turbot were marketed as live fish for domestic consumption in large cities near the east coast of China, such as Guangzhou, Shenzhen, Hongkong, Shanghai, and Dalian. Markets in the smaller cities are not developed yet. At present, its market has been distributed all around China with the increasing of production and dropping of price (see Figure 11.4). Further development of turbot farming in China will result in a broad domestic consumption of turbot with the more rational price and better recognition by consumers of this exotic species.

11.5.6 Production economics

In China, most turbot farms are located in three provinces: Shandong, Liaoning, and Hebei, and one municipality directly under the Central Government, Tianjin, all being located along Bohai bay. Among these locations, Shandong province is the origin of turbot farming, and hence has far more farms than any other province. According to government statistics, the greenhouse area in flatfish production in Shandong province was 5,300 thousand m^2 in 2006, which included 4,240 thousand m^2 for turbot. The yearly production of turbot is 40 thousand tonnes in Shandong province, which was 70% of the total turbot production in China in 2006 (see Figure 11.5). The estimated production of other flatfish in Shandong was about 19 thousand tonnes, as the production of flatfish was not estimated by species definitely. The second farming center of turbot is Liaoning province. According to the 2006 statistic of Huludao city in Liaoning province, there is 400 thousand m^2 turbot farming area in this city with a production of 10 thousand tonnes, which contribute to 20% of the national production. According to 2005 China fishery statistics, the flatfish farming area of the three provinces is about 8,060 thousand m^2, providing 65.3 thousand tonnes production, which includes over 50 thousand tons of turbot. According to FAO (Fishstat plus) fishery statistic, the total production of flatfish reached its maximum of 83 thousand tons in 2005 (see Figure 11.3). However, the finding of prohibited nitrofuran metabolites in sample fish taken from Shanghai markets in 2006 caused a wide distrust of the customers to the farmed turbot, and resulted in quick shrinking of the turbot market and production. Close cooperation among the industry, research institutes, and relative governmental administration is needed to improve and guarantee the safety of farmed turbot.

The direct farm cost (e.g., stock, feed, labor, maintenance, transportation, etc.) for farmed turbot is estimated at USD 5.85/kg (Lei 2006) in a typical turbot farm. However, if the indirect cost (e.g., insurance, depreciation, etc.) is included, the total cost of farmed turbot is estimated about USD 7.0/kg in China. Further research works on dry pellet formulation, disease prevention and control are needed to produce turbot at a lower cost.

11.6 Summary: industry constraints and future expectations

Turbot farming has developed into a very important mariculture industry in China in less than 20 years. This rapid development is the result of a combination of a suitable culture species, favorable natural resources, and proper market strategy. However, there are still obstacles impeding the further development of this industry. Problems, such as the availability of well water, pollution of the coastal environment, and safety of market product, are becoming more and more evident. Therefore, we need to improve constantly the existing culture method, upgrading equipment and technology, seize opportunities, and accelerate the construction of semiclosed or closed recirculation farming systems along the northern Chinese coast.

To achieve the above objectives, a key national collaborative project by 15 research institutions and farming companies, is aiming to improve the techniques in five major areas of turbot farming including selective breeding, nutrition and feed, disease prevention, recirculation culture systems, and product safety. It can be anticipated that turbot farming will further develop soon into a more sustainable and healthy mariculture industry in China.

Literature cited

Chang, Q., Liang, M.Q., Wang, Y.G., Gao, C.R., and Wang, J.L. 2005. Effect of dietary vitamin C on non-specific immunity of turbot (*Scophthalmus maximus*). *Marine Fisheries Research* 26(5):22–26 (in Chinese with English abstract).

Chen, M.Y., Zhang, X.M., and Lian, J.H. 2005. Development of some digestive enzymes and alkaline phosphatase activities in turbot *Scophthalmus maximus* larvae and juveniles. *Periodical of ocean university of China* 35(3):483–486 (in Chinese with English abstract).

Fu, X.H., Sun, M., and Sun, S.C. 2005. Activity of digestive enzymes in *Scophthalmus maximus*. *Journal of Fishery Sciences of China* 12(1):26–32 (in Chinese with English abstract).

Jiang, K.Y., Li, Y., Li, J., Wang, L., and Wang, Y.J. 2005. Eco-nutrition requirement of protein for juvenile turbot (*Scophthalmus maximus* L.). *Marine Sciences* 29(9):65–70 (in Chinese with English abstract).

Lei, J.L. 2005. *Turbot Farming Techniques*, 2nd ed. Shanghai Scientific and Technical Publishers, Shanghai (in Chinese).

Lei, J.L. 2006. Outlook of the marine fish culture industry in China. *Marine Fisheries Research* 27(2):1–9 (in Chinese with English abstract).

Lei, J.L., and Liu, X.F. 1995. Preliminary study on the introduction and culture of turbot (*Scophthalmus maximus*). *Modern Fisheries Information* 10(11):1–3 (in Chinese with English abstract).

Lei, J.L., Ma, A.J., Chen, C., and Zhuang, Z.M. 2005b. The present status and sustainable development of turbot (*Scophthalmus maximus* L.) culture in China. *Engineering Science* 7(5):30–34 (in Chinese with English abstract).

Lei, J.L., Ma, A.J., and Liu, X.F. 2005a. Turbot (*Scophthalmus maximus*) culture techniques. In: Lei, J.L. (ed.) *Marine Fish Culture Theory and Techniques*. Chinese Agriculture Press, Beijing, pp. 524–591 (in Chinese).

Lei, J.L., Men Q., and Wang, Y.G. 2002. Review of "green house+ deep well seawater" industrialized culture pattern of turbot (*Scophthalmus maximus*). *Marine Fisheries Research* 23(4):1–7 (in Chinese with English abstract).

Lei, J.L., and Zhang, X.L. 2001. Experiment of rearing *Scophthalmus maximus* Linnaeus by using deep well seawater in the farming plant. *Modern Fisheries Information* 16(3):10–12 (in Chinese with English abstract).

Liang, M.Q., Chang, Q., Wang, Y.G., Gao, C.R., and Wang, J.L. 2005. Influences of vitamin E and lipid sources on non-specific immunity of turbot (*Scophthalmus maximus* Linnaeus). *Marine Fisheries Research* 26(5):15–21 (in Chinese with English abstract).

Ma, A.J., Chen, C., Lei, J.L., Chen S.Q., and Zhuang, Z.M. 2005a. The effect of protein and n-3 HUFA on the reproduction of turbot (*Scophthalmus maximus*). *Marine Fisheries Research* 26(1):7–12 (in Chinese with English abstract).

Ma, A.J., Chen, C., Lei, J.L., Chen, S.Q., Zhuang, Z.M., and Wang, Y.G. 2005b. Effect of stocking density on growth, feed conversion and pigmentation improvement of turbot (*Scophthalmus maximus* L.). *Oceanologia et Limnologia Sinica* 36(3):207–212 (in Chinese with English abstract).

Ma, A.J., Chen, S.Q., Lei, J.L., and Wang, Y.G. 2004. Study on effect of vitamin and mineral mixture on the turbot pigmentation improvement. *Marine Fisheries Research* 25(5):26–30 (in Chinese with English abstract).

Ma, C.H., Chen, D.G., and Shen, W.Q. 2003. Food consumption amount and evacuation rate of turbot *Scophthalmus maximus*. *Fisheries Science* 22(5):5–8 (in Chinese with English abstract).

Men, Q. 2002. Overview on turbot *Scophthalmus maximus* (Linnaeus) introduced to China for ten years. *Modern Fisheries Information* 17(9):14–17 (in Chinese with English abstract).

Men, Q., Lei, J.L., and Wang, Y.G. 2004. Biology and critical breeding techniques of turbot *Scophthalmus maximus*. *Marine Sciences* 28(3):1–4 (in Chinese with English abstract).

Salvesen, I., Øie, G., and Vadstein, O. 1997. Surface disinfection of Atlantic halibut and turbot eggs with glutaraldehyde: evaluation of concentrations and contact times. *Aquaculture International* 5(3):249–258.

Su, K., Zhang, H.S., Xiao, B.Q., and Ma, X.N. 2003. Turbot culture under high density in a sea water closed recirculation system. *Modern Fisheries Information* 18(5):9–13 (in Chinese).

Sun, Z.Z., and Yan, Y.X. 2003. Experiment on industrial turbot (*Scophthalmus maximus*) farming. *Marine Fisheries Research* 24(1):6–10 (in Chinese with English abstract).

Wang, P., Jiang, X.L., Jiang, Y.H., Cao, J.M., and Yang, X.S. 2006. Immunoregulatory effect of alginate oligosaccharides on *Scophthalmus maximus*. *Marine Sciences* 30(8):6–9 (in Chinese with English abstract).

Zeng, Q.M., Lin, W.L., Wu, L.F., Zhou, C., and Lin, Y.J. 2006. The techniques of cultivating turbot (*Scophthalmus maximus*) in Southern China. *Marine Sciences* 30(1):1–4 (in Chinese with English abstract).

Zhang, X.L., Jiang, S.C., Wang, Z.M., Zou D., Cong, Y.Q., and Wang, X.B. 2006. A Study of the similarities and differences in the breeding of *Scophthalmus maximus* and *Paralichthys olivaceus*. *Transactions of Oceanology and Limnology* 3(3):105–110 (in Chinese with English abstract).

Zhu, J.X., Wang, Y.G., Liu, H., Liu, Y.G., and Lei, J.L. 2004. Status and problems of turbot seed production in China (I). *Scientific Fish Farming* 9: 1–2 (in Chinese).

Section 4

North and South America Stock Enhancement

Chapter 12

Stock enhancement of southern and summer flounder

John M. Miller, Robert Vega, and Yoh Yamashita

12.1 Introduction

Traditional management, which is by catch restriction, has not worked well for most exploited species of flatfish. In the case of southern flounder in North Carolina, United States, stocks are now less than 6% of the virgin stock (NC DMF 2005). In Texas, declines in the southern flounder's relative abundance have prompted Texas Parks and Wildlife Department to explore other management options, including stock enhancement, to improve the statewide populations. For one reason or another, usually pressure from commercial interests, North Carolina managers will not restrict catch enough for the stocks to rebuild themselves. In fact, hardly any wild species of commercially exploited animal has been able to withstand the economic pressures to overexploit it, be it terrestrial or aquatic. This situation, coupled with habitat degradation and increasing population pressure, means that even if a species could be restored to some former level of abundance by restricting exploitation, it would not likely remain that way for long.

Two exceptions to this natural supply and demand imbalance are as follows: (1) where sportsfishers practice catch and release; and (2) where animals are reared and stocked in their natural habitat. In the latter case, it is noteworthy that the only species in Japan which have maintained their former abundance in the face of heavy exploitation are species which are intensively stocked, one of which is the Japanese flounder (Yamashita and Aritake, this volume). Thus, it is possible to at least envision sustainable catches of marine fish species with more proactive supply-side management. Unlike freshwater fish, until recently, spawning and rearing marine species could not be done, at least in quantities sufficient to make substantive additions to wild stocks. But in the past 2 or 3 decades, great strides in aquaculture of marine species have been made, and now it is only a question of whether a stocking program should be initiated. In the cases of southern and summer flounder, both can be spawned and reared in hatcheries (Burke *et al.* 1999; Daniels and Watanabe 2002; Daniels *et al.* 2007,), thus making it reasonable to consider the more proactive management approach, stocking.

The perspective of this chapter is flounder stocking efforts in Texas (recently begun) and in North Carolina, where stocking of marine fish has not begun. In fact, in the southern flounder management plan, the NC Division of Marine fishes wrote, "Do not endorse funding for pilot research on the feasibility of southern flounder stock enhancement at this time." (NC DMF 2005). Apparently, despite abundant evidence to the contrary, the belief that southern flounder can be managed by catch restriction still persists.

Thus, this chapter cannot take a retrospective data-based approach. Rather, we rely on principles derived from other flatfish research to discuss the feasibility and potential benefits of stocking southern and summer flounder.

12.2 Previous work

With the exception of recent southern flounder rearing and stocking efforts in Texas, the only substantive efforts to stock flounder in the United States were some very limited pilot releases of summer flounder conducted as part of a PhD. dissertation in North Carolina (Kellison 2000), some parts of which have been published. Kellison released small numbers (hundreds) of juvenile summer flounder at four sites near Beaufort, North Carolina, and attempted to follow them for about 2 months to get estimates of growth and survival rates (Kellison *et al.* 2003). Recapture rates of the hatchery-reared fish were so low that growth estimates were not possible. Kellison suggested that the low recapture rates were due to predation, but it was not possible to track the released fish to know how much of the attrition was due to emigration. Hatchery-reared fish were more susceptible to predation by blue crabs (*Callinectes sapidus*) in the laboratory than wild fish (Kellison *et al.* 2000). The Japanese have shown that many more fish must be released to detect any effect (Kitada and Kishino 2006).

12.3 Rationale for stocking

There are four fundamental reasons for considering stocking. First, to overcome recruitment limitation, which is where natural recruitment is insufficient to reach a species' carrying capacity of an ecosystem. This strategy is termed "stock enhancement." Second, to restore the spawning stock biomass of overfished species to the point where they can sustain higher exploitation or to restore the ecological balance of the fish communities. This strategy is termed "restocking." The third strategy, "sea ranching," is where natural recruitment is low, but there is no intention of augmenting spawning. Larger individuals are stocked in a "put and take" operation, and harvested before they spawn. A fourth strategy is "ecological," which is used to restore the community structure or to conduct ecological experiments, e.g., to determine carrying capacity or investigate species interactions (Miller and Walters 2004). Both the goals and the evaluation measures are different in the four cases.

Since both summer and southern flounder occupy estuaries as juveniles and usually migrate into coastal waters before the age of 3 years (Miller *et al.* 1984) when most are legal sized, sea ranching would seem to be likely to succeed if fish were stocked in the estuaries. But either restocking or stock enhancement could also be successful in the case of flounders, since the Standing Stock Biomass (SSB) of each species is far below that desired and both species are recruitment limited.

12.4 Likelihood stocking would increase production

The necessary condition for stock enhancement to work is that the natural rate of recruitment is below what the ecosystem can support. In the case of both species of flounder in North Carolina and southern flounder in Texas, spawning occurs offshore (Figure 12.1) in winter, and the larvae must reach and negotiate narrow inlets (or passes) to reach juvenile nursery areas in the estuaries. Since larvae are not strong enough to swim from the spawning areas to the inlets (Miller *et al.* 1985), nor can they swim fast enough to oppose normal currents

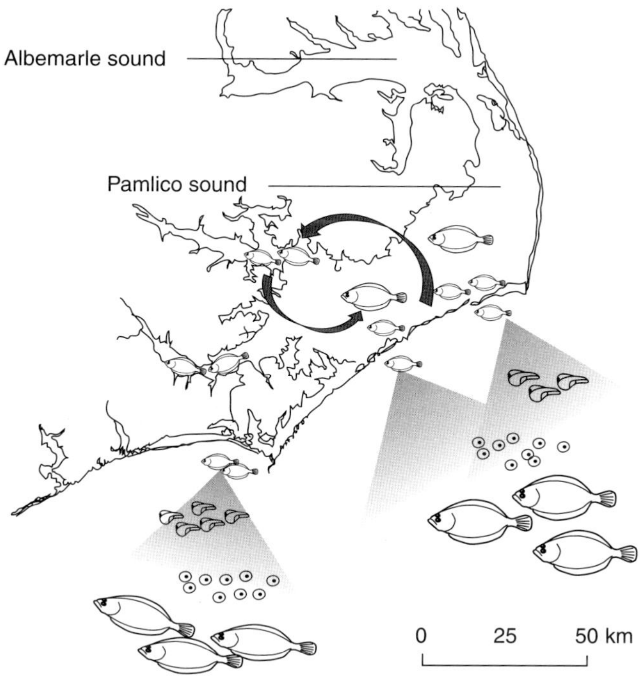

Figure 12.1 The life cycle of southern and summer flounder showing adult spawning area (offshore), egg and larval migration to inlets, juvenile migration to nursery areas, and 1- to 3-year residence time in the open sounds (arrows) before returning offshore to spawn.

in inlets, they are subject to ocean currents for at least the first few weeks of life. In North Carolina, the interannual variability in these currents during the winter–spring migration phase results in a tenfold variation in the numbers of southern flounder juveniles reaching the nursery areas (Taylor *et al.* 2009). Taking the maximum level of recruitment as a minimum estimate of carrying capacity, since there is no evidence for density-dependent growth limitation in southern flounder (Guindon and Miller 1995), means stocking juveniles in estuarine nurseries could potentially increase recruitment at least up to tenfold in the poorest years. Considering even the years of highest recruitment are probably well below historical levels due to overfishing and depressed SSB, now at about 6% of virgin stock (NC DMF 2005), substantially greater gains seem plausible. Additional evidence to support this comes from published papers showing the lack of food limitation (Guindon and Miller 1995). In addition, Taylor *et al.* (2009) found that in different years, the colonization of individual bays varied up to 20-fold. This suggests the possibility that stocking those same bays that are undercolonized in certain years could increase production. Strictly speaking, the carrying capacity of none of the estuaries or nursery bays is known, and no doubt it has been reduced by environmental degradation; but this unknown is itself an argument for conducting pilot enhancement studies to determine this. A final reason to suppose stocking would increase production is the fact that good and weak year classes of southern flounder in North Carolina persist for up to at least 3 years (Taylor *et al.* 2009). This is interpreted as evidence that events preceding recruitment to the nurseries are the main determinants of year class strength. The corollary to this is the idea that stocking flounder at a stage later than that which is responsible for interannual variability, i.e., stocking older juveniles would likely be successful.

As mentioned earlier, it is clear that stocking can at least help maintain populations in the face of heavy exploitation (Yamashita and Aritake, this volume). In the cases of both Texas and North Carolina, it is even more likely to be successful. The coastline of Japan, where Japanese flounder are stocked, is an open coastline, whereas in the case of both North Carolina and Texas, southern flounder inhabit semienclosed bays, or lagoons (Figure 12.1), for at least their first 2 years (Taylor *et al.* 2009). By that time, they are legal size. It is well known that the most successful stocking is in semienclosed bays (Yamashita and Yamada 1999). In the case of Albemarle Sound, stocking southern flounder would likely be particularly successful. Albemarle Sound is relatively difficult to colonize owing to the fact that winds which favor transport of larvae through inlets also create pressure gradients in Eastern Albemarle Sound which oppose currents carrying larvae from entering the Sound through Croatan Sound (Weisberg and Pietrafesa 1983). This means that though Albemarle Sound is a favorable habitat for southern flounder (they grow well there), it is poorly colonized by species which spawn offshore in winter, such as southern flounder. Thus, stocking (sea ranching) Albemarle Sound could be particularly fruitful. Similarly, and particularly in the case of southern flounder, which thrive in low-salinity habitats, stocking them in the upper reaches of the Neuse and Pamlico Rivers, i.e., upstream from the limit of larval transport by currents could also be particularly successful (Pietrafesa *et al.* 1986).

12.5 Management changes to support stocking efforts

At least two changes in the current management practices are important, if not necessary, to help ensure success of any large-scale stocking. First of all, in the case of restocking (to increase SSB), a short-term moratorium on fishing should be implemented to speed up the recovery of adult populations. Second, in North Carolina, a ban on inshore trawling should be implemented to protect released as well as wild juveniles. Nearly 80% of the shrimp harvested in North Carolina comes from trawlers working inside the Sounds. The bycatch, which may be as high as 18 kg of finfish per kg of shrimp, comprise 67% flatfish, of which 73% is juvenile southern flounder and 26% juvenile summer flounder (Logothetis and McCuiston 2006). Unfortunately, the inshore shrimping grounds in North Carolina include the habitats and migratory pathways of flounder in summer and fall. Texas, like most other coastal states in the southeastern United States, has already prohibited inshore netting.

12.6 Potential risks and rewards of stocking

Potential adverse impacts of stocking include the following: genetic dilution, disease introduction, depression of wild stocks, etc.. These have already been discussed many times in the stock enhancement literature (Howell *et al.* 1999; Leber *et al.* 2004), so we will not reiterate them here (but see VII, below). But one impact of large-scale stocking could be to alter the community structure of the ecosystem. Whether or not this would be bad depends on the value attached to the target (stocked) species versus the other members of the community. In freshwater systems, hardly anyone seems to be concerned about the displacement of so-called "minor" species. In estuaries, the abundance of a few "minor" species (Ross and Epperly 1985) could be altered. For sure, the level of secondary production in most systems seems dependent on diversity. But who would object to more southern flounder, e.g., in the estuary—even if it were at the expense of pin fish, lizard fish, or most other species that might be somewhat reduced? Since most target species for fishing (or stocking) are at or near the top of the food chain, and these have generally been severely depressed (Pauly *et al.* 1998), worries about depressing other target species would seem largely unfounded. On the other hand, if stocked southern flounder depressed numbers of red drum, e.g., this would likely be perceived as a negative effect. The problem is that we do not understand the trophic and other links among species to predict how the community structure would change with stocking. Such changes are better known in freshwater systems, partly because stocking has been practiced for decades. Again, the only way to find out what changes would likely occur, if any, would be to stock a target species on a meaningful scale in a limited area and measure the community changes. On the other hand, if a species is being considered for stocking, it probably is because its numbers have been depressed, by overfishing, e.g., so stocking would be more likely to restore a former ecological balance than to upset it. All of these considerations argue that serious discussions among stakeholders should precede any large-scale stocking, and that pilot releases should be undertaken to learn more about the risks and rewards of stocking (Miller and Walters 2004).

In southern flounder, like winter and Japanese flounder and other flatfish (and, probably, summer flounder), sex is determined by the rearing temperature at certain susceptible stages (Goto *et al.* 1999, 2000; Yamamoto, E. 1999; Lukenbach *et al.* 2003, Howell *et al.* 2005). It is possible that phenotypic males could be stocked that are XX, i.e., genetic females, by manipulating the juvenile rearing temperature. If so, the sex ratio of their offspring could be highly skewed toward females. North Carolina, being at the northern limit of the range of southern flounder, makes it more likely that in certain years, the sex ratio of the wild stock is skewed in unusually cold winters. In principle, the summer flounder, being at the southern edge of its range in North Carolina could be skewed in the opposite direction in warm years. The extent of neither is known, but if it proves to be the case, then releases of a preponderance of a particular genotype, made possible by manipulating the hatchery environment, could be used to help restore a genetic balance.

12.7 Issues that need resolution before stocking is implemented

1. What is the carrying capacity of the system to be stocked? This question would be best answered with fish released at an ecologically meaningful scale—50–100 thousand/year, as the Japanese have shown (Atushi and Masuda 2004; Howell and Yamashita 2005). As numbers are increased, signs of density-dependent reduction in growth or survival would signal the carrying capacity being approached. Also, other species' abundance should be monitored, since competitors and predators also determine the carrying capacity.
2. What is the genetic profile of the wild stock? This needs to be determined so that any changes in genetic structure because of releases of hatchery fish can be detected (see Seikai *et al.*, this volume).
3. What would be the impact of released fish on the community structure in the release habitat? Possibly, the only way to determine this is through pilot releases on a limited scale, where the community structure is measured before and after releases.
4. What are the common diseases of hatchery and wild fish, and how are they detected? A protocol for screening hatchery fish needs to be developed before any releases.

12.8 Hatchery and stocking protocols to increase success

That fish reared in hatcheries are often ill-equipped to survive in nature is a well known fact (Olla *et al.* 1994; Kellison *et al.* 2003; Arai *et al.* 2007; Sparrevohn and Stottrup 2007). The lack of experience with predators makes hatchery fish particularly vulnerable to predation. Furthermore, being fed pellets from above usually results in hatchery fish spending more time swimming up in the water column looking for food instead of burying in the substrate and waiting for food (Furuta 1998). Visually keyed predators, notably blue crab (*C. sapidus*), are attracted by movement. Furuta *et al.* (1998) found the same thing in hatchery-reared Japanese flounder, which were preyed upon by fish. In some Japanese prefectures, hatchery fish are kept in cages in the field prior to release to acclimate them to wild food. At Texas Parks and Wildlife Marine Development Center, juvenile fish spend some time in rearing ponds before release, no doubt increasing their ability to recognize and feed on wild food. Also there, artificial structures in ponds are being evaluated that simulate more natural habitat conditions.

Crowding in the hatchery has different effects, depending on the species and stage. Certain species, such as Japanese flounder, respond with higher growth rates at low densities and lower growth rates when fish area is over twice the tank area (Jeon *et al.* 1993). It is not known how southern or summer flounder respond to density in the hatchery.

Besides a good habitat, there is an optimal size and season for releases. In flatfish, this seems to be 9 or 10 cm (Yamashita *et al.* 1994). Optimal in this case means the size at which cost is least per gram of net production by the released fish. Releasing smaller fish is cheaper, but survival is poorer (Yamashita and Yamada 1999).

In principle, rearing fish in bacteria-rich water improves their immunocompetence, preparing them for life in natural waters. Fish can also be fed probiotics to improve resistance to disease.

To help prevent genetic dilution, Texas broodstock is replaced with at least 25% new wild broodstock each year, and broodstock are not kept in the hatchery for more than 4 years. Molecular markers are being developed to characterize the genotype of broodstock.

Clearly, stocking fish in better habitats results in greater returns, but also stocking in semienclosed bays is more effective in terms of retention rate than stocking open coastal areas (Howell and Yamashita 2005).

The quality of the seed, or released fish, can also be dependent on how the broodstock are spawned (Waters 1999). In general, offspring from forced spawning, either physical or with hormones, are often poorer in quality than those from parents who are induced to spawn in the hatchery by temperature and daylength manipulation. Unfortunately, some species cannot be induced to spawn and must be stripped of gametes.

One of the primary bottlenecks to the Texas hatchery program has been adult male broodstock viability. The viability of captive males seems to wane after one season, and the timely collection of new male broodstock requires both good fortune and planning.

12.9 Socioeconomic aspects

Despite many clear cases where releasing flounder results in increased catches (Howell and Yamashita 2005), it is not generally profitable to stock flounder for commercial harvest at present. However, stocking flounder for sportsfishing may be profitable, since sportsfishermen are willing to pay much more to catch them. For example, every dollar invested in fisheries in Florida returns much more to the State (R. Taylor, unpublished data).

Mentioned frequently in the literature is the fact that when stakeholders are engaged in a stocking program, they behave more responsibly toward the resource (Copeland *et al.* 1998; Kitada and Kishino 2006). The fish become "theirs," rather than some common property resource, and there follows a reduced incidence of taking undersized fish or overexploitation.

Success measures usually involve measuring the market return rate (MRR)—the number (or weight) of returned fish divided by the number of stocked fish. About the only place this has been measured for flatfish is in Japan, where, depending upon species, the MRR ranges from less than 1% to over 23% (Yamashita and Aritake, this volume). MRR would be appropriate for sea ranching efforts, but for stock enhancement or restocking, some measure of increased abundance of recruits or adults would be appropriate.

In order to determine the return rate, released fish need to be marked to distinguish them from the wild stock. There are many marking methods, depending on the size of the released fish, e.g., coded wire tags, colored latex injections, etc. But in the case of many hatchery-reared flatfish, pigment on the abocular (blind) side occurs. This is true for southern and summer flounder. So the released fish can be easily distinguished from wild stock, which have no pigment on the blind side. Of course, if the released fish need to be further distinguished, e.g., where or when they were released, some conventional marking method must be used. In some cases, genetic markers can be inserted to distinguish hatchery-reared fish.

12.10 Who should pay?

In general, the ones who utilize the resource, commercial or sportsfishers, should pay. In the case of sportsfishers, this can be done through license fees. In the case of commercial fishers, some cost-recovery method, such as that of harvesting salmon in Alaska by the hatcheries is probably necessary. Some additional cost-recovery is accomplished in Japan by selling some of the juveniles produced to private growout operations. However, the initial cost of a hatchery and the preliminary research to determine if stocking would work and should be done is certainly beyond the means of fishers. Therefore, when viewing stocking as a tool for managing public resources, at least the initial costs should be borne by the State. The model for flounder should be no different than that for stocking freshwater species—in which cases the hatcheries are state- or federally owned and operated.

12.11 Conclusion

The era of expecting sustainable fishing to be supported by natural reproduction of diminishing stocks has passed. Target fish stocks, as well as their habitats, have already been altered by fishing and development and pollution to the point where a return to former, "natural," states is impossible. It is time to recognize that fishing pressures plus habitat losses and degradation now exceed Mother Nature's ability to compensate. Fortunately, it is now possible, with responsible application of hatchery and release technology, to supplement wild stocks. Such measures should be used to augment, not replace, conventional management. It is only a question of how, not if, we proceed to an era of manipulated ecosystems.

NC estuarine and coastal ecosystems formerly supported nearly 20 times the present stock size of southern flounder, plus several other species whose stocks have been severely depleted by overexploitation. Notwithstanding the habitat degradation that has no doubt occurred, this suggests that considerable excess productive capacity exists that could be utilized by judicious stocking of additional fish. Provided managers would take steps to reduce bycatch of juveniles and support the necessary pilot research to guide stocking efforts, such as stocking undercolonized nurseries, it is possible that production and harvest of southern flounder could be increased manyfold. Albemarle Sound represents an especially productive potential new habitat for flounder. With such efforts plus the development of production aquaculture, it is now possible to envision a sustainable supply of southern flounder for the foreseeable future.

Literature cited

Arai, T., Tominaga, O., Seikai, T., and Masuda, R. 2007. Observational learning improves predator avoidance in hatchery-reared Japanese flounder *Paralichthys olivaceous* juveniles. *Journal of Sea Research* 58:59–64.

Atushi, S., and Masuda, Y. 2004. Effectiveness of the release of hatchery-produced stock of Japanese flounder *Paralichthys olivaceous* in Kagoshima Bay, southern Japan. *Nippon Suisan Gakkaishi* 70(6):910–921 (in Japanese with English summary).

Burke, J.S., Sekai, T., Tanaka, Y., and Tanaka, M. 1999. Experimental intensive culture of summer flounder, *Paralichthys dentatus*. *Aquaculture* 176:135–144.

Copeland, B.J., Miller, J.M., and Waters, E.B. 1998. *The Potential for Flounder and Red Drum Stock Enhancement in North Carolina: Summary of a Workshop*. UNC Sea Grant Publication. Raleigh, NC.

Daniels, H.V., and Watanabe, W.O. 2002. *A Practical Hatchery Manual: Production of Southern Flounder Fingerlings*. North Carolina Sea Grant, Raleigh, NC, 40 p.

Daniels, H.V., Watanabe, W.O., and Murashige, R. 2007. Southern flounder growout manual. Proceedings of the Southern Flounder Workshop, Wallace, NC. September 27, 2007. NC State University Cooperative Extension, Raleigh, NC.

Furuta, S. 1998. Comparison of feeding behavior of wild and hatchery-reared Japanese flounder, *Paralichthys olivaceous*, juveniles by laboratory experiments. *Nippon Suisan Gakkaishi* 64:393–397.

Furuta, S., Watanabe, T., and Yamada, H. 1998. Predation by fishes on hatchery-reared Japanese flounder, *Paralichthys olivaceous*, juveniles released in the coastal area of Tottori Prefecture. *Nippon Suisan Gakkaishi* 64:1–7.

Goto, R., Mori, T., Kawamata, K., Matsubara, T., Mizuno, S., Adachi, S., and Yamauchi, K. 1999. Effects of temperature on gonadal sex determination in barfin flounder *Verasper moseri*. *Fisheries Science* 65:884–887.

Goto, R., Kayaba, T., Adachi, S., and Yamauchi, K. 2000. Effects of temperature on sex determination in marbled sole *Limanda yokohamae*. *Fisheries Science* 66:400–402.

Guindon, K.Y., and Miller, J.M. 1995. Growth potential of juvenile southern flounder (*Paralichtys lethostigma*) in low salinity nursery areas of Pamlico Sound, North Carolina. *Netherlands Journal of Sea Research* 34(1/3):89–100.

Howell, B.R., Fairchild, E.A., and Rennels, N. 2005. Skewed sex ratios in cultured winter flounder, *Pseudopleuronectes americanus*: implications for stock enhancement. Aqua 2006. The annual meeting of the European Aquaculture Society. August 9, 2005, Trondheim, Norway.

Howell, B.R., Moksness, E., and Svasand, T. (eds) 1999. *Stock Enhancement and Sea Ranching*. Fishing News Books, Oxford, 606 pp.

Howell, B.R., and Yamashita, Y. 2005. Aquaculture and stock enhancement. In: Gibson, R.N. (ed.) *Flatfishes: Biology and Exploitation*. Blackwell Science Limited., Oxford, pp. 347–371, 391p.

Kellison, G.T. 2000. Evaluation of stock enhancement potential for summer flounder (*Paralichthys dentatus*): an integrated laboratory, field and modeling study. PhD. Dissertation. North Carolina State University, Raleigh, NC.

Kellison, G.T., Eggleston, D.B., and Burke, J.S. 2000. Comparative behavior and survival of hatchery-reared versus wild summer flounder (*Paralichthys dentatus*). *Canadian Journal of Fisheries and Aquatic Sciences* 57(9):1870–1877.

Kellison, G.T., Eggleston, D.B., Taylor, J.C., Burke, J.S., and Osborne, J.A. 2003. Pilot evaluation of summer flounder stock enhancement potential using experimental ecology. *Marine Ecology Progress Series* 250:263–278.

Kitada, S., and Kishino, H. 2006. Lessons learned from Japanese marine finfish stock enhancement programmes. *Fisheries Research* 80:101–112.

Jeon, I.G., Min, K.S., Lee, J.M., Kim, K.S., and Son, M.H. 1993. Optimal stocking density for olive flounder, *Paralichthys olivaceous*, rearing in tanks. *Bulletin of the National Fisheries Research Development Agency of Korea* 48:57–70.

Leber, K.M., Kitada, S., Blankenship, H.L., and Svasand, T. 2004. *Stock Enhancement and Sea Ranching: Developments, Pitfalls and Opportunities*, 2nd edn. Blackwell Publishing, Oxford, 562 pp.

Logothetis, E., and McCuiston, D. 2006. An assessment of the bycatch captured in the inside commercial shrimp fishery in southeastern North Carolina. Sea Grant FRG # 04-EP-01, 05-EP-03. Poster.

Lukenbach, J.A., Godwin, J., Daniels, H.V., and Borksi, R.J. 2003. Gonadal differentiation and effects of temperature on sex determination in southern flounder (*Paralichthys lethostigma*). *Aquaculture* 216:315–327.

Miller, J.M., Crowder, L.B., and Moser, M.L. 1985. Migration and utilization of estuarine nurseries by juvenile fishes: an evolutionary perspective. *Contributions in Marine Science* 27:338–342.

Miller, J.M., Reed, J.P., and Pietrafesa, L.J. 1984. Patterns, mechanisms and approaches to the study of migrations of estuarine-dependent fish larvae and juveniles. pp. 209–225 In: McCleave, J.D., Arnold, G.P., Dodson, J.J., and Neill, W.H. (eds) *Mechanisms of Migration in Fishes*. Plenum Press, New York, 574 p.

Miller, J.M., and Walters, C.J. 2004. Experimental ecological tests with stocked marine fish. In: Leber, K.M., Kitada, S., Blankenship, H.L., and Svasand, T. (eds) *Stock Enhancement and Sea Ranching: Development, Pitfalls and Opportunities*. Proceedings of the 2nd International Symposium on Stock Enhancement and Sea Ranching. Blackwell Publishing, Oxford, pp. 142–152.

NC DMF 2005. North Carolina Fishery Management Plan. Southern Flounder (*Paralichthys lethostigma*). North Carolina Division of Marine Fisheries, Morehead City, NC, 335 p.

Olla, B.L., Davis, M.W., and Ryer, C.H. 1994. Behavioral deficits in hatchery-reared fish: potential effects on survival following release. *Aquaculture and Fisheries Management* 25(Supplement 1):19–34.

Pauly, D., Christensen, V., Dalsgaard, J., Froese, R., and Torres, F., Jr. 1998. Fishing down marine food webs. *Science* 279(5352):860–863.

Pietrafesa, L.J., Janowitz, G.S., Miller, J.M., Noble, E.B., Ross, S.W., and Epperly, S.P. 1986. Abiotic factors influencing the spatial and temporal variability of juvenile fish in Pamlico Sound, North Carolina. In: Wolfe, D.A. (ed.) *Estuarine Variability*. Academic Press, New York, pp. 341–353.

Ross, S.W., and Epperly, S.P. 1985. Utilization of shallow estuarine nursery areas by fishes in Pamlico Sound and adjacent tributaries. In: Yanez-Arancibia (ed.) *Fish Community Ecology in Estuaries and Coastal Lagoons: Towards an Ecosystem Integration*. UNAM Press, Mexico, pp. 207–232, 654 p.

Seikai, T., Kikuchi, K., and Fujinami, U. (this volume). Culture of Japanese flounder.

Sparrevohn, C., and Stottrrup, J.G. 2007. Post-release survival and feeding in reared turbot. *Journal of Sea Research* 57:151–161.

Taylor, J.C., Miller, J.M., Pietrafesa, L.J., Dickey, D.A., and Ross, S.W. (2009). Winter winds and river discharge determine juvenile southern flounder (*Paralichthys lethostigma*) recruitment and distribution in North Carolina estuaries. *Journal of Sea Research*.

Waters, E.B. 1999. *Flounder Aquaculture and Stock Enhancement in North Carolina: Issues, Opportunities and Recommendations*. UNC Sea Grant Publication UNC-SG-99-02. Raleigh, NC, 24 p.

Weisberg, R.H., and Pietrafesa, L.J. 1983. Kinematics and correlation of the surface wind field in the South Atlantic Bight. *Journal of Geophysical Research* 88(8):4593–4610.

Yamamoto, E. 1999. Studies on sex-manipulation and production of cloned populations in hirame, *Paralichthys olivaceous* (Temminick et Schlegel). *Aquaculture* 173:235–246.

Yamashita, Y., and Aritaki, M. (this volume) Stock Enhancement of Japanese Flounder in Japan.

Yamashita, Y., Nagahora, S., Yamada, H., and Kitigawa, D. 1994. Effects of release size on survival and growth of Japanese flounder *Paralichthys olivaceous* in coastal waters off Iwate Prefecture, northern Japan. *Marine Ecology Progress Series* 105(3):269–276.

Yamashita, Y., and Yamada, H. 1999. Release strategy for Japanese flounder fry in stock enhancement programmes. In: Howell, B.R., Moksness, E., and Svasand, T. (eds) *Stock Enhancement and Sea Ranching*. Fishing News Books, Oxford, pp. 191–204.

Section 5

Europe Stock Enhancement

Chapter 13

Stock enhancement Europe: turbot *Psetta maxima*

Josianne G. Støttrup and C. R. Sparrevohn

13.1 Introduction

13.1.1 Turbot biology and ecology

Turbot *Psetta maxima* is a flatfish belonging to Scophthalmidae, distributed throughout Europe from the Mediterranean and Black Sea to the Baltic Sea and Norwegian coast (Muus *et al.* 1998). Genetic differences between North Sea and Baltic Sea turbot and between eastern and western Mediterranean turbot have been reported (Nielsen *et al.* 2004; Suzuki *et al.* 2004). Florin and Höglund (2007) found no genetic differences between Baltic Sea and Kattegat turbot, but this could be due to the Kattegat being a strong transition zone with high numbers of both North Sea and Baltic Sea stocks. Differences in hemoglobin genotypes were found between turbot from Iceland and west Norway, Skagerrak, Kattegat, and the Baltic Sea (Imsland *et al.* 2003).

Throughout its life cycle, turbot occupies the coastal zone (Neumann and Píriz 2000). Turbot spawn at depths from 10–40 m, with no apparent concentration of adults for spawning (Rae and Devlin 1972; Van Der Land 1991). Turbot are summer spawners and produce pelagic eggs. In the Baltic, they are adapted to the low-saline environment producing demersal eggs that survive salinities as low as 7 psu (Nissling and Westin 2004). Eggs and larvae are otherwise planktonic and are driven by currents toward the coasts, where they settle as juveniles (Figure 13.1). The younger age groups concentrate in sandy or mixed bottom, shallow waters with a narrow depth distribution from the shoreline to around 2 m (Riley *et al.* 1981), <3 m in Galician bays (Iglesias 1981), to 4 m in British estuaries (Jones 1973) and in Skagerrak/Kattegat (Pihl 1989; Støttrup *et al.* 2002), or to depths of 8 m in the eastern Baltic (Scherbich 1998). In one exposed sandy beach in Kattegat, turbot juveniles occupied a broader depth range (Sparrevohn and Støttrup 2008), and the larger fish (age 2 and older) showed a diel migration pattern moving to shallower waters (<4 m) at night, possibly to feed (Støttrup *et al.* 2002).

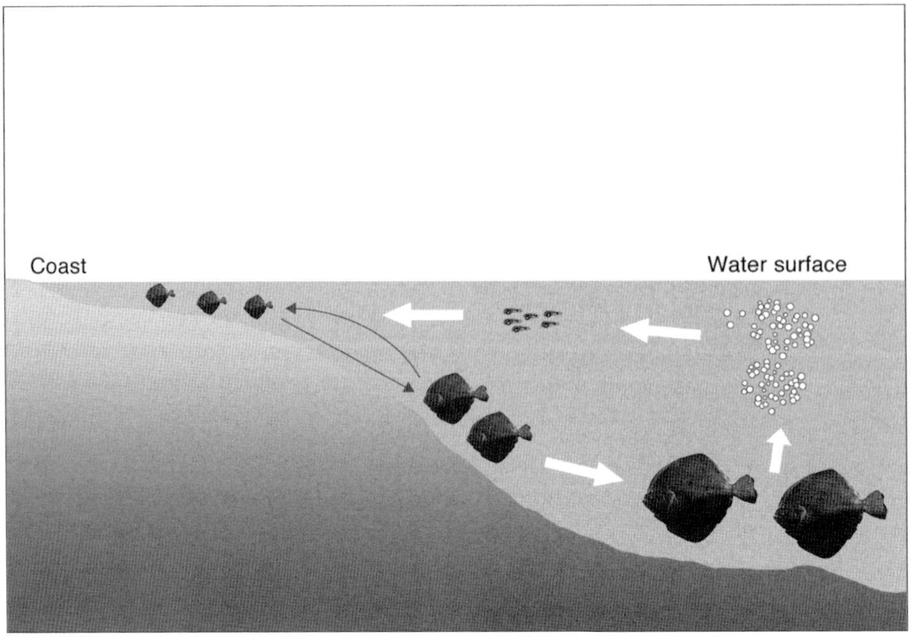

Figure 13.1 Schematic illustration of the life cycle of turbot *Psetta maxima*. See text for description. (Drawn by Peter Waldorph.)

13.2 Turbot production

13.2.1 Commercial capture of turbot in Europe

Turbot capture in Europe is around 7,000 metric tonnes but has varied from around 4,000 tonnes in mid-1980s to 9,000 tonnes in the early 1990s (FAO 2006). In the Baltic, turbot is mainly caught as by-catch in the cod fishery. The total fishery for this species in this area peaked in 1996 with 1,210 tonnes but declined to 310 tonnes in 2006 (ICES 2007). In the Black Sea, turbot are caught primarily by Turkey, followed by former USSR, Bulgaria, and Romania (Zengin *et al.* 2006). Due to heavy misreporting, catch statistics are not very reliable with an estimated total mean of around 2,800 tonnes.

13.2.2 Turbot aquaculture in Europe

Turbot are also produced commercially in several European countries including Spain, Portugal, France, Norway, and Iceland (FAO 2008). Denmark produces mainly juveniles, which are sold to on-growing farms in southern Europe. European production of turbot for consumption almost equals total capture of turbot and is around 5,000 tonnes. About half of this production comes from Spain (Danancher and Garcia-Vazquez 2007). In Turkey, turbot juveniles are produced for stocking purposes (Zengin *et al.* 2007).

In most hatcheries, turbot juveniles are produced on two types of live-prey; rotifers *Brachionus plicatilis* during first feeding, switching to the first naupliar stages of *Artemia* sp. (Støttrup and McEvoy 2003). Once first-feeding is established and the larvae are sufficiently large to capture and ingest *Artemia* nauplii, these are co-fed for a couple of days until the larvae have weaned onto the larger prey items, *Artemia* nauplii. Since these prey types are nutritionally inadequate for turbot larvae, they are enriched with an emulsion containing lipids, proteins, vitamins, and minerals considered essential for normal development and growth of turbot larvae. A nutritionally inadequate larval diet has not only short-term consequences resulting in deformed or malpigmented juveniles, but may impact later life stages. In juvenile sole *Solea solea*, the juvenile's ability to survive cold winters was compromised because of an inadequate diet during the larval stage (Howell 1994).

Larval nutrition may therefore play an important role in stocking in determining the quality of the fry and affecting the success of the releases. Copepods are the normal live prey for most marine fish larvae in nature and are nutritionally superior even to enriched rotifers and *Artemia* nauplii (Støttrup and McEvoy 2003). In Denmark, fish for releases are generally produced in semi-extensive systems and fed copepods during the larval stage (Urup 1994; Engell-Sørensen *et al.* 2004).

13.3 Turbot stocking

13.3.1 Aim of stocking turbot in Europe

Stocking of turbot in Europe is still at an experimental stage, chosen primarily because of its availability in production, and generally used as a model species to test the potential for enhancing local stocks. One exception is the Turkish restocking of Black Sea turbot (Zengin *et al.* 2007). Table 13.1 provides an overview of number and size of turbot released in different countries over time and Figure 13.2 of the localities for released turbot. Experiments with turbot releases have been conducted primarily in Danish coastal waters since 1998, in Spanish bays (Iglesias and Rodrıguez-Ojea 1994; Iglesias *et al.* 2003; Balsa *et al.* 2006), off the Belgian coast (Delbare and De Clerck 2000), and in the Black Sea (Maslova 2002).

In Denmark, the releases are conducted through the National Coastal Fisheries Management Program, financed through fishing licenses for anglers and recreational fisheries, which by law must have a valid license to fish in coastal waters. Turbot restocking in Turkish coast of the Black Sea is still continued in Turkey, after completion of a case study conducted during 1999–2002 (M. Zengin, personal communication; Zengin *et al.* 2007). In other countries, turbot releases have been short-term experiments lasting from a single to a few number of years and discontinued.

13.4 Rationale for turbot stocking

The rationale for releasing fish has been debated over several decades and several attempts have been made to empirically prove or disprove the theories. Støttrup

Table 13.1 Overview on turbot releases in Europe since 1989.

Denmark

Year	Locality	Total numbers released	Size at release (average cm ± SD or size range)
1989	Limfjord (1)	5,678	—
	W. Baltic (10)	1,198	—
1990	Limfjord (1)	6,121	—
	W. Baltic (10)	1,203	—
1991	Limfjord (1)	100,000	—
	North Zealand (4)	7,581	11 ± 1
1992	Limfjord (1)	7,498	—
	North Zealand (4)	7,953	12 ± 2
	Odense Fjord (8)	4,538	13 ± 2
1993	Limfjord (1)	45,000	3–5
	North Zealand (4)	89,939	11 ± 1
	Vejle Fjord (6)	65,500	6–10
	North Fyn (7)	60,800	5–6
1994	Limfjord (1)	60,000	5–18
	Ålborg Bay (2)	70,942	5–12
	North Zealand (4)	162,762	5–13
	Sejerø Bay (5)	37,300	5–12
	Vejle Fjord (6)	20,000	9 ± 1
1995	Limfjord (1)	59,400	3–5
	Ålborg Bay (2)	85,628	10 cm
	Århus Bay (3)	12,500	10 cm
	North Zealand (4)	101,072	5–19
1996	North Zealand (4)	132,830	6–21
1997	North Zealand (4)	31,479	11 ± 2
	Vejle Fjord (6)	57,085	3–6
1998	Near Limfjord (1)	5,000	15–20
	Ålborg Bay (2)	33,839	19 ± 3
	North Zealand (4)	19,641	8–19
1999	Ålborg Bay (2)	10,707	9 ± 1
2000	Århus Bay (3)	13,000	10
2001	Århus Bay (3)	18,596	8 ± 1
2002	Limfjord (1)	105,587	1–10
	Århus Bay (3)	10,700	10.5
2003	Århus Bay (3)	9,657	12 ± 1
			17 ± 2
2004	Ålborg Bay (2)	15,053	8–12
	Århus Bay (3)	44,345	3–10
	North Zealand (4)	21,747	8–10
	Vejle Fjord (6)	5,179	15–22
	North Fyn (7)	3,734	15–22
2005	Århus Bay (3)	20,000	3–5
	Århus Bay (3)	4,500	—
2006	Århus Bay (3)	7,500	4–11
	Vejle Fjord (6)	40920	7–10
2007	Vejle Fjord (6)	7,967	9–16
	Sound, Kattegat (9)	22,000	5–7
2008	Limfjord (1)	2,000	—

Table 13.1 (*Continued*)

Spain				
Year	**Release locality**	**Numbers released**	**Size**	**Reference**
1992	Vigo Bay	2,981	19 cm	Iglesias and Rodrıguez-Ojea 1994
1996	Vigo Bay	594	4–6 cm	Iglesias *et al.* 2003
2005	W Galicia (Finisterre)	7,500	13 cm	J.C. Marino (personal communication)
2006	W and SW Galicia (Finisterre and Arousa Bay)	10,000	11 cm	J.C. Marino (personal communication)
2007	SW Galicia (Arousa Bay)	5,000	12 cm	J.C. Marino (personal communication)
2008	SW Galicia (Arousa Bay)	8,000	0.5–15 cm	J.C. Marino (personal communication)
Russia				
1992–1994	Black Sea	15,000	—	Maslova 2002
1995	Black Sea	50,000	—	Maslova 2002
1996	Black Sea	50,000	—	Maslova 2002
1997	Black Sea	50,000	—	Maslova 2002
Belgium				
1998	Off Belgian coast	1,962	22 cm	Delbare

Number in brackets indicates map locality in Figure 13.1. Size at release is given either as average, average ± SD or size range in cm.

et al. (2008) provided a rationale for releasing cod *Gadus morhua* larvae in the highly unique ecosystem of the eastern Baltic. This was possible due to the abundant and high quality information available on the ecosystem dynamics, trophic interactions, and the population dynamics of the species. For turbot, the approach has been different partly because of the lack of information on this species, or the highly complicated dynamics of the system where the work has been conducted. Thus far, work on turbot stock enhancement has been based on empirical studies conducted on released fish in a variety of semienclosed or open coastal systems.

The main aim for the turbot releases conducted in Europe has been stock enhancement. In Turkey, turbot releases were aimed at restocking the Black Sea turbot population. However, few studies have set up well-defined criteria against which to measure success of their releases, one of the recommendations for a responsible approach to stocking put forward by Blankenship and Leber (1995). Criteria for flatfish stocking were put forward in Støttrup *et al.* (2002) and Støttrup and Sparrevohn (2007) and include considerations about population regulating factors, and parameters that affect survival after their release.

In the following, different aspects related to the success criteria of stock enhancement will be dealt with. These include techniques for recognizing released fish, methods for release, choice of release site/habitat, and postrelease mortality.

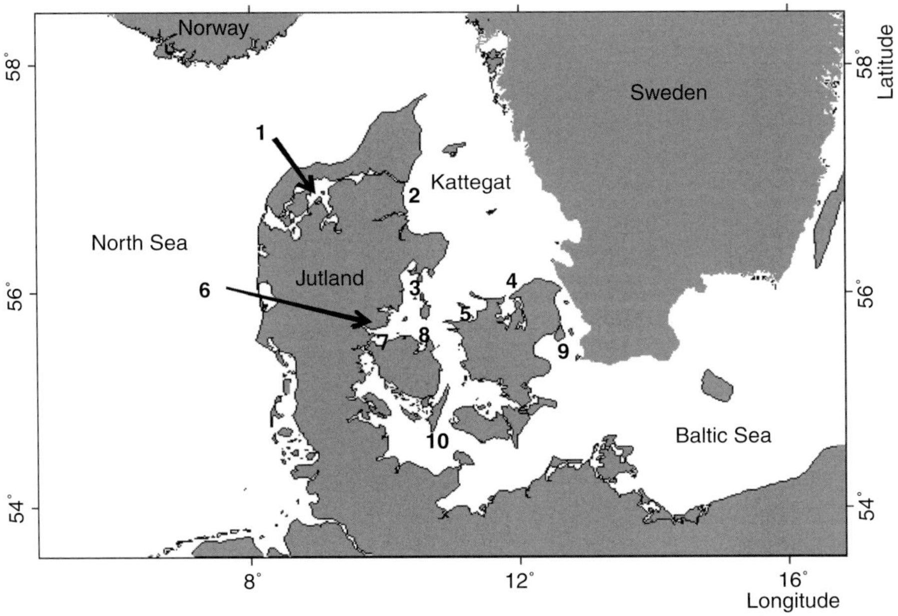

Figure 13.2 Map of Denmark showing the sites for releases of turbot. Limfjord = 1, Aalborg Bay = 2, Aarhus Bay = 3, North Zealand = 4, Sejerø Bay = 5, Vejle Fjord = 6, North of Fynen = 7, Odense Fjord = 8, the Sound = 9, and Western Baltic = 10.

13.5 Origin of fish for stocking

Common practice in stocking, especially in experimental work, is the use of fish derived from a local broodstock. In Denmark, many of the releases were in fjords or coastal shores within Kattegat, which is a transient sea between the high saline Skagerrak and North Sea and low-saline Baltic Sea. In the Kattegat proper, turbot originating from Skagerrak have been used for the releases; in the Belt Sea or further south in the western Baltic, only turbot originating from that area have been released (Table 13.1, Figure 13.1). In the Belgian study, fish were obtained from a French turbot-rearing farm, with no mention of origin. In Spain and the Black Sea, fish for releases were reared from juveniles from broodstock of local origin. Black Sea turbot reared for stocking originated from local broodstock (Zengin *et al.* 2007).

13.6 Marking and tagging techniques

External T-bar tags have been used for tagging fish >8 cm in both Spain and Denmark (Iglesias and Rodrıguez-Ojea 1994; Støttrup *et al.* 2002). These tags are easy to apply and easily recognized by commercial and recreational fishermen. There is sufficient place on each tag for the return-address and because the tags are numbered consecutively, they provide information on individual fish. Local fishermen may be encouraged to return tags with information on catch date, catch position, and size of fish, in return for small rewards (Støttrup *et al.* 2002). Disadvantages include tag losses and lesions around the area where the tag is inserted due to continuous abrasion. This was especially due to tag fouling by algae and mussels and was shown to negatively affect growth (Støttrup *et al.* 2002). The sturdier Petersen disc was used by Delbare and De Clerck in Belgium (2000). This disc is 14 mm in diameter and is fixed to the fish using titanium wire. The tagging process is much slower than T-bar tagging, but tag losses do not occur. Also, the tag is more securely tied to the individual fish reducing the risk of lesions around the insertion points. This tag type has the same advantages as the T-bar tag. Iglesias *et al.* (2003) used acrylic paint injected subcutaneously with a 0.5 mL needle on the ventral surface of the fish to identify the released fish. The tagging rate is possibly somewhere in between the T-bar and disc tagging. Tag losses are considered low with this method, but the fish cannot be identified individually and it is unlikely that commercial or recreational recaptures would be recognized or reported.

For tagging larger numbers of fish, Støttrup *et al.* (2002) used alizarin complexone to mark the otoliths. Marking marine fish with alizarin complexone or alizarin red is well described in the literature (Tsukamoto 1988; Iglesias and Rodrıguez-Ojea 1997; Lagardère *et al.* 2000). The fish are immersed in a bath for 24 hours, after which the mark is washed out. Applying the correct dose is important both for avoiding marking mortality and for subsequently being able to recognize the mark. In turbot juveniles 4–6 cm in size, an alizarin dose of 50 mg/L for 24 hours was found optimal and could be applied with no fish losses (J.G. Støttrup, unpublished data). With this dosage, marked fish could be recognized at least 3 years after release and it was not necessary to pretreat the otolith before observations under the UV microscope (Støttrup *et al.* 2002). Fish growth did not seem to be affected by this mark (Paulsen and Støttrup 2004). Disadvantages include the inability to differentiate among individuals and mark recognition, a labor-intensive procedure of extracting otoliths and examining these under a UV-light microscope.

13.7 Release procedures

Turbot juveniles have been released directly from shore (Sparrevohn *et al.* 2002; Iglesias *et al.* 2003) or from a boat in slightly deeper waters (Delbare and De Clerck 2000; Støttrup *et al.* 2002). When released from shores, this has been done either through a pipe connected directly to the transport truck, or in cases where the road was too far from the sea, with the help of local volunteer recreational fishermen who helped transport the fish in buckets to shallow waters. In deeper waters, fish have been released from tanks aboard ships, but as close to the sandy bottom as possible (Delbare and De Clerck 2000). In Denmark, the fish were released directly through a pipe from the transport truck on board a ferry to the water surface at around 6 m depth (Støttrup *et al.* 2002). When released during daytime, the fish swam directly to the bottom where they remained generally inactive on the sandy bottom. During twilight, they spent more time swimming at the surface (J.G. Støttrup, personal observations) and may have increased their vulnerability to pelagic predators. A study was conducted to examine the effects of transport and release procedures on turbot growth and survival in laboratory simulation experiments (Paulsen and Støttrup 2004). This species was shown to be highly robust and relatively unaffected by handling procedures.

13.8 Choice of release site/habitat

An appropriate release habitat/area is important for successful stock enhancement. Density-dependent and independent biomass regulation may negatively affect stocking results. Density-dependent regulation of the biomass should be minimal as this counteracts the aim of a stock enhancement, which is to increase the biomass by adding individuals. An artificial increase in fish biomass has the potential to impose intraspecific competition and in the worst case, the carrying capacity of the area may be exceeded resulting in a net population biomass growth of zero (MacCall 1990). In such a case, there is a high risk of released individuals displacing wild individuals. It is equally important to select an area with low-density independent regulation of biomass, i.e., selecting an area with suitable hydrographic conditions for growth (salinity/temperature), a proper degree of exposure to wave energy, good oxygen conditions, the right sediment type, and low predation mortality. Choosing an inappropriate area in terms of density-independent factors is just as critical as releasing fish into an area with high-density dependent regulation and can result in complete failure. This was demonstrated with the release of mullet *Mugil cephalus* into an area not suitable as nursery grounds for this species (Leber and Arce 1996). No recaptures were obtained, probably due to a very high mortality. Density-independent and dependent regulations are typically inversely correlated where only density dependent factors will change with varying release magnitude. Therefore, while selecting release habitat, evaluating density-dependent and independent biomass regulation only makes sense if the magnitude of the biomass released is taken into account. This is rarely done and typically an area with an already existing wild population will be preferred as area for release because an already existing wild population indicates the absence of critical density-independent regulation. Density-dependent regulation is most often ignored, even though the downside of selecting an area with a naturally occurring population is an increased risk of density-dependent regulation of the biomass.

In the Danish stock enhancement program releases of artificially reared juvenile turbot were carried out at three different locations in Kattegat to examine area suitability as release habitat. Growth performance of the released individuals differed significantly between release habitats (Sparrevohn and Støttrup 2008), but also no growth differences between wild and artificially reared individuals within an area was present (Paulsen and Støttrup 2004). The three habitats differed in their degree of wind exposure, where high exposure areas were believed to be the best habitat for turbot (Gibson 1973; Riley *et al.* 1981). As predicted, the highest abundance of turbot was observed in the area with the highest exposure but growth was found to be inferior compared to the more sheltered areas. The slower growth observed was not necessarily a result of competition for food since the frequency of empty stomachs was similar across areas. Instead, food quality seemed to be responsible for the different performance. An analysis of the diet revealed that in the area where turbot switched from exoskeleton carrying prey items such as crustaceans to fish (e.g., sand gobies, *Pomatoschistus minutus*) at smallest length, turbot juveniles had the highest growth. The conclusion was that there is a potential in choosing areas not normally considered to be optimal nurseries. Firstly, because these areas provide optimal density-independent conditions, in this case, unexploited prey items of better quality than in nurseries, normally highly recruited by juveniles, and secondly, choosing an area with a limited natural population will minimize the risk of intraspecific competition.

Other factors however may also be important in determining optimal release habitats. Migration distance and the rapidity by which released fish disperse from the release site vary between release sites. Size of fish released may impact migration patterns of released fish. A limited migration was observed for larger turbot released in central Kattegat (Bagge 1987) and the smaller turbot released at North Zealand were first observed to migrate offshore at the age of 3 and 4 years (Støttrup *et al.* 2002). This was interpreted as a spawning migration toward more offshore spawning grounds.

Low recapture rates, or lack of recaptures can be a major problem in stocking experiments. Turbot releases in the Limfjord, a fjord located in North Jutland (Figure 13.2 site 1), showed lower and highly scattered recaptures in comparison to releases in other areas, indicating either a higher postrelease mortality or rapid migration out of the Limfjord (Støttrup *et al.* 1998). Since the low recaptures could also have been due to a lower fishing intensity in that area, the results are difficult to interpret.

13.9 Release strategy and magnitude of release

There has been some debate on whether fish should be released in one area (concentrated release) or scattered. Yamashita and Yamada (1999) argued that concentrated releases were more effective as this would reduce risk of exposure to predators, whereas food availability was ample as long as the magnitude was within the carrying capacity of that area. Most of the turbot releases were concentrated releases and at North Zealand (Figure 13.2 area 4) were equivalent to the supplement of 1 individual/1000^2 (age-1 fish) to the release area. Since the releases contributed an average of 25% to age-1 local stock (Støttrup *et al.* 2002), resulting densities were within those found in natural flatfish populations elsewhere. For example Jones (1973) caught juvenile turbot at densities of 4.9 individual/1000^2. Thus, it was presumed that the releases at North Zealand did not exceed the carrying capacity in that area. In a study conducted in 1995, more recaptures were obtained by spreading the release over five areas as compared to concentrating the release in one area (Støttrup *et al.* 2002). It was suggested that an increased exposure to food for a size of fish where size refuge from most predators has been reached would not negatively affect survival. Indeed, the higher recaptures indicate increased survival. The slower growth in these turbot released in North Zealand, was attributed to prey quality in this site rather than density-dependent interactions as discussed in the previous section.

On the other hand, an increased depth distribution resulting from concentrated age-0 turbot releases conducted from 1993 to 1995 was interpreted as a spillover effect to less favorable habitat (MacCall's basin model; MacCall 1990) (Sparrevohn and Støttrup 2007). Between 80,000 and 155,000 age-0 turbot were released each year resulting in around 25% addition to the indigenous year-class and an area-based contribution of 14–20 individual/1,000 (Støttrup *et al.* 2002; Sparrevohn and Støttrup 2007).

Turbot releases are generally conducted during summer or autumn, reflecting the spawning season and time needed to reach a certain size in a hatchery. In Denmark, age-0 group were released around September, by which time they had reached a minimum size of 5 g in weight and 5 cm in total length. In the following spring, age-1 turbot reached a minimum size of 10 cm under rearing conditions. They were generally released as soon as water temperatures had exceeded 10°C to ensure good feeding conditions. In the Black Sea, optimum period for release was found to be October–November, by which time reared turbot had reached an average of 11 cm in length (Zengin *et al.* 2007). This period for release was considered optimal to decrease vulnerability to commercial nets and prevent incurring further hatchery costs for their extra maintenance.

13.10 Postrelease mortality and conditioning

Several laboratory and enclosure studies have shown that the relative survival of artificially bred individuals is lower than for wild counterparts (Kellison *et al.* 2000; Hossain *et al.* 2002; Fairchild and Howell 2004). Unfortunately, reliable quantitative estimates of postrelease mortality from field experiments are difficult to obtain because of the methodology constrains caused by the changing spatial distribution of the released fish with time after release. This problem was recognized by Leber *et al.* (1996) and attempts to solve it by applying various spatially based sampling approaches have been tried (e.g., Furuta *et al.* 1997).

For turbot, postrelease mortality was investigated, assuming that movement of turbot away from the release location could be described as a diffusion process (Sparrevohn *et al.* 2002). The diffusion model is not generic and can only be applied to the dispersal of certain species, since two assumptions should be satisfied. First assumption is that the released individuals move independently of each other. This assumption seems to be fulfilled in the case of turbot but may not be applicable to schooling species such as herring *Clupea clupea* or cod. The second assumption is that distances moved by each single fish within a given time follows the central limit theorem. In other words, the distance moved must not be infinite and each single fish should have the same distribution of movements (directions and distances) from which it can choose a movement at random. If two or more types of dispersal strategies are present in the population, the resulting density distribution of fish (assuming no constrains on the movement such as a physical barrier) will not be normal but leptokurtic, a result found for some fish species and related to their individual boldness (Fraser *et al.* 2001).

Table 13.2 Experiments carried out to investigate postrelease mortality.

Year	Size (cm ± SD)	Number released	Type	Days monitored	Release date
2001	7.5 ± 0.6	3,529	Naïve	9	May 14
2002	10.5	10,649	Naïve	3[a]	May 10
2003	11.8 ± 0.9	5,000	Naïve	7	June 3
2003	17.1 ± 1.7	5,000	Naïve	7	June 3
2004	9.8 ± 0.9	2,500	Naïve	8	May 4
2004	9.8 ± 0.9	2,500	Conditioned	8	May 4
2005	3–5	10,000	Naïve[b]	6	June 10
2005	3–5	10,000	Conditioned[b]	6	June 10
2006	4–5	2,500	Conditioned	5	May 5
2006	11 ± 1.0	2,500	Conditioned	5	May 5

[a]Sampling was ended since the recaptures was very small. During the 2 days, 507 tags were found lying on the beach.
[b]Due to complications in detection of the alizarin marked otoliths of the naïve fish they could not be separated from the conditioned turbot.
Naïve fish are fish that were transported and released directly from the hatchery, whereas conditioned fish were encaged at the release location prior to actual release.

A series of experiments was carried out between 2003 and 2006 in Denmark in order to examine the effect of size-at-release and the effect of conditioning fish on postrelease mortality. Table 13.2 provides an overview of these releases.

From the model (Sparrevohn *et al.* 2002), the fraction of released fish still present in the area (called the "recatchable part") could be estimated daily. This fraction will theoretically be less than one due to mortality and the fact that for most gears the catchability (efficiency) will be lower than 100%. The decline in recatchable part was interpreted as a high initial mortality lasting a few days and decreasing gradually until a natural level was reached (Figure 13.3). Postrelease mortality was therefore estimated from the initial steep decline. Duration of the postrelease mortality was interpreted as the duration of the decline in these plots. Postrelease mortality of approximately 34% per day was estimated for small (7.5 cm) turbot released in 2001 (Figure 13.3a; Sparrevohn *et al.* 2002; Sparrevohn and Støttrup 2007). Using slightly larger (10.5 cm) turbot, the experiment was repeated the following year, but it was impossible to estimate mortality due to very low recaptures (Table 13.2). During this release, a high seagull (great black-backed gull *Larus marinus* and herring gull *Larus argentatus*) activity was witnessed and on several occasions, the gulls were observed to capture released turbot, which led to later conclusions on high potential avian postrelease predation in specific coastal areas (Støttrup and Sparrevohn 2007).

In 2003, two larger size groups (11.8 and 17.1 cm) of turbot were released. Postrelease mortality was estimated at 50% per day and 7% per day for these two groups, respectively (Figures 13.3b and 13.3c; Sparrevohn and Støttrup 2007). Presumably, the largest fish were too large to be eaten by the gulls. This assumption was supported by visual observations of gulls having caught large turbot without being able to handle and swallow the specimen. A size functional response was thus demonstrated for seagull predation on released turbot and postrelease mortality could be minimized by releasing the appropriate size of fish relative to the local predominant predator.

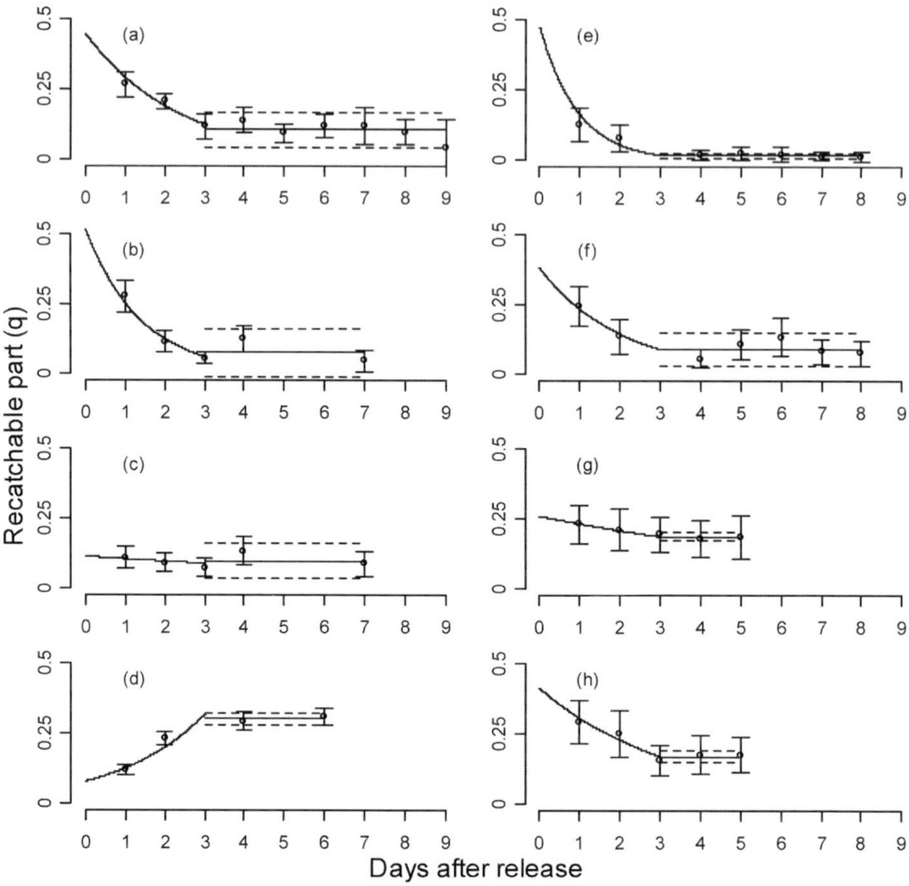

Figure 13.3 Recatchable part of turbot together with 95% confidence intervals relative to days after release. (a) 7.5 cm naïve turbot released in 2001; (b) 11.8 cm naïve turbot released in 2003; (c) 17.1 cm naïve turbot released in 2003; (d) conditioned and naïve turbot of size 3–5 cm released simultaneously in 2004; (e) 9.8 cm naïve turbot released in 2004 simultaneously with the 9.8 cm conditioned released in 2004 shown in figure (f); (g) 11 cm conditioned turbot released in 2006; (h) 4–5 cm conditioned turbot released in 2006. The solid line until day 3 shows the decrease in numbers and the slope is the postrelease mortality with the intercept at the *y*-axis being the catchability. The horizontal lines indicate the average recatchable part of the released fish (solid) and 95% confidence limit (dashed) from day 3 onwards. (Figures 13.3 a–c, e, and f reprinted from *Journal of Sea Research*, 57 (2–3), Sparrevohn C.R., and Støttrup J.G., Postrelease survival and feeding in reared turbot, 151–161, 2007, with permission from Elsevier. Figures 13.3d, g, and h redrawn from data in Sparrevohn and Støttrup 2008.)

An alternative approach to reduce postrelease mortality was to condition the released fish before release. It has been shown possible to improve the behavioral skill of reared fish to potentially improve their postrelease survival (Howell 1994; Hossian *et al.* 2002). The conditioning procedure applied was rather simple. Turbot intended for release were kept in cages in their release habitat for a period (3–6 days) prior to release (Figure 13.4). Depending on size of the individual fish, between 250 and 2,500 fish were kept in each cage. Postrelease mortality of conditioned fish was almost half that estimated for naïve turbot released (Figure 13.3e and 13.3f; Sparrevohn and Støttrup 2007). However,

Figure 13.4 Cages designed for flatfish conditioning before release. The cages have no bottom to allow direct contact with the bottom sediment. A coarse net is mounted on top to avoid avian predation. (Photo: Claus R. Sparrevohn.)

postrelease mortality in conditioned fish was still high (estimated at 38% per day) and it was proposed that the naïve fish had attracted a high number of predators, which influenced mortality on the conditioned fish released in the same area. This suggestion was confirmed in a subsequent experiment. In 2006, only conditioned fish were released and postrelease mortality was very low as shown in Figure 13.3g.

Similar experiments with small fish (4–5 cm) proved more problematic. Figure 13.3d shows a "positive mortality", indicating both a low mortality but also flaws in the experimental setup using this fish size. There were problems in identifying the naïve from the conditioned fish. The fish had been marked with alizarin complexone as in previous experiments, but for reasons unknown, the mark was not clearly distinguishable. In a subsequent experiment, however, a relatively high postrelease mortality was shown for this size group (Figure 13.3h) and was higher than that in the 11 cm size group released at the same time (Figure 13.3g).

The success of conditioning reared fish may be due to a number of factors including color adaptation or improved cryptic behavior. Wild turbot adapt their skin color to the surrounding sediment, presumably to reduce predation (Lanzing 1977). In the hatcheries, the fish are maintained in a constant light and background environment and it has been suggested that the hatchery environment may impair the ability of reared flatfish to color adapt or bury (Howell 1994). Reared sole were demonstrated to be slow in color adaptation (Ellis *et al.* 1997), taking 4–7 days to adapt to a lighter background (sand) but much longer time to color adapt. Winter flounder *Pseudopleuronectes americanus* required a minimum of 90 days to color adapt to match the sediment (Fairchild and Howell 2004). Color adaptation in turbot has not been studied in detail. Reared turbot, aged 1 year or older, used in the Danish experiments seemed generally slow to color adapt, and reared turbot were often easily recognizable from wild turbot in the catches conducted over a number of days after release.

In the wild, flatfish bury in the surrounding sediment for smaller or larger periods. Burial behavior in flatfish is thought to provide protection from predators (Kruuk 1963). In turbot, burial behavior has been also suggested to provide the ability to lie in wait for, and ambush, mobile prey (Holmes and Gibson 1983). Reared sole maintained on hard substrata in the absence of sand have been observed to attempt to bury (Howell and Canario 1987) and to bury as frequently as wild sole (Ellis *et al.* 1997) indicating innate behavior. This could also be true for turbot since reared turbot immediately buried after release (Iglesias and Rodrıguez-Ojea 1994). However, burial efficiency was lower in the initial attempts of reared fish as compared to wild fish, indicating a need to learn techniques or improve muscular efficiency (Ellis *et al.* 1997). In sole, the "learning" period took up to 12 days (Ellis *et al.* 1997), whereas in winter flounder this only took a couple of days (Fairchild and Howell 2004). Preliminary studies have shown that turbot also need some time to learn how to bury efficiently (Støttrup and Nielsen 1998), but more work needs to be conducted on cryptic behavior in turbot and how this may affect postrelease survival in reared turbot.

13.11 Cost–benefit of the releases

So far, no in-depth analysis on cost–benefits for stocking turbot has been conducted in Europe. The Danish program is funded through licenses from the recreational fisheries and the releases supported by both recreational and commercial fishermen. A simplified model for economic viability was presented by Støttrup and Sparrevohn (2007) whereby 8% turbot recaptures would be sufficient to even out the costs of release based on commercial values of fish. This is because turbot is commercially a highly valued fish fetching high market prices. Sparrevohn (2008) estimated the commercial value of the releases looking at the value of growth increment (weight) and survival after release and found that 10 cm turbot fry intended for stocking should have a maximum price of 1.5 to 2.0 € to cover production costs. However, for recreational fisheries to take place, it is not necessary that the costs are covered by the expenses incurred for the releases. Furthermore, in countries such as Denmark, recreational fishermen are not allowed to sell their catch. Thus, new models need to be put forward to describe and estimate cost–benefits for releases of fish for recreational purposes.

13.12 Perspectives

Stock enhancement of turbot is a viable tool for enhancing local stocks for recreational or commercial fishery. The first priority, however, should be to implement the responsible approach sensu (Blankenship and Leber 1995). This should ensure healthy fry of local origin, a breeding program that ensures a genetic variability similar to that of the wild local species, proper objectives, criteria, and appropriate measures for success. Most importantly, the releases should be monitored to ensure the success of the stocking taking into consideration a constantly changing environment. Also, monitoring should ensure that no adverse environmental changes or impacts occur.

Research emphasis should be directed toward improving field techniques for monitoring, for valuating sites for releases, and improving knowledge on the ecology of the species.

Development of appropriate models for estimating cost–benefits of release including recreational value is required.

Stocking should be included in regional fisheries management programs, including recovery programs for specific species where appropriate. Before embarking on stocking, the rationale for stocking should be reviewed including an initial review of the ecosystem dynamics and species population dynamics to ensure that the criteria for stock enhancement or restocking are fulfilled.

13.13 Acknowledgments

We would like to thank the Danish Coastal Fisheries Management Program for funding most of the release work with turbot. Thanks are due to José Carlos Mariño for providing information on stocking in Spain.

Literature cited

Bagge, O. 1987. Tagging of turbot and brill in the Kattegat 1965–1970. *ICES, Demersal Fish Committee. C.M.* G:10, 3 pp.

Balsa, J.C.M., Linares, F., and Luis, R.-J. 2006. Stock enhancement project of turbot *Psetta maxima* in Galicia (NW Spain). First Experience. 2006 AQUA Conference, Fortezza da Basso Convention Centre, Firenze, Italy, May 9–13, 2006.

Blankenship, H.L., and Leber, K.M. 1995. A responsible approach to marine stock enhancement. *American Fisheries Society Symposium* 15:165–175.

Danancher, D., and Garcia-Vazquez, E. 2007. Turbot – *Scophthalmus maximus*. Biology, ecology and genetics. Genimpact final scientific report, University of Oviedo, Oviedo, Spain, pp. 55–61.

Delbare, D., and De Clerck, R. 2000. Release of reared turbot in Belgian coastal waters as a tool for stock enhancement. *ICES CM 2000* O:(02), 13 pp.

Ellis, T., Howell, B.R., and Hughes, R.N. 1997. The cryptic responses of hatchery-reared sole to a natural sand substratum. *Journal of Fish Biology* 51:389–401

Engell-Sørensen, K., Støttrup, J.G., and Holmstrup, M. 2004. Rearing of flounder (*Platichthys flesus*) juveniles in semi-extensive systems. *Aquaculture* 230:475–491.

Fairchild, E.A., and Howell, W.H. 2004. Factors affecting the post-release survival of cultured juvenile *Pseudopleuronectes americanus*. *Journal of Fish Biology* 65(A):69–87.

FAO 2010a. http://www.fao.org/fishery/statistics/global-capture-production/en.

FAO 2010b. http://www.fao.org/fishery/statistics/global-aquaculture-production/en.

Florin, A.-B. and Höglund, J. 2007. Absence of population structure of turbot (*Psetta maxima*) in the Baltic Sea. *Molecular Biology* 16:115–126.

Fraser, D.F., Gilliam, J.F., Daley, M.J., Le, A.N., and Skalski, G.T. 2001. Explaining leptokurtic movement distributions: intrapopulation variation in boldness and exploration. *American Naturalist* 158:124–135.

Furuta, S., Watanabe, T., Yamada, H., Nishida, T., and Miyanaga, T. 1997. Changes in distribution, growth and abundance of hatchery-reared Japanese flounder *Paralichthys olivaceus* released in the coastal area of Tottori Prefecture. *Nippon Suisan Gakkaishi* 63:877–885.

Gibson, R.N. 1973. The intertidal movements and distribution of young fish on a sandy beach with special reference to the plaice (*Pleuronectes platessa* L.). *Journal of Experimental Marine Biology and Ecology* 12:79–102.

Holmes, R.A., and Gibson, R.N. 1983. A comparison of the predatory behaviour in flatfish. *Animal Behaviour* 31:1244–1255.

Hossain, M.A.R., Tanaka, M., and Masuda, R. 2002. Predator-prey interaction between hatchery-reared Japanese flounder juvenile, *Paralichthys olivaceus*, and sandy shore crab, *Matuta lunaris*: daily rhythms, anti-predator conditioning and starvation. *Journal of Experimental Marine Biology and Ecology* 267:1–14.

Howell, B.R. 1994. Fitness of hatchery-reared fish for survival in the sea. *Aquaculture and Fisheries Management* 25(1):3–17.

Howell, B.R., and Canario, A.V.M. 1987. The influence of sand on the estimation of resting metabolic rate of juvenile sole, *Solea solea* (L.). *Journal of Fish Biology* 31:277–280.

ICES 2007. Report of the ICES Advisory Committee on Fishery Management, Advisory Committee on the Marine Environment and Advisory Committee on Ecosystems, 2007. ICES Advice, Books 8, 147 pp.

Iglesias, J. 1981. Spatial and temporal changes in the demersal fish community of the Ria de Arosa (NW Spain). *Journal of Marine Biology* 65:199–208.

Iglesias, J., and Rodrıguez-Ojea, G. 1994. Fitness of hatchery reared turbot, *Scophthalmus maximus* L., for survival in the sea: first year results on feeding, growth and distribution. *Aquaculture and Fisheries Management* 25(Suppl. 1):179–188.

Iglesias, J., Ojea, G., Otero, J.J., Fuentes, L., and Ellis, T. 2003. Comparison of mortality of wild and released reared 0-group turbot, *Scophthalmus maximus*, on an exposed beach (Ría da Vigo, NW Spain) and a study on the population dynamics and ecology of the natural population. *Fisheries Management and Ecology* 10:51–59.

Iglesias, J., and Rodrıguez-Ojea, G. 1997. The use of alizarin complexone for immersion marking of the otoliths of embryos and larvae of the turbot *Scophthalmus maximus* (L): dosage and treatment time. *Fisheries Management and Ecology* 4:405–417.

Imsland, A.K., Scanu, G., and Naevdal, G. 2003. New variants of the haemoglobins of turbot (*Scophthalmus maximus*): possible use in population genetics studies and aquaculture. *Sarsia* 88(1):55–64.

Jones, A. 1973. The ecology of young turbot, *Scophthalmus maximus* (L.), at Borth, Cardiganshire, Wales. *Journal of Fish Biology* 5:367–383.

Kellison, G.T., Eggleston, D.B., and Burke, J.S. 2000. Comparative behaviour and survival of hatchery-reared versus wild summer flounder (*Paralichthys dentatus*). *Canadian Journal of Fisheries and Aquatic Sciences* 57:1870–1877.

Kruuk, H. 1963. Diurnal periodicity in the activity of the common sole, *Solea vulgaris* Quensel. *Netherlands Journal of Sea Research* 2:1–28.

Lagardère, F., Thibaudeau, K., and Bégout Anras, M.L. 2000. Feasibility of otolith markings in large juvenile turbot, *Scophthalmus maximus*, using immersion in alizarin-red S solutions. *ICES Journal of Marine Science* 57(4):1175–1181.

Lanzing, W.J.R. 1977. Reassessment of chromatophore pattern regulation in two species of flatfish (*Scophthalmus maximus*; *Pleuronectes platessa*). *Netherlands Journal of Sea Research* 11:213–222.

Leber, K.M., and Arce, S.M. 1996. Stock enhancement in a commercial mullet, *Mugil cephalus* L., fishery in Hawaii. *Fisheries Management and Ecology* 3:261–278.

Leber, K.M., Arce, S.M., Sterritt, D.A., and Brennan, N.P. 1996. Marine stock-enhancement potential in nursery habitats of striped mullet, *Mugil cephalus*, in Hawaii. *Fishery Bulletin* 94:452–471.

MacCall, A.D. 1990. *Dynamic Geography of Marine Fish Populations*. University of Washington Press, Seattle, WA, 153 pp.

Maslova, O.N. 2002. Problems and achievements in seed production of the Black Sea turbot in Russia. *Turkish Journal of Fisheries and Aquatic Sciences* 2:23–27.

Muus, B.J., Nielsen, J.G., Dahlstrøm, P., and Nyström, B.O. 1998. *Havfisk og fiskeri*. Gads Forlag, Copenhagen, Denmark, pp. 338.

Neumann, E., and Píriz, L. 2000. Svenskt småskaligt kustfiske – problem och mögligheter. *Fiskeriverket Rapport* 2:3–40.

Nielsen, E.E., Nielsen, P.H., Meldrup, D., and Hansen, M.M. 2004. Genetic population structure of turbot (*Scophthalmus maximus* L.) supports the presence of multiple hybrid zones for marine fishes in the transition zone between the Baltic Sea and the North Sea. *Molecular Ecology* 13:585–595.

Nissling, A., and Westin, L. 2004. Salinity requirements for successful spawning of Baltic and Belt Sea cod and the potential for cod stock interactions in the Baltic Sea. *Marine Ecology Progress Series* 152:261–271.

Paulsen, H., and Støttrup, J.G. 2004. Growth rate and nutritional status of wild and released reared juvenile turbot in southern Kattegat, Denmark. *Journal of Fish Biology* 65:210–230.

Pihl, L. 1989. Abundance, biomass and production of juvenile flatfish in southeastern Kattegat. *Netherlands Journal of Sea Research* 24(1):69–81.

Rae, B.B., and Devlin, S.D.E. 1972. The turbot, its fishery and biology in the Scottish area. *Marine Research* 1:27.

Riley, J.D., Symonds, D.J., and Woolner, L. 1981. On the factors influencing the distribution of 0-group demersal fish in coastal waters. The early life history of fish: recent studies. *Rapports et Procès-verbaux des Réunions du Conseil International pour l'Exploration de la Mer.* 178:223–228.

Scherbich, L.V. 1998. Daily age of juvenile turbot *Psetta maxima* (Pleuronectiformes, Scophthalmidae) and some peculiarities of juvenile and adult fishes distribution in the coastal area of the Baltic Sea, adjacent to Kaliningrad region. *ICES C.M.* CC:(6), 10 pp.

Sparrevohn, C.R. 2008. Evaluating and optimizing stock enhancement of a natural flatfish stock. PhD thesis, Wageningen University, Wageningen IMARES, Holland.

Sparrevohn, C.R., Nielsen, A., and Støttrup, J.G. 2002. Diffusion of fish from a single release point. *Canadian Journal of Fish and Aquatic Sciences* 59(5):844–853.

Sparrevohn, C.R., and Støttrup, J.G. 2007. Post release survival and feeding in reared turbot. *Journal of Sea Research* 57:151–161.

Sparrevohn, C.R., and Støttrup, J.G. 2008. Diet, abundance and distribution as indices of turbot (*Psetta maxima* L.) release habitat suitability. *Reviews in Fisheries Science* 16:338–347.

Støttrup, J.G., Lehmann, K., and Nicolajsen, H. 1998. Turbot, *Scophthalmus maximus*, stocking in Danish coastal waters. In: Cowx, I.G. (ed.) *Stocking and Introduction of Fish*. Fishing News Books, Oxford, UK, pp. 301–318.

Støttrup, J.G., and McEvoy Lesley, A. (eds) 2003. *Live Feeds in Marine Aquaculture*. Blackwell Science, Oxford, U.K., 318 pp.

Støttrup, J.G., and Nielsen, B. 1998. How does cryptic behavior of reared turbot *Scophthalmus maximus* compare with that of their wild counterparts. In: Grizel, H., and Kestemont, P. (eds) *Aquaculture and Water: Fish Culture, Shellfish Culture and Water Usage*, Special Publication no. 26. European Aquaculture Society, Oostende, pp. 257–258.

Støttrup, J.G., Overton, J.L., Paulsen, H., Möllmann, C., Tomkiewicz, J., Pedersen, P.B., and Lauesen, P. 2008. Rationale for restocking the eastern Baltic. *Reviews in Fisheries Science* 16:58–64.

Støttrup, J.G. and Sparrevohn, C.R. 2007. Can stock enhancement enhance stocks? *Journal of Sea Research* 57:104–113.

Støttrup, J.G., Sparrevohn, C.R., Modin, J., and Lehmann, K. 2002. The use of releases of reared fish to enhance natural populations. A case study on turbot *Psetta maxima* (Linné, 1758). *Fisheries Research* 59(1–2):161–180.

Suzuki, N., Nishida, M., Yoseda, K., ÜstÜndag, C., Sahin, T., and Amaoka K. 2004. Phylogeographic relationships within the Mediterranean turbot inferred by mitochondrial DNA haplotype variation. *Journal of Fish Biology* 65(2):580–585.

Tsukamoto, K. 1988. Otolith tagging of ayu embryo with fluorescent substances. *Nippon Suisan Gakkaishi* 54:1289–1295.

Urup, B. 1994. Methods for the production of turbot fry using copepods as food. In: Lavens, P., and Remmerswaal, R.A.M. (eds) *Turbot Culture: Problems and Prospects*, Special Publication no. 22. European Aquaculture Society, Ghent, Belgium, pp. 47–53.

Van Der Land, M.A. 1991. Distribution of flatfish eggs in the 1989 egg survey in the southeastern North Sea, and mortality of plaice and sole eggs. *Netherlands Journal of Sea Research* 27:277–286.

Yamashita, Y. and Yamada, H. 1999. Release strategy for Japanese flounder fry in stock enhancement programmes. In: Howell, B.R., Moksness, E., and Svåsand, T. (eds) *First International Symposium on Stock Enhancement and Sea Ranching*. Fishing News Books, Blackwell Scientific Publications, Oxford, pp. 191–204.

Zengin, M., Gümüs, A., and Bostanci, D. 2006. Age and growth of the Black Sea turbot, *Psetta maxima* (Linneaus, 1858) (Pisces: Scophthalmidae), estimated by reading otoliths and by back-calculation. *Journal of Applied Ichthyology* 22:374–381.

Zengin, M., Polaat, H., Kutlu, S., and Gümüş, A. 2007. An investigation on recruitment of hatchery-reared Black Sea turbot juveniles to natural stocks and its bioecological characteristics. Project Number: TAGEM/HAYSÜD/2000/17/03/010, Final Report. Central Fisheries Research Institute, Trabzon, Turkey, 193 pp.

Section 6

Asia Stock Enhancement

Chapter 14

Stock enhancement of Japanese flounder in Japan

Yoh Yamashita and Masato Aritaki

14.1 Background

Due to the rise in world population and general shift to higher food levels from grain to meat in developing countries, the state of food insecurity is becoming more serious (FAO 2006). The total capture fishery production of the world has been static around 90–95 million metric tons (MT) during recent years (FAO 2007). However, Watson and Pauly (2001) even reported that global capture fishery landings have started to decrease since the late 1980s. In contrast, aquaculture production reached 47.8 million MT in 2005, which is an increase of 3.5 times above the level in 1990, indicating the increasing importance of the role of aquaculture in fishery food production. However it is misleading to consider that aquaculture contributes to global net fish production, because it requires large inputs of wild fish for feed (raw or as pellets), and reduces wild fish supplies through habitat degradation and wild seedstock collection (Naylor *et al.* 2000). Stock enhancement provides a link between the capture fisheries and aquaculture, and its present situation and perspectives are worth evaluating.

In Japan, there has been a significant decline in the catch of coastal fisheries after a peak in 1985 (Figure 14.1). If we exclude catch of Pacific sardine (*Sardinops melanostictus*), which was tremendously abundant from the late 1970s to the early 1990s, the coastal catch has been continuously decreasing from the late 1960s (Figure 15.3 in Howell and Yamashita 2005). This decline is mainly attributable to overfishing and habitat degradation. However, mitigation measures including fishing regulation and conservation/regeneration of the habitat have not been effective in fully restoring the living resources of these coastal areas. A third way to replenish depleted stocks is by stock enhancement (Howell and Yamashita 2005). A national stock enhancement program was initiated in 1963 (Imamura 1999) and approximately 70 species (half of them are fish) are currently released in Japan (Fisheries Agency *et al.* 2005–2008).

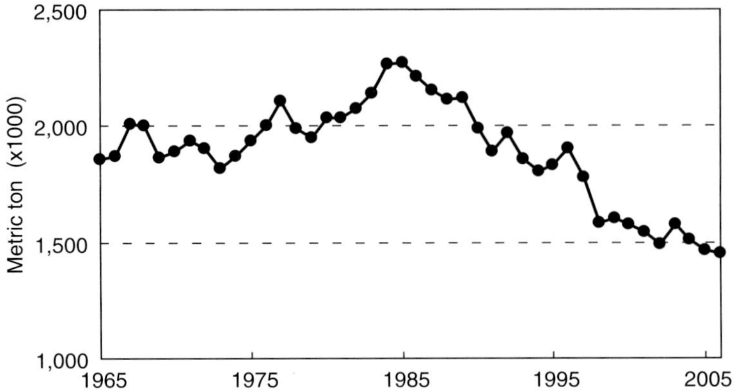

Figure 14.1 Annual catch from coastal fisheries in Japan (including sardine). (Ministry of Agriculture, Forestry and Fisheries 1978–2007.)

14.2 Summary of catch and stock enhancement data for Japanese flounder

Japanese flounder (*Paralichthys olivaceus*; referred to as flounder below) is distributed in coastal waters of eastern Asian countries, Japan, China, Korea, and Russia. Fishery catch (excluding aquaculture production) of this species in Japan has fluctuated between 5,500 and 9,000 MT since 1965 (Figure 14.2). This species is the most important finfish in the marine stock enhancement program in Japan. The release of juvenile hatchery flounder is carried out by each prefecture under the responsibility of each prefectural government supported by the national hatcheries of the Fisheries Research Agency. Reliable statistical data on flounder stock enhancement activities in Japan have been available since 1977. The total number of flounder juveniles released each year has increased from 260,000 in 1977 to 30.4 million in 1999. Currently, around 25 million are being released in almost all coastal prefectures from Kagoshima (31°N) to Hokkaido (45°N) (Figures 14.2 and 14.3).

If we adopt 11.7% (Kitada and Kishino 2006) as the mean contribution rate (the incidence of hatchery fish in the total number of fish of the same species landed at markets) in Japan, a total fishery catch of released flounder is calculated to be around 800 MT in the recent years. Aquaculture production of flounder shows similar levels to the total fishery catch, from 6,000 to 8,500 MT in the 1990s, but it has declined to 4,591 MT in 2005 probably because of the increase of cheaper imported farmed flounder from China and Korea.

Application of release practices of hatchery-cultured juveniles has been divided into three categories: "restocking" to restore severely depleted spawning biomass, "stock enhancement" to augment the natural supply of juveniles and optimize harvest by overcoming recruitment limitations, and "sea farming" releasing to unenclosed marine and estuarine environments for harvest at a large size (put, grow, and take) (Bell *et al.* 2008). In Japan, the release of the flounder is categorized as stock enhancement. In this chapter, release technologies for efficient stocking and practical effectiveness in flounder stock enhancement are summarized and future perspectives for the management of the flounder stocks are discussed.

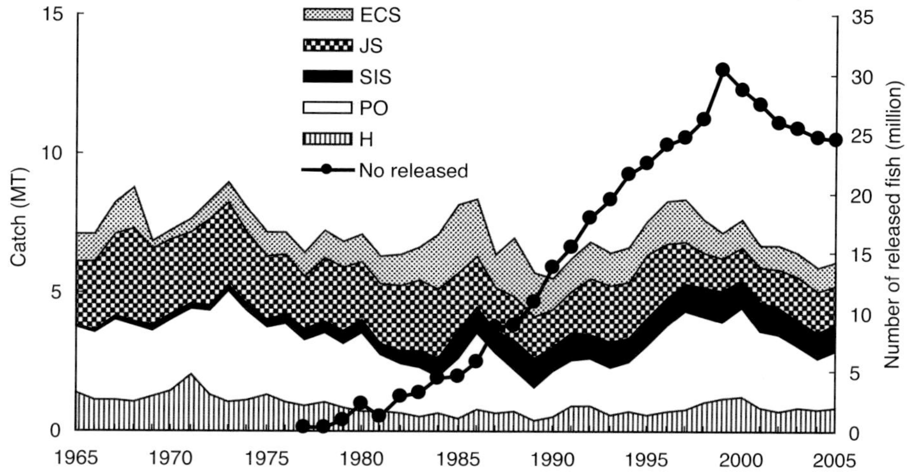

Figure 14.2 Change in catches of Japanese flounder for five regions in Japan and total number of released hatchery-cultured juveniles. ECS, East China Sea; JS, Japan Sea; SIS, Seto Inland Sea; PO, Pacific Ocean; H, Hokkaido (see Fig. 15.3). (Ministry of Agriculture, Forestry and Fisheries 1978–2007; Fisheries Agency and Japan Sea Farming Association 1980–2003; Fisheries Agency and Fisheries Research Agency 2004; Fisheries Agency *et al.* 2005–2008.)

14.3 Release strategy

In fish market surveys, almost all released flounder (>95%) could be clearly distinguished from wild flounder by the occurrence of permanent black pigments (hypermelanosis) on the blind (abocular) side (Tominaga and Watanabe 1998), which wild fish are lacking. However, the incidence of juvenile hatchery flounder which have no such pigments on the blind side has recently increased (see Chapter 8), so the contribution rate may be underestimated if only blind side pigments is used as a sole marker of hatchery flounder. An accurate estimate of the percentage of hatchery flounder having black pigments on the blind side is needed to correct contribution rates. In addition, if we need to identify release information such as size, time, or place of release, hatchery flounder are tagged by fin cuts, chemical or fluorescent dyes, anchor tags, or genetic tags.

Evaluation of the following criteria is considered to be important to enhance the efficiency and success of release technology. In addition to these key criteria, genetic diversity, fitness, and quality of seed must be taken into consideration for responsible stock enhancement (see Chapter 8). These criteria are as follows: (1) size at release; (2) release habitat; (3) release timing; (4) release magnitude; and (5) release method and conditioning.

14.3.1 Size at release

Mortality of released juvenile flounder is caused mainly by predation with high predation mortality occurring within 1–2 weeks of release (Furuta *et al.* 1998; Yamashita and Yamada 1999). Because predation is strongly affected by the predator–prey size relationship, a large size at release is important to minimize

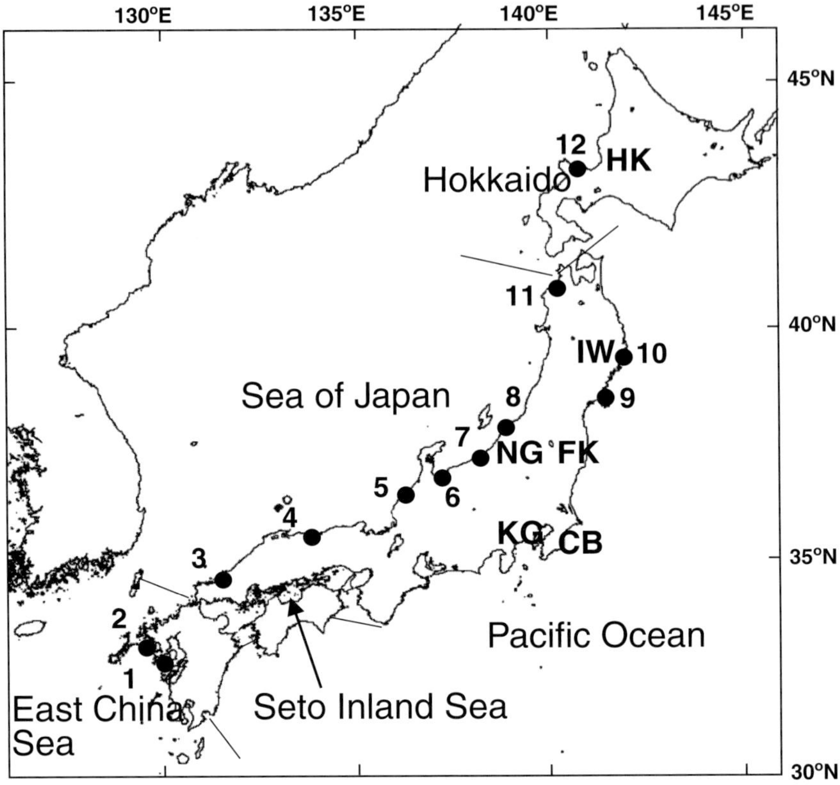

Figure 14.3 Map showing areas in Japan and sampling stations of wild juvenile flounder shown in Figure 14.6. KG indicates Kanagawa, CB Chiba, FK Fukushima, NG Niigata, IW Iwate, and HK Hokkaido prefectures which show the highest market return rate.

postrelease mortality (Yamashita *et al.* 1994). In addition to piscivorous fishes (Furuta *et al.* 1998; Yamashita and Yamada 1999), nocturnal predation by crabs and cuttlefish has been recognized as a source of mortality of released juvenile flounder (Hossain *et al.* 2002; Saitoh *et al.* 2003).

The market return rate (MRR, defined as the incidence of the number of returned fish at market/number of released fish) of hatchery flounder with TL at release, <200 mm and released from 1983 to 2004 ranged between 0.01 and 32.0% (Figure 14.4). The MRR tended to increase with increasing size at release. The number of cases with an MRR of more than 5% was 54 out of 171 when the TL at release was less than 100 mm and 34 out of 43 cases when the TL at release was >100 mm. However, the release of large-sized fish does not always ensure a high MRR. The highest MRR values are reported for fish between 80 and 100 mm TL at release. This is consistent with the results from a large-scale release-recapture experiment in which juvenile flounder of a broad size range (40–150 mm) were released simultaneously after being marked with alizarin complexone on the otoliths to identify individual release length (Yamashita *et al.* 1994). Generally, as fish size increases, juvenile fish become less vulnerable to predation and have a greater ability to recover from stress (Masuda and

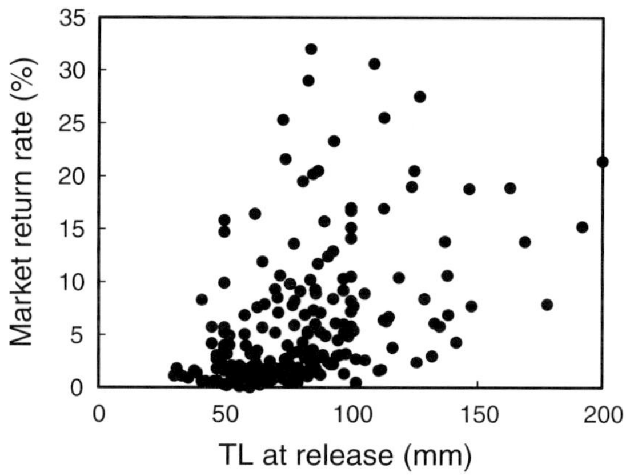

Figure 14.4 Relationship between the TL at release and the market return rate of commercial sized Japanese flounder ($N = 213$). (Fisheries Agency *et al.* 1990–2007; Nakamura 1996; Furuta 1998; Okouchi *et al.* 1999; Hiyama and Kimura 2000; Atsuchi and Masuda 2004; National Association for the Promotion of Productive Seas 2007, 2008.)

Ziemann 2000). However, Makino *et al.* (2006) reported that the relationship between size and learning capability in prey capture could be represented as a quadratic curve with the maximum learning capability at 70 mm in standard length (SL) for striped knifejaw (*Oplegnathus fasciatus*). Because the learning capability of hatchery fish to adapt to the new natural environment of release site is thought to affect postrelease survival, the highest MRR around 80–100 mm TL found in the released flounder can be partly explained by the size and learning capability relationship.

Size at release is also closely associated with the cost to culture juveniles in the hatchery (Leber 1995; Munro and Bell 1997; Yamashita and Yamada 1999). In a given size of culture tank, costs and production period are lower and the number of cultured fish for release is higher with smaller size at release. This means the best size at release for economic efficiency (net income/hatchery production and release costs) is sometimes different from the ecologically optimal size at release (Yamashita and Yamada 1999). The average size at release in the flounder stock enhancement program in Japan has increased from approximately 50 mm TL during the early 1980s to 80 mm in recent years (Figure 14.5). However, there are large differences in TL at release among prefectures depending on the prefecture's release strategy. For example, in 2006, release sizes ranged from 43 to 116 mm TL.

14.3.2 Release habitat

Wild flounder nursery grounds are found in shallow sandy areas with 0.1–1 mm diameter sediments from 0.5 to 10 m in depth (Tanaka *et al.* 1989; Yamamoto *et al.* 2004a). The flounder predominantly consume copepods at settling stage and mysids from settlement stage, around 10 mm TL, to postsettlement juvenile stages up to 150 mm TL (Minami 1982; Yamada *et al.* 1998; Tanaka *et al.* 2005,

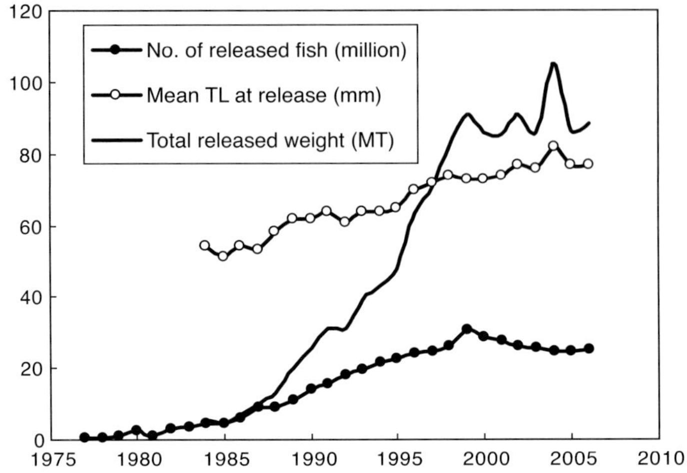

Figure 14.5 Annual change of the number of released hatchery-cultured flounder, average TL at release and estimated total weight of released juveniles. (Fisheries Agency and Japan Sea Farming Association 1980–2003; Japan Sea Farming Association 2004; Fisheries Agency *et al.* 2005–2008.)

2006). Juveniles prey on crangonid shrimp and gammarid amphipods when mysids are not sufficient (Yamamoto *et al.* 2004b; Yamamoto and Tominaga 2007). Juvenile flounder also are able to catch small fish as prey soon after settlement, and a piscivorous feeding habit becomes evident beginning around 50 mm TL if prey fishes such as anchovy, sand lance, and gobies are available (Yamada *et al.* 1998; Yamashita and Yamada 1999). Feeding conditions of wild juvenile flounder collected from 12 nursery grounds from southwestern to northeastern Japan in 2000 are compared in Figure 14.6 (recalculated from Tanaka *et al.* 2006). Relative stomach fullness (RSF; defined as the percentage ratio of the individual stomach content weight to the size-dependent maximum stomach content weight which is the same as the stomach content index in Tanaka *et al.* 2005) can be used to represent prey availability of each nursery ground. Stations 8, 10, and 12 in Figure 14.6 seem to have ideal prey conditions with abundant mysids and small prey fishes.

Hatchery-cultured juvenile flounder are generally released into nursery grounds that are appropriate for wild juveniles. Feeding incidence (percentage of released fish having food in their stomachs, i.e., the reciprocal of the percentage of empty stomach) of hatchery fish has been monitored as a simple index of habitat prey availability and degree of adaptation. Reported feeding incidence of hatchery flounder one week after release has ranged from 15 to 95% (Fisheries Agency *et al.* 1990–2007). Habitat suitability for stocking should be judged using these feeding condition indices. In addition, release sites should be selected from the viewpoint of predation refuge (Furuta 1996). However, there has been no systematic research on the relationship between release habitat and effectiveness of stock enhancement.

Sparrevohn and Støttrup (2008) suggested that an area that provides high survivorship but has low accessibility for the wild populations is an area with a high stock enhancement potential for turbot (*Psetta maxima*). For flounder, the

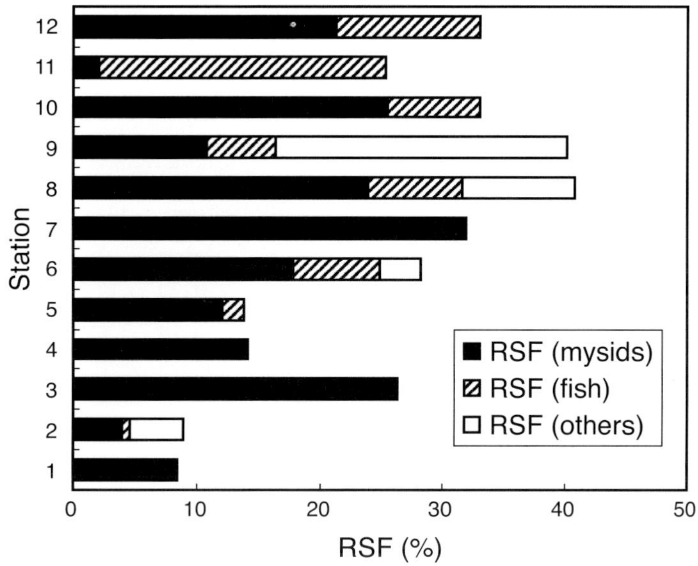

Figure 14.6 Regional changes in relative stomach fullness (RSF) of wild Japanese flounder juveniles collected from the 12 nursery areas shown in Figure 14.3. (Recalculated from Tanaka *et al.* (2006).) RSF is defined as the ratio of the percentage of individual stomach content weight to the size-dependent maximum stomach content weight.

area of western Tokyo Bay with no mysids and abundant small fishes, where wild juveniles cannot recruit and survive due to the lack of mysids, is applicable for this case. Juveniles (>70 mm TL) released into this area did not require mysids and began to consume abundant small prey fishes after release resulting in high growth (300 mm in the first year) and MRR (25–32%) (Nakamura 1996).

14.3.3 Release season

Flounder spawning season varies widely depending on latitude and extends from January in the south to August in northern Japan (Minami 1997). Although, the abundance of mysids, which is the most dominant prey for juvenile flounder, also shows considerable seasonal fluctuations, settlement of wild juvenile flounder shows close synchrony with the peak season of mysids (Fujii and Noguchi 1996; Tanaka *et al.* 1998; Yamada *et al.* 1998). Predators of juvenile flounder, such as piscivorous fish, cuttlefish, and crabs migrate to the shallow areas seasonally. Consequently, hatchery-cultured flounder should be released when prey are most abundant and predators the lowest.

There is only one published research paper on the relationship between release season and stocking effectiveness (Furuta 1998). In the inshore waters of Tottori Prefecture, mysids are abundant in April and May, and then rapidly decrease to a very low level by mid- or late June. The two major predatory fishes, large (>age +1) flounder and bartail flathead (*Platycephalus indicus*), migrate into shallow coastal waters in June. Before 1992, juvenile flounder were released from

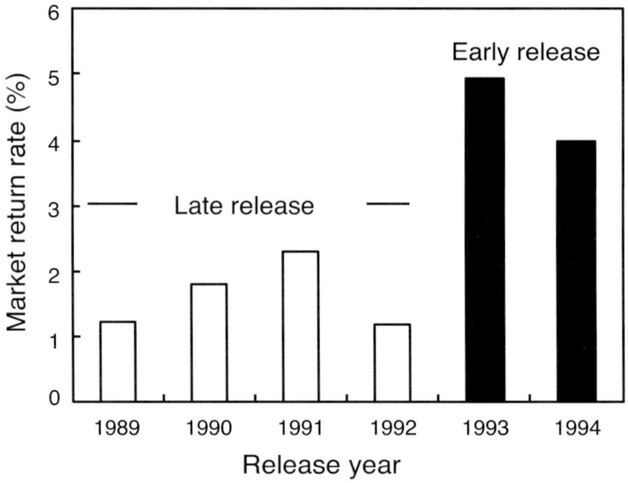

Figure 14.7 Comparison in the market return rate of hatchery flounder between late (mid-June to July in 1989–1992) release and early (mostly May in 1993 and 1994) release in Tottori Prefecture. (Redrawn from Furuta (1998).)

mid-June to July and consequently, a large number of the released flounder starved and/or were preyed on, resulting in high mortality rates (Furuta 1996; Furuta *et al.* 1998). Due to those results, the release season was changed to mostly May in 1993 and 1994 while maintaining the same size at release. The MRR approximately tripled from 1.6 ± 0.5% (SD) in 1989–1992 to 4.9% (1993) and 4.0% (1994), presumably because of higher survival of released juveniles (Figure 14.7).

14.3.4 Release magnitude

Release magnitude of hatchery fish must be determined on the basis of carrying capacity. Although MacCall (1990) defined it as the population biomass at which per capita population growth is zero, the concept of carrying capacity is generally used ambiguously. In the context of stock enhancement, the carrying capacity can be considered as the surplus productivity available for hatchery fish. The purpose of stock enhancement is to augment the target stock by using excess trophic resources. If released fish compete for limited productivity against wild conspecifics or other commercially important fish species and reduce the survival of wild populations, so-called replacement occurs (Kitada and Kishino 2006). Hilborn (1999) cautioned that stock enhancement programs need to ensure that cultured fish do not simply displace wild fish without any net increase in total production. Replacement does not occur only as density-dependent mortality in terms of number, but also displacement in weight is likely in the field when competition reduces the growth rate of wild fish instead of increasing the mortality rate.

Most flatfish species settle and concentrate into the two-dimensional nursery habitat at the juvenile stage (Rijnsdorp *et al.* 1992; Gibson 1994; Van der Veer *et al.* 2000) and consequent density-dependent mechanisms may easily occur during this stage. The stomach fullness (Figure 14.6) and growth rate (Tanaka *et al.* 1998) of wild juvenile flounder are generally higher in northeastern areas than southwestern areas of Japan. In addition, all the highest MRR (>20%, 12 cases out of 213 cases in Figure 14.4) were reported from northeastern coasts such as Kanagawa (KG), Chiba (CB), Fukushima (FK), Niigata (NG), Iwate (IW), and Hokkaido (HK) prefectures (Figure 14.3) suggesting that shallow coastal areas in northeastern Japan generally have higher productivity as nursery grounds for the flounder.

Growth rate of wild fish can be an index of the carrying capacity and habitat suitability of a nursery ground. Because the specific growth rate of the juvenile flounder is predominantly determined by prey availability and temperature (Yamashita *et al.* 2001), the maximum growth rate is temperature-dependent when prey is sufficient. In a 6-year study (2001–2006) in Sendai Bay and the Joban area off the northeastern Pacific coast, growth rates of wild juveniles were reported to be near the maximum value estimated from temperature regardless of juvenile density (Uehara *et al.* 2008). In particular, the growth rate of the dominant year class that occurred in 2005 when juvenile density was more than 10 times higher than densities in the lowest years, was close to the predicted maximum value as well. This indicates that these areas have sufficient productivity to support large fluctuations of wild flounder recruitment and possibly the addition of hatchery fish. In contrast, Kitada and Kishino (2006) reported the possible density-dependent mortality of released fish in two local areas.

Significantly lower growth rates of wild juveniles than the predicted maximum value suggests that there is no unutilized productivity for stock enhancement. Numerical modeling will allow quantitative determination of the optimal magnitude of hatchery seed release (Taylor and Suthers 2008).

14.3.5 Release method and conditioning

Healthy hatchery fish that can immediately adapt to natural conditions must be produced for stocking (Howell and Yamashita 2005). Physical handling for transport and release, low dissolved oxygen (DO), and rapid temperature change during transportation may damage hatchery fish and lead to postrelease mortality.

Released hatchery juveniles are more susceptible to predation than wild juveniles (Olla *et al.* 1994; Furuta *et al.* 1998; Kellison *et al.* 2000). The main cause of this susceptibility to predation for juvenile hatchery flounder is the difference in feeding behavior (Furuta 1996; Tanaka *et al.* 1998). Cultured juvenile flounder fed on commercial pellets have three specific feeding behaviors which differ from wild fish: (1) longer swimming time off the bottom; (2) returning to the bottom far from their initial position; and (3) a low occurrence of burrowing behavior. Cultured flounder could be trained to exhibit more natural behaviors by rearing at low density, with sandy substratum or use of a diet of live mysids (Tanaka *et al.* 1998). In addition, juvenile flounder are reported to be capable of predator conditioning through predator-exposure learning processes in the laboratory (Arai *et al.* 2007). However, the effectiveness of such conditioning for flounder has not been assessed postrelease. In other flatfish species, it has been demonstrated that conditioning improves the ability to bury in the sediment for winter flounder (*Pseudopleuronectes americanus*) juveniles in the laboratory (Fairchild and Howell 2004) and that conditioning before release improved the postrelease survival of turbot in the field (Sparrevohn and Støttrup 2007).

14.4 Evaluation of the effectiveness of the stock enhancement

14.4.1 Effectiveness of the flounder stock enhancement

To provide indices of the effectiveness of stock enhancement, contribution rate (the number of returned hatchery-fish/the total number of fish of the same species landed at market), MRR (previously defined), economic efficiency (previously defined), stocking efficiency (survival rate to recruitment, defined as the number of 1-year-old recruited hatchery fish/number of released hatchery fish), and yield per release (grams of hatchery fish caught/individual released fish) have been used.

The contribution rate of hatchery-cultured flounder has been reported to range from 0.1 to 57.4% depending on year and release area (Kitada and Kishino 2006). Contribution rate can be used as an index of the stocking impact on the wild population, but not as a measure of the contribution to production increase. MRR is the most commonly used index of stocking efficiency (e.g., Figures 14.4 and 14.7). The average MRR reported for all of Japan (from 1983 to 2004 year class) shown in Figure 14.4 was 5.99 ± 6.48% ($N = 213$) with 39 cases of MRR >10% in 11 areas. The problem of this index is that values increase with the increasing ratio of younger fish in the catch. Economic efficiency was reported from Kagoshima Bay (1.06 ± 0.1; 1993–1995) (Atsuchi and Masuda 2004), Miyako Bay (1.64 ± 0.32; 1989–1992) (Okouchi *et al.* 1999), Fukushima Prefecture (0.85 ± 0.36; 1994–2002) (Tomiyama *et al.* 2008), and Inland Sea side of Yamaguchi Prefecture (9.10 ± 4.20; 1992–1994) (Hiyama and Kimura 2000). In this index, generally cost does not include construction and maintenance of hatchery facilities. Income consists of only market sales and excludes sales relating to recreational flounder fishing which has become increasingly popular. Stocking efficiency and yield per release seem useful to evaluate the release efficiency; however, these data are not available for most areas. Presently, the stocking efficiency data are only available in Kagoshima Bay where the survival rate to recruitment ranged from 1.1 to 5.1% (Atsuchi and Masuda 2004). Miyako Bay and Kagoshima Bay both report yield per released individual; the average from these bays is 46.2 ± 22.9 g (Kitada and Kishino 2006). In addition, 80–100 MT of hatchery juveniles are currently released annually (Figure 14.5) and contribute to approximately 800 MT production (previously mentioned) as fishery catch each year.

As Kitada and Kishino (2006) emphasized, the effectiveness of stock enhancement should be evaluated by the contribution to the net increase in harvest or abundance of the target species. We analyzed the correlation between the number of juvenile hatchery flounder released and the commercial fishery catch

Table 14.1 Relationship between the number of hatchery juvenile flounder released and the annual fishery catch two years postrelease in 40 areas (35 prefectures and 5 areas of Hokkaido).

Region	Regression coefficient			
		Significant		Significant
	+	+	−	−
East China Sea	2	1	3	1
Japan Sea	0	0	10	5
Seto Inland Sea	7	5	2	0
Pacific Ocean	8	5	3	0
Hokkaido	4	2	1	0
Total	21	13	19	6

[a]Statistics from the Fisheries Agency, the Fisheries Research Agency, the Japan Sea Farming Association, and the National Association for the Promotion of Productive Seas in 1977–2003 for release data and the Ministry of Agriculture, Forestry and Fisheries in 1977–2005 for catch data.
[b]The values in the table indicate the number of areas (see Figure 14.3 for the regions). Prefectures abutting two regions are assigned to the one with the higher catch.

2 years later in 40 areas (35 prefectures and 5 areas of Hokkaido) under the assumption that 2-year old fish constitute the majority of the catch. The year of onset of release is different by each prefecture or area. The results of the analysis showed that 21 areas had a positive regression coefficient and 13 of these areas were statistically significant ($P < 0.05$). Nineteen areas had negative coefficients with six of those areas being significant ($P < 0.05$) (Table 14.1). There were areal trends that the Pacific Ocean and Seto Inland Sea sides had positive coefficients and Japan Sea side except Hokkaido had negative coefficients. Generally, fishery catches have tended to decline in the Japan Sea and the East China Sea since the late 1960s and mid-1980s, respectively, while catches have risen in the Seto Inland Sea, the Pacific, and Hokkaido coasts since the 1960s, the mid-1980s, and the early 1990s, respectively (Figure 14.2). This fluctuation is thought to affect the results of the regression analysis. Effectiveness of the flounder stock enhancement can be detected in small local areas. However, the effects of stock enhancement may be masked by the magnitude of natural recruitment in large stocks (Kitada and Kishino 2006).

Although almost all coastal prefectures have carried out stock enhancement programs for flounder for a long period, detailed analyses of the effectiveness for each area are limited. All prefectural governments responsible for this program have an obligation to conduct postrelease monitoring and research on the effectiveness because stocking is conducted using money from taxes from citizens and sales income of fishermen.

14.4.2 Comparison with the stock condition of other coastal commercial fishes

The fundamental purpose of flounder stock enhancement is to stabilize the catch and increase productivity. To elucidate the effectiveness of the stocking of the flounder from these two view points, we analyzed the coefficient of variance (CV)

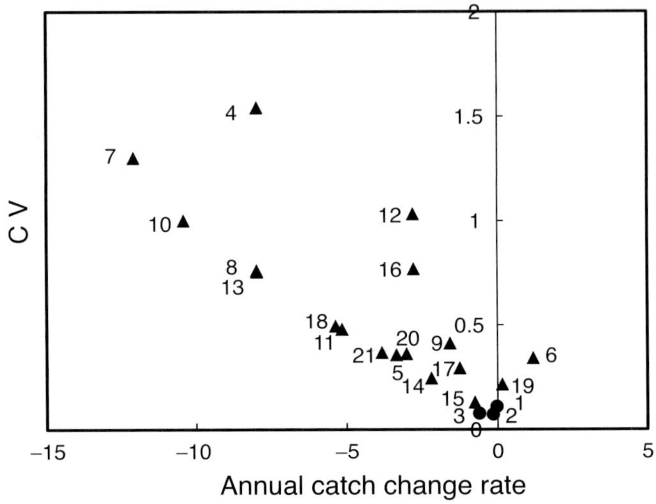

Figure 14.8 Relationship between the coefficient of variance and the annual catch change rate (linear regression coefficient/average annual catch) for 30 years (in 1976–2005) in 21 coastal taxa. Closed circles indicate the three intensively stocked species. (1) Japanese flounder, (2) red sea bream (*Pagrus major*), (3) black sea bream (*Acanthopagrus schlegeli*, including *Sparus sarba*), (4) Pacific herring (*Clupea pallasii*), (5) Pacific cod (*Gadus macrocephalus*), (6) Okhostk Atka mackerel (*Pleurogrammus azonus*), (7) rock fishes (*Sebastes* spp.), (8) Broadbanded thornyhead (*Sebastolobus macrochir*), (9) sand fish (*Arctoscopus japonicus*), (10) Scianidae, (11) Synodontidae, (12) melon seed (*Psenopsis anomala*), (13) Daggertooth pike conger (*Muraenesox cinereus*), (14) Largehead hairtail (*Trichiurus lepturus*), (15) sea breams (*Evynnis japonica, Dentex tumifrons*), (16) Spanish mackerel (*Scomberomorus niphonius*), (17) Exocoetidae, (18) Mugilidae, (19) sea bass (mainly *Lateolabrax japonicus*), (20) sand lance (*Ammodytes personatus*), and (21) coastal flatfish (excluding Japanese flounder).

and the change rate of annual catch (linear regression coefficient was divided by average catch for standardization) for 30 years (in 1976–2005) in 21 coastal taxa including flounder (Ministry of Agriculture, Forestry and Fisheries 1978–2007) (Figure 14.8). Only two species, sea bass (*Lateolabrax japonicus*) and Atka mackerel (*Pleurogrammus azonus*), showed significant positive linear regression coefficients. Although the other 19 taxa showed negative trends for 30 years, the decline in the flounder, red sea bream (*Pagrus major*), and black sea bream (*Acanthopagrus schlegeli*) catches was the lowest. The coefficient of variance of fishery catch was also the lowest in the order of black sea bream, red sea bream, and flounder (Figure 14.8), clearly indicating that the market landings of these three species have been stable for the last 30 years. These three species have been the most important targets of long-term stock enhancement programs in Japan, with a total number of fish released at 50 billion red sea bream, 43 billion flounder, and 14 billion black sea bream from 1977 to 2005. In contrast, fishery catch of less- or non-targeted coastal fishes, except sea bass and Atka mackerel, have fluctuated and declined since the mid-1970s. Although the landings of the three intensively stocked species did not increase after the commencement of stocking, the stable catch during the last 30 years may indicate that the mass release of juveniles has contributed to sustaining recruitment to the commercial fisheries.

14.5 Future perspectives

Stock enhancement is a management tool to augment limited and/or fluctuating marine resources by supplementing the stocks (Munro and Bell 1997; Blaxter 2000; Bell *et al.* 2008). The following key points should be considered for the future promotion of the flounder stock enhancement program in Japan.

1. Healthy, high quality, and genetically diverse hatchery fish should be released.
2. The stocking program has been implemented as a unit of prefectural government. However, because adult flounder can migrate over several hundreds of km and the range of a local stock as a management unit may extend over several prefectures (Nishida *et al.* 1997), stock management plans for this species including stock enhancement and fishing regulations should be considered under the collaboration of interprefectural (regional) unit.
3. Suitability and carrying capacity of the release site predominantly affect postrelease survival and are significant factors determining the effectiveness of stock enhancement. Ecological studies of the release site and postrelease monitoring of the community, including the released flounder, should be conducted to determine the optimum release strategy and evaluate the impact of stocking. In particular, advancement of technology to estimate the optimal release magnitude in relation to the carrying capacity of the release site is required.
4. Effectiveness and economic efficiency should be analyzed for each stock enhancement program. The purposes of hatchery releases may be different at each local area, such as increase of catch, stabilize catch fluctuation, create new stocks for commercial and/or recreational fishing, and so on. Therefore, depending on purposes, all stakeholders must be involved in the decision-making on the future aspects and the evaluation of effectiveness.

14.6 Acknowledgments

We are most grateful to Drs. Y. Tsuruta, National Association for the Promotion of Productive Seas, E. A. Fairchild, University of New Hampshire, and J. M. Miller, North Carolina State University, for their fruitful comments on the manuscript.

Literature cited

Arai, T., Tominaga, O., Seikai, T., and Masuda, R. 2007. Observational learning improves predator avoidance in hatchery-reared Japanese flounder *Paralichthys olivaceus* juveniles. *Journal of Sea Research* 58:59–64.

Atsuchi, S., and Masuda, Y. 2004. Effectiveness of the release of hatchery-produced stock of Japanese flounder *Paralichthys olivaceus* in Kagoshima Bay, southern Japan. *Nippon Suisan Gakkaishi* 70(6):910–921 (in Japanese with English summary).

Bell, J.D., Leber, K.M., Blankenship, H.L., Loneragan, N.R., and Masuda, R. 2008. A new era for restocking, stock enhancement and sea ranching of coastal fisheries resources. *Reviews in Fisheries Science* 16(1–3):1–9.

Blaxter, J.H.S. 2000. The enhancement of marine fish stocks. *Advances in Marine Biology* 38:1–54.

Fairchild, E.A., and Howell, W.H. 2004. Factors affecting the post-release survival of cultural juvenile *Pseudopleuronectes americanus*. *Journal of Fish Biology* 65:1–19.

FAO 2006. *The State of Food Insecurity in the World 2006*. Food and Agriculture Organization of the United Nations, Rome.

FAO 2007. *The State of World Fisheries and Aquaculture 2006*. Food and Agriculture Organization of the United Nations, Fisheries and Aquaculture Department, Rome.

Fisheries Agency and Japan Sea Farming Association 1980–2003. *Annual Statistics of Seed Production and Release in 1977–2001*. Japan Sea Farming Association, Tokyo (in Japanese).

Fisheries Agency and Fisheries Research Agency 2004. *Annual Statistics of Seed Production and Release in 2002*. Fisheries Research Agency, Yokohama (in Japanese).

Fisheries Agency, Fisheries Research Agency and National Association for the Promotion of Productive Seas 2005–2008. *Annual Statistics of Seed Production and Release in 2003–2006*. National Association for the Promotion of Productive Seas, Tokyo (in Japanese).

Fisheries Agency, Fisheries Research Agency, Japan Sea Farming Association, National Association for the Promotion of Productive Seas and Prefectural Governments 1990–2007. *Annual Reports on the Stock Enhancement Technology Development Projects*. Fisheries Agency Japan, Tokyo (in Japanese).

Fujii, T., and Noguchi, M. 1996. Feeding and growth of Japanese flounder (*Paralichthys olivaceus*) in the nursery ground. In: Watanabe, Y., Yamashita, Y., and Oozeki, Y. (eds) *Survival Strategies in Early Life Stages of Marine Resources*. A. A. Balkema, Rotterdam, pp. 141–151.

Furuta, S. 1996. Predation on juvenile Japanese flounder (*Paralichthys olivaceus*) by diurnal piscivorous fish: Field observations and laboratory experiments. In: Watanabe, Y., Yamashita, Y., and Oozeki, Y. (eds) *Survival Strategies in Early Life Stages of Marine Resources*. A. A. Balkema, Rotterdam, pp. 285–294.

Furuta, S. 1998. Behavioral and ecological studies on release techniques of hatchery-reared Japanese flounder *Paralichthys olivaceus*. *Bulletin of the Tottori Prefectural Fisheries Experimental Station* 35:1–76 (in Japanese with English summary).

Furuta, S., Watanabe, T., and Yamada, H. 1998. Predation by fishes on hatchery-reared Japanese flounder *Paralichthys olivaceus* juveniles released in the coastal area of Tottori Prefecture. *Nippon Suisan Gakkaishi* 64(1):1–7.

Gibson, R.N. 1994. Impact of habitat quality and quantity on the recruitment of juvenile flatfishes. *Netherlands Journal of Sea Research* 32:191–206.

Hilborn, R. 1999. Confessions of a reformed hatchery basher. *Fisheries* 24:30–31.

Hiyama, S. and Kimura, H. 2000. On stocking effectiveness of flounder *Paralichthys olivaceus* larva release in the Seto-Inland Sea off Yamaguchi Prefecture. *Bulletin of the Yamaguchi Inland-Sea Prefectural Fisheries Experimental Station* 29:1–8 (in Japanese).

Hossain, M.A.R., Tanaka, M., and Masuda, R. 2002. Predator-prey interaction between hatchery-reared Japanese flounder juvenile, *Paralichthys olivaceus*, and sandy shore crab, *Mututa lunaris*: daily rhythms, anti-predator conditioning and starvation. *Journal of Experimental Marine Biology and Ecology* 267: 1–14.

Howell, B.R., and Yamashita, Y. 2005. Aquaculture and stock enhancement. In: Gibson, R.N. (ed.) *Flatfishes: Biology and Exploitation*. Blackwell Science, Oxford, pp. 347–371.

Imamura, K. 1999. The organization and development of sea farming in Japan. In: Howell, B.R., Moksness, E., and Svåsand, T. (eds) *Stock Enhancement and Sea Ranching*. Fishing News Books, Oxford, pp. 91–102.

Kellison, G.T., Eggleston, D.B., and Burke, J.S. 2000. Comparative behaviour and survival of hatchery-reared versus wild summer flounder (*Paralichthys dentatus*). *Canadian Journal of Fisheries and Aquatic Science* 57:1870–1877.

Kitada, S., and Kishino, H. 2006. Lessons learned from Japanese marine finfish stock enhancement programmes. *Fisheries Research* 80:101–112.

Leber, K.M. 1995. Significance of fish size-at-release on enhancement of striped mullet fisheries in Hawaii. *Journal of the World Aquaculture Society* 26:143–153.

MacCall, A.D. 1990. *Dynamic Geography of Marine Fish Populations*. University of Washington Press, Seattle, WA.

Makino, H., Masuda, R., and Tanaka, M. 2006. Ontogenetic changes of learning capability under reward conditioning in striped knifejaw *Oplegnathus fasciatus* juveniles. *Fisheries Science* 72:1177–1182.

Masuda, R., and Ziemann, D.A. 2000. Ontogenetic changes of learning capability and stress recovery in Pacific threadfin juveniles. *Journal of Fish Biology* 56: 1239–1247.

Minami, T. 1982. The early life history of flounder *Paralichthys olivaceus*. *Bulletin of the Japanese Society of Scientific Fisheries* 48(11):1581–1588 (in Japanese with English summary).

Minami, T. 1997. Ecological aspects, life history. In: Minami, T., and Tanaka, M. (eds) *Biology and Stock Enhancement of Japanese Flounder*. Koseisha-Koseikaku, Tokyo, pp. 1–24 (in Japanese).

Ministry of Agriculture, Forestry and Fisheries. 1978–2007. *Annual Statistics of Fisheries and Aquaculture Production in 1976–2005*, Association of Agriculture and Forestry Statistics, Tokyo (in Japanese).

Munro, J.L., and Bell, J.D. 1997. Enhancement of marine fisheries resources. *Reviews in Fisheries Science* 5:185–222.

Nakamura, R. 1996. Fisheries and stocking of flatfishes. *Bulletin of Japanese Society of Fisheries Oceanography* 60:271–275 (in Japanese).

National Association for the Promotion of Productive Seas 2007, 2008. *Annual Reports on the Restoration Project of Coastal Stocks by Stock Enhancement*. National Association for the Promotion of Productive Seas, Tokyo (in Japanese).

Naylor, R.L., Goldberg, R.J., Primavera, J.H., Kautsky, N., Beveridge, M.C.M., Clay, J., Folke, C., Lubchenco, J., Mooney, H., and Max, T. 2000. Effects of aquaculture on world fish supplies. *Nature* 405:1017–1024.

Nishida, M., Ohkawa, T., and Fujii, T. 1997. Ecological aspects, population structure. In: Minami, T. and Tanaka, M. (eds) *Biology and Stock Enhancement of Japanese Flounder*. Koseisha-Koseikaku, Tokyo, pp. 41–51 (in Japanese).

Okouchi, H., Kitada, S., Tsuzaki, T., Fukunaga, T., and Iwamoto, A. 1999. Numbers of returns and economic return rates of hatchery-released flounder *Paralichthys olivaceus* in Miyako Bay – evaluation by fish market census. In: Howell, B.R., Moksness, E., and Svåsand, T. (eds) *Stock Enhancement and Sea Ranching*. Fishing News Books, Oxford, pp. 573–582.

Olla, B., Davis, M.W., and Ryer, C.H. 1994. Behavioural deficits in hatchery-reared fish: potential effects on survival following release. *Aquaculture and Fisheries Management* 25(Suppl. 1):19–34.

Rijnsdorp, A.D., van Beek, F.A., Flatman, S., Millner, R.M., Riley, J.D., Giret, M., and De Clerck, R. 1992. Recruitment in sole stocks, *Solea solea* (L.) in the northeast Atlantic. *Netherlands Journal of Sea Research* 29:173–192.

Saitoh, K., Takagaki, M., and Yamashita, Y. 2003. Detection of Japanese flounder-specific DNA from gut contents of potential predators in the field. *Fisheries Science* 69:473–477.

Sparrevohn, C.R., and Støttrup, J.G. 2007. Post-release survival and feeding in reared turbot. *Journal of Sea Research* 57:151–161.

Sparrevohn, C.R., and Støttrup, J.G. 2008. Diet, abundance and distribution as indices of turbot (*Psetta maxima* L.) release habitat suitability. *Reviews in Fisheries Science* 16(1–3):1–10.

Tanaka, M., Goto, T., Tomiyama, M., and Sudo, H. 1989. Immigration, settlement and mortality of flounder (*Paralichthys olivaceus*) larvae and juveniles in a nursery ground, Shijiki Bay, Japan. *Netherlands Journal of Sea Research* 24(1):57–67.

Tanaka, M., Seikai, T., Yamamoto, E., and Furuta, S. 1998. Significance of larval and juvenile ecophysiology for stock enhancement of the Japanese founder, *Paralichthys olivaceus*. *Bulletin of Marine Science* 62(2):551–571.

Tanaka, Y., Yamaguchi, H., Gwak, W.S., Tominaga, O., Tsusaki, T., and Tanaka, M. 2005. Influence of mass release of hatchery-reared Japanese flounder on the feeding and growth of wild juveniles in a nursery ground in the Japan Sea. *Journal of Experimental Marine Biology and Ecology* 314:137–147.

Tanaka, Y., Ohkawa, T., Yamashita, Y., and Tanaka, M. 2006. Geographical differences in stomach contents and feeding intensity of juvenile Japanese flounder *Paralichthys olivaceus*. Nippon Suisan Gakkaishi 72(1):50–57 (in Japanese with English summary).

Taylor, M.D., and Suthers, I.M. 2008. A predatory impact model and targeted stock enhancement approach for optimal release of mulloway (*Argyrosomus japonicus*). *Reviews in Fisheries Science* 16(1–3):125–134.

Tominaga, O., and Watanabe, Y. 1998. Geographical dispersal and optimum release size of hatchery-reared Japanese flounder *Paralichthys olivaceus* released in Ishikari Bay Hokkaido, Japan. *Journal of Sea Research* 40:73–81.

Tomiyama, T., Watanabe, M., and Fujita, T. 2008. Community-based stock enhancement and fisheries management of the Japanese flounder in Fukushima, Japan. *Reviews in Fisheries Science* 16(1–3):146–153.

Uehara, S., Kurita, Y., Tomiyama, T., Yoneda, M., Oshima, M., and Yamashita, Y. 2008. Pre- and post-settlement processes in determining year-class strength of Japanese flounder off the Pacific coast of northern Japan. In: *Book of Abstract, The 7th International Flatfish Symposium*, University of Lisbon, Lisbon.

Van der Veer, H.W., Berghahn, R., Miller, J.M., and Rijnsdorp, A.D. 2000. Recruitment in flatfishes, with special emphasis on North Atlantic species: Progress made by the Flatfish Symposia. *ICES Journal of Marine Science* 57:202–215.

Watson, R., and Pauly, D. 2001. Systematic distortions in world fisheries catch trends. *Nature* 414:534–536.

Yamada, H., Sato, K., Nagahora, S., Kumagai, A., and Yamashita, Y. 1998. Feeding habits of the Japanese flounder *Paralichthys olivaceus* in Pacific coastal waters of Tohoku district, northeastern Japan. *Nippon Suisan Gakkaishi* 64:247–256 (in Japanese with English summary).

Yamamoto, M., Makino, H., Kagawa, T., and Tominaga, O. 2004a. Occurrence and distribution of larval and juvenile Japanese flounder *Paralichthys olivaceus* at sandy beaches in eastern Hiuchi-Nada, central Seto Inland Sea, Japan. *Fisheries Science* 70:1089–1097.

Yamamoto, M., Makino, H., Kobayashi, J., and Tominaga, O. 2004b. Food organisms and feeding habits of larval and juvenile Japanese flounder *Paralichthys olivaceus* at Ohama Beach Hiuchi-Nada, the central Seto Inland Sea, Japan. *Fisheries Science* 70:1098–1105.

Yamamoto, M., and Tominaga, O. 2007. Daily rations and food availability of Japanese flounder *Paralichthys olivaceus*, small flounder *Tarphops oligolepis* and sandy goby *Favonigobius gymnauchen* at a sandy beach in the central Seto Inland Sea, Japan. *Fisheries Science* 73:314–323.

Yamashita, Y., Nagahora, S., Yamada, H., and Kitagawa, D. 1994. Effects of release size on survival and growth of Japanese flounder *Paralichthys olivaceus* in coastal waters off Iwate Prefecture, northeastern Japan. *Marine Ecology Progress Series* 105:269–276.

Yamashita, Y., and Yamada, H. 1999. Release strategy for Japanese flounder fry in stock enhancement programs. In: Howell, B.R., Moksness, E., and Svåsand, T. (eds) *Stock Enhancement and Sea Ranching*. Fishing News Books, Oxford, pp. 191–204.

Yamashita, Y., Tanaka, M., and Miller, J.M. 2001. Ecophysiology of juvenile flatfish in nursery grounds. *Journal of Sea Research* 45:205–218.

Section 7

Flatfish Worldwide

Chapter 15

Disease diagnosis and treatment

Edward J. Noga, Stephen A. Smith, and Oddvar H. Ottesen

Disease is almost always one of the most serious economic and management problems that are faced by fish farmers and it is very important that the farmer be aware of the severe damage that can be caused by disease, as well as implementing ways of controlling it. As flatfish culture has expanded, so has the number of reported diseases and the recognition of disease as a major impediment to successful propagation. In flatfish species that are routinely cultured, such as Japanese flounder, turbot, and Atlantic halibut, many diseases have been described, and managing them as much as possible is essential for a successful operation. While the number of diseases that are known to affect many other flatfish are few, this merely reflects the fact that those fish have only been cultured on a small scale. This chapter will cover some of the most basic aspects of disease management, as well as summarizing the diseases that have been reported in flatfish.

15.1 General signs of disease

A successful farmer must know when his fish are "doing well" or "having problems." This mainly comes with experience because each fish species varies in its normal behavior and appearance. It is important to gain a good grasp of what is "normal" because only by knowing this can one recognize when a problem arises. This is especially important in aquaculture because catching problems early, before they become a catastrophe, is a key to a successful operation.

Some of the most common signs of a problem in a cultured flatfish population include:

- Decreased feed consumption—Fish often may stop eating altogether when they are sick. However, even fish that have a low-grade infection (e.g., a mild parasite infestation) may exhibit decreased feeding rate or a lower feed conversion ratio (FCR). This may not be apparent without a close examination of carefully maintained records.

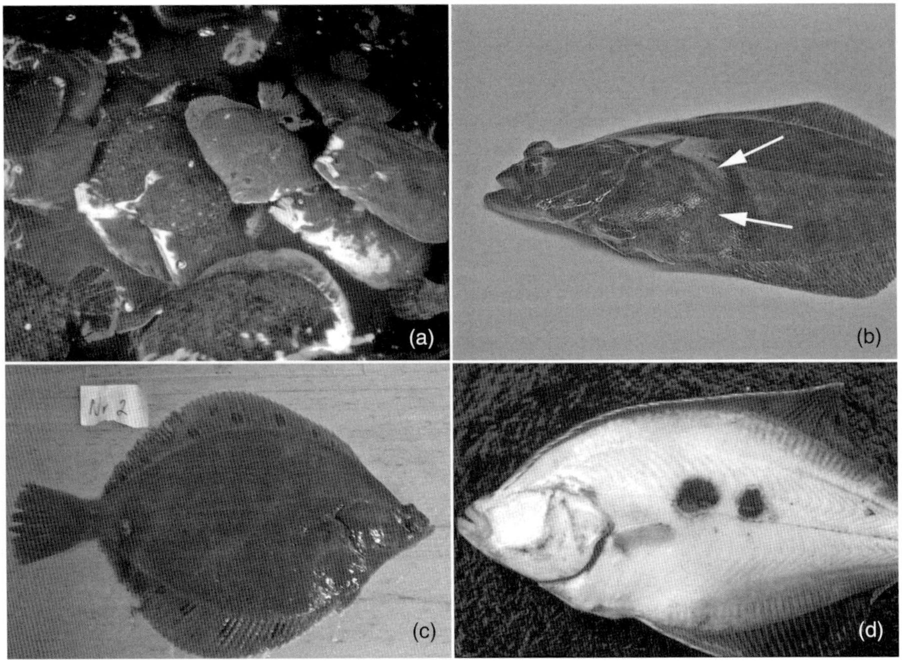

Figure 15.1 Common gross lesions in flatfish. (a) Depigmentation (white areas on up side); (b) abdominal swelling (arrows); (c) skin ulcers with hemorrhage; and (d) erosion and hemorrhage on the fins. (Photograph courtesy of H. Moller)

- Poor growth—This problem goes hand in hand with decreased feed consumption or low FCR.
- Unexplained losses—All commercial aquaculture operations will experience some losses during production. The amount of losses will vary with the farm, as well as the life stage (e.g., often highest during the hatchery stage). Any losses above this "norm" should be closely investigated for cause.
- Abnormal behavior and/or appearance—These are referred to as "clinical signs." Changes in level of activity (more or less active), response to presence of feed, locomotion, among others, can give clues to certain diseases. One should also be aware of the normal color pattern and the appearance of the flatfish species being cultured, in order to discern changes that may occur with disease. Some common changes include presence of abnormal pigmentation patterns, or red or white areas on the body. Other changes include fin erosion, ulcers, swollen eyes, or a swollen abdomen (Figure 15.1).

15.2 Viral diseases

Viral diseases are one of the most serious disease threats to aquacultured fish and a steadily increasing number of viral infections have been reported in flatfish (Table 15.1).

Table 15.1 Viral pathogens reported in flatfish species.

Virus	Flatfish species													
	Turbot	Greenland halibut	Atlantic halibut	Plaice	Greenback flounder	Japanese flounder	Southern flounder	Summer flounder	Barfin flounder	Winter flounder	European flounder	Dab	English sole	Dover sole
Nodavirus	D(C)		D(C)			D(C)			D(C)	A(W)				
VHS	D(C)	A(W)	D(C)[a]	A(W)		A(W)					A(W)	A(W)	A(W)	
Hirame rhabdovirus						D(C)								
Reo-like virus			D(C)					D(C)						
Lymphocystis				D(W)		D(C)					D(W)	D(W)		
Herpesvirus scophthalmi	D(C)													
Japanese flounder herpesvirus						D(C)								
Aquabirnavirus	D(C)	A(W)	D(C)	A(W)	A(W)	A(W) D(C)	A(W)			A(W)	A(W)	A(W) D(C)		A(W)
Turbot Iridovirus	D(C)					A(C)[a]								
Red Sea Bream Iridovirus						D(C)								

D, causes disease; A, asymptomatic; W, wild fish; C, cultured fish.
[a]Experimental infection.

Viruses are microscopic, intracellular pathogens that replicate (reproduce) by penetrating a host cell and then commanding the cell to produce its progeny. Viruses tend to be host-specific (usually affecting only one species or a closely related group of species), although some have a very broad host range. The severity of a particular viral infection is typically temperature-dependent (i.e., the virus causes the most severe disease within a certain temperature range). Viral infections usually cause the most severe disease in young fish. While older fish may not develop disease, they often can become carriers (i.e., harbor the virus for long periods and transmit it to others).

Diagnosis of a viral disease is based on the history (i.e., the circumstances surrounding the outbreak) and observation of clinical signs that match the suspected disease. This must also include identification of the virus. Viral identification requires specialized techniques, and specimens should be submitted to a competent laboratory. Identification techniques that might be used include specific antibodies or use of a gene probe on tissues or after virus isolation (in cell culture) from infected tissues. Some viruses may be present in low numbers without causing disease, requiring quantification to determine whether they are actually causing the clinical signs and mortalities. Some viruses may be shed from asymptomatic carriers, especially during spawning time. In such cases, the shedders usually do not appear sick but they can transmit the virus to their offspring.

There are no drugs available to treat viral infections in any fish, nor are there any commercially available vaccines for viral diseases of any flatfish species. Thus, control relies on environmental management and biosecurity (see Health Management in Flatfish Aquaculture). Especially important in other cultured fish is the use of eggs and broodstock known to be free of certain viruses. However, this is not yet feasible in flatfish aquaculture because there is a lack of domesticated broodstock, requiring that wild-caught fish with an unknown history of virus exposure be used for propagation.

15.2.1 *Nodaviral diseases* (viral nervous necrosis, VNN, viral encephalopathy and retinopathy, VER, nervous necrosis virus, NNV, fish encephalitis)

Nodaviruses that affect fish are in the genus *Betanodavirus*, family Nodaviridae (Munday *et al.* 2002). They cause acute to chronic disease in at least 30 species of marine fish and are present worldwide, except Africa. Various viruses have caused epidemics in fish in Japan, Europe (Norway, Mediterranean Sea, and probably the Irish Sea and the Isle of Man), the Caribbean Sea (Martinique), and much of the southern tropical Pacific Ocean.

Clinical disease is usually seen in larvae and less commonly in juveniles. The earliest onset of disease also varies among viruses, but may occur as soon as one-day posthatch. Lesions are usually more severe and mortalities highest in younger fish, but some nodavirus diseases can affect even market-size fish. Older fish may have a predilection to develop retinal damage.

Nodaviruses cause disease in several flatfish species (Table 15.1). Nodaviral infection in larval and juvenile halibut is a major obstacle to culture in Norway and may be the most serious impediment to halibut culture (Bergh *et al.* 2001). Halibut appear to only develop clinical signs in early juvenile stages, when their immune system is not yet developed (Grotmol *et al.* 1997). Clinical signs occur mostly at the feeding and weaning stage (40–100 days posthatch), but yolk sac fry may also be affected. However, older fish can become carriers. There are also some reports of large halibut (several kilograms) showing clinical signs of VER (Korsnes *et al.* 2005). There is likely both vertical (broodstock to offspring) and horizontal (cohort to cohort) transmission. High concentrations of virus have been detected in seawater of rearing units and this virus appears to originate from the infected fish in the hatchery (Nerland *et al.* 2007). Recovering halibut can carry the virus as a subclinical infection for over one year. Nodavirus is very stable and can survive in seawater for a long time. Nodavirus also causes disease in barfin flounder, turbot, and Japanese flounder (Tanaka *et al.* 2003). Nodavirus has also been isolated from clinically normal wild winter flounder (Gagne *et al.* 2004).

When high mortalities occur in susceptible larval or juvenile fish in hatcheries without the presence of detectable pathogens (e.g., parasites, bacteria) in the clinical workup, nodavirus infection should be ruled out. For virological examination, whole larvae or small juveniles should be sampled. From larger fish, the brain, spinal cord, and eyes should be sampled. Disinfection and quarantine is the only proven means of controlling most nodaviral epidemics. VNN in striped jack was successfully controlled by ozonation of fertilized eggs combined with detection and elimination of virus-carrying broodstock (Mushiake *et al.* 1994). Ozonation is also experimentally successful in treating Atlantic halibut eggs (Bergh *et al.* 2001). Selection of nodavirus-free spawners using an immunoassay to detect the virus has successfully reduced the incidence of VER in juvenile barfin flounder (Watanabe *et al.* 2000) and other fish. Atlantic salmon are experimentally susceptible to nodavirus from Atlantic halibut, so these two fish should not be cultured in close proximity. Temperature might be more important than host specificity for the distribution of the various nodavirus subtypes (Korsnes *et al.* 2005), suggesting that virus transfer between unrelated fish species may be common.

15.2.2 Viral hemorrhagic septicemia

Viral hemorrhagic septicemia (VHS), caused by a rhabdovirus of the genus *Novirhabdovirus*, is a major cause of mortality in freshwater salmonids in freshwater, and has been increasingly found in nonsalmonid fish. Several European marine isolates of viral hemorrhagic septicemia virus (VHSV) have caused significant losses to turbot fry in aquaculture (King *et al.* 2001). There have been three VHS outbreaks on turbot farms since 1989: one each in Germany, Scotland, and Ireland. Clinical signs and pathology of VHS in flatfish are similar to those in salmonids. Turbot exhibit the classical clinical signs of VHS seen in salmonids (i.e., swollen abdomen with fluid and exophthalmos ["popeye"], as well as hemorrhages [reddening] in the skin, eyes, muscles, and on the surfaces of internal organs). Cultured Japanese flounder in Japan and Korea also suffer high mortalities. Even market-size fish (1 kg) have had mass mortalities. The disease is very similar to hirame rhabdovirus disease (see below) but Japanese flounder with VHS have more prominent fluid accumulation in the peritoneal and pericardial cavities (Isshiki *et al.* 2001). Atlantic halibut are experimentally susceptible (Skall *et al.* 2005).

VHSV has also been isolated from asymptomatic, wild caught English sole, Greenland halibut, Japanese flounder, plaice, dab, and European flounder (Skall *et al.* 2005; Anonymous 2007). There is circumstantial evidence that nonvirulent VHSV isolates can become pathogenic. Data also suggest that VHSV might have originated in the marine environment and may constitute a potential risk for mariculture. Many isolates cultured from wild fish in European waters are pathogenic to turbot, suggesting that exposure to wild fish (and possibly raw seawater) might be a significant risk factor for disease outbreaks in turbot. It might be possible for flatfish to become infected by eating VHSV-infected fish that die and fall to the bottom.

In salmonids, where VHS is a very serious problem, it is probably spread mainly via transport of infected farmed fish. But, VHSV infection from the marine environment is a constant threat to control programs for VHS in cultured salmonids and thus cocultivation of rainbow trout and flatfish in mariculture should be avoided. The introduction of farmed fish from seawater into freshwater (except for nonsusceptible species) should also be avoided (Skall *et al.* 2005).

15.2.3 Miscellaneous viral diseases

Hirame rhabdovirus disease caused mass mortalities in Japanese flounder in Japan during the 1980s but has been much less prevalent in recent years (Isshiki *et al.* 2001). Clinical signs are very similar to VHS, but fish usually display more hemorrhage in the lateral muscles and viscera.

Aquabirnavirus infection, caused by marine aquabirnavirus (MABV), has been identified in Japanese flounder and spotted halibut in Japan (Isshiki *et al.* 2004). Co-occuring bacterial infection worsens the outcome of the disease (Pakingking *et al.* 2003). Aquabirnavirus is also associated with mortalities in juvenile Atlantic halibut (Bergh *et al.* 2001) and has been isolated from several clinically healthy wild flatfish (Table 15.1). Feeding birnavirus-infected bivalves to flatfish increases the infection levels of the flatfish (Skall *et al.* 2000).

Reo-like virus causes liver damage accompanied by high mortality in juvenile, farmed Atlantic halibut (Ferguson *et al.* 2003) and summer flounder (Wada *et al.* 2009). Turbot iridovirus (a megalocytivirus) has caused mass mortality of cultured turbot in Korea (Oh *et al.* 2006a). It can propagate in Japanese flounder but does not cause disease in that species. However, another megalocytivirus, red sea bream iridovirus, can cause significant mortality in Japanese flounder (Do *et al.* 2005).

Lymphocystis is a very common viral infection in freshwater and marine fish, causing tumor-like growths on the skin, fins, and mouth. It is mainly a problem because the disfigurement of the fish reduces carcass value. However, in some cases, proliferations around the mouth interfere with feeding. It has caused major losses in Japanese flounder (Iwamoto *et al.* 2002).

Herpesvirus scophthalmi infection, identified in cultured turbot in Europe, causes a characteristic posture, with the head and tail of the fish raised while lying on the bottom (Liewes 1984). An unclassified herpesvirus, responsible for a disease called viral epidermal necrosis (viral epidermal hyperplasia), has caused mass mortalities in Japanese flounder larvae (Miyazaki 2005) but has recently been a much less serious problem.

15.3 Bacterial diseases

A number of bacteria are important pathogens in both wild and cultured flatfish and are responsible for serious economic losses (Table 15.2). Infections are often precipitated by some stress that upsets the natural defenses against these agents (e.g., overcrowding, low dissolved oxygen, high ammonia, transport, high temperature). Some may cause primarily external (skin/gill) infection; most can cause internal (systemic) disease. Pathogens that may present as only skin infections include flavobacteria, aeromonads, and vibrios. Fish may present with fin "rot," an imprecise general term for ulcerative, necrotic lesions that affect the fins. Various bacteria are often present in fin rot lesions, but some stress is considered to be the primary cause. The fin rot syndrome includes several diseases and idiopathic responses (see Noninfectious Diseases).

Bacterial skin infections can advance to become internal (systemic), leading to much greater and more acute mortality. In other cases, fish may develop systemic bacterial infections with no skin involvement, or may later show skin damage as a consequence of systemic infection. The classical signs associated with systemic (internal) bacterial infection are reflective of the various toxins produced by the bacteria and include hemorrhage of internal organs, especially those involved in filtering blood (spleen, kidney). Kidney and/or spleen are often enlarged. External signs may include skin ulcers, fin necrosis, or hemorrhages on the body and fins. Fish may have eye damage and/or fluid accumulation in the abdomen. Not all bacteria that cause systemic disease produce the above clinical signs, but these signs are common. Fish-pathogenic bacteria may reside in the environment, sometimes indefinitely, or on/in apparently normal fish (latent carriers). Also, culture of a pathogen does not prove it is the cause. For example, *Vibrio ichthyoenteri* and *Photobacterium damselae* subsp. *damselae* (also in the family Vibrionaceae), known fish pathogens, have been isolated from summer flounder experiencing chronic mortalities, but their role in causing disease in that species is unclear (Gauger *et al.* 2006).

Most bacterial infections are susceptible to some type of antibiotic but the particular antibiotic that is effective varies greatly among bacterial species and even strains of bacteria. Thus, the bacterium must usually be cultured from the fish and its antibiotic sensitivity determined by a competent laboratory before a proper treatment can be started. In addition, there are very strict regulations on the use of antibiotics, and these vary greatly among countries. There are a few commercially available vaccines for bacterial diseases of flatfish that have been primarily developed for other commercial species (e.g., salmonids). Control must also include appropriate environmental management and biosecurity (see Health Management in Flatfish Aquaculture), but a number of bacteria may reside naturally in the environment, making their exclusion a challenge.

Table 15.2 Bacterial pathogens reported in flatfish species.

	Flatfish species										
Bacterium	**Turbot**	**Atlantic halibut**	**Plaice**	**Senegalese sole**	**Japanese flounder**	**European flounder**	**Dab**	**Summer flounder**	**English sole**	**Dover sole**	**Shotted halibut**
Vibrio sp.[a]	D(C)								D(C)	D(C)	
Vibrio ichthyoenteri					D(C)			D(C)[b]			
Vibrio harveyi				D(C)	D(C)			D(C)			
V. qinhuangdoara											
Vibrio parahaemolyticus					D(C)						
Photobacterium damsela subsp. *piscicida*	D(C)				D(C)			D(C)[b]			
Moritella viscosa	D(C)	D(C)	D(C)								
Streptococcus iniae					D(C)						
Streptococcus parauberis	D(C)				D(C)						
Edwardsiella tarda					D(C)						
Mycobacterium								D(C)			
Aeromonas salmonicida	D(C)	D(W)	D(W)		D(C)	D(W)	D(W)				D(C)
Tenacibaculum maritimum	D(C)				D(C)					D(C)	D(C)
Tenacibaculum ovolyticum		D(C)									

D, causes disease; A, asymptomatic; W, wild fish; C, cultured fish.
[a]Most if not all flatfish are probably susceptible to at least one vibrio.
[b]Role in disease uncertain.

15.3.1 External infections

So-called "atypical" strains of *Aeromonas salmonicida* have been isolated from skin ulcers of several wild flatfish (Table 15.2), including Atlantic halibut, which are also experimentally susceptible (Gudmundsdóttir *et al.* 2003). Vibriosis (see below) can also commonly occur as skin ulcers (red spot). *Tenacibaculum maritimum* (formerly *Flexibacter maritimus*) has caused skin and gill erosions in Dover sole (black patch necrosis) and Japanese flounder, especially juveniles (Austin and Austin 2007). *Tenacibaculum ovolyticum* (formerly *Flexibacter ovolyticus*) has caused mortality of Atlantic halibut eggs and larvae, with puncturing of the egg leading to death (Hansen *et al.* 1992).

15.3.2 Internal infections

Vibrios are the most common bacterial infection in marine fish and are usually more pathogenic at higher temperatures, so are mostly a problem in summer. Vibriosis has been reported in turbot, English sole, and Dover sole (Liewes 1984), but most if not all flatfish are probably susceptible. Diseased fish may be very dark with a distended abdomen and are often severely anemic. The most common vibrio associated with fish disease is *Vibrio anguillarum* (also named *Listonella anguillarum*). *Vibrio harveyi*, which now includes *V. carchariae*, has caused mass mortalities in Japanese flounder (Oh *et al.* 2006b) and enteritis and stunting in cultured summer flounder in a disease called flounder infectious necrotizing enteritis (Soffientino *et al.* 1999). *Vibrio ichthyoenteri* causes larval enteritis, a serious disease in larval Japanese flounder (Muroga 2001). Other vibrios reported to cause disease in cultured flatfish include *V. parahaemolyticus* (Li *et al.* 2005) and *V. qinhuangdoara*. Vibrio infections have also been observed in wild winter flounder; Soffientino *et al.* 1999). Different geographic strains of Atlantic halibut vary in their susceptibility to *Vibrio anguillarum* (Hoare *et al.* 2002). Immersion vaccine provides good protection against *V. anguillarum* in Atlantic halibut (Bowden *et al.* 2002).

Edwardsiella tarda, the cause of edwardsiellosis, is one of the most serious threats to Japanese flounder culture (Zheng *et al.* 2005). It is present on flounder farms even when the disease is not occurring, but dies quickly in seawater when not infecting fish, suggesting that terrestrial runoff might be an important source of infection (Mamnur *et al.* 1994). Streptococcosis, which includes diseases caused by *Streptococcus* and *Lactococcus*, are important problems in cultured Japanese flounder. *Streptococcus iniae*, as well as *S. parauberis* and *Lactococcus garviae*, have caused in disease in Japan and Korea when the temperature is high (summer). *S. iniae* is resident in the sediment and water of farms with infected fish (Nguyen *et al.* 2002). Edwardsiellosis occurs at the same time of the year and simultaneous infections can occur. *Streptococcus parauberis* has also caused disease in juvenile and adult turbot and Japanese flounder (Kim *et al.* 2006). *Mycobacterium*, a slow-growing bacterium causing chronic infections in many marine fish, has been observed in cultured summer flounder in an intensive recirculating aquaculture system (Hughes and Smith 2002). This pathogen is of considerable potential concern due to the lack of any effective methods for its treatment or prevention.

15.4 Parasitic and other eukaryotic diseases

Parasitic infestations and infections are one of the most common diseases of cultured flatfish (Table 15.3). Almost all of the external and internal parasites that have been reported from cultured flatfish can have detrimental health consequences when present in high numbers within a population (Schram and Haug 1988); however, the majority of the parasites causing disease in culture can be easily controlled with good management practices, regular monitoring, and appropriate therapeutic intervention (Svendsen and Haug 1991). However, parasites that can cause internal infections (i.e., not on the surface of the skin or gills) can be very difficult to treat.

15.4.1 Protozoan parasites

The most serious protozoan parasites of cultured flatfish are the opportunistic histophagous ciliates that cause scuticociliatosis. Ciliates in the genera *Uronema*, *Philasterides*, *Miamiensis*, and *Pseudocohnilembus* have been recorded from cultured Japanese flounder, turbot, and plaice (Ototake and Matsusato 1986; Inglesias *et al.* 2001; Kim *et al.* 2004). Scuticociliates have a direct life cycle (i.e., no intermediate host, the fish is the only host) and also do not need a fish host to survive. They initially infest the surface of the skin and gills but often proceed systemically to the internal organs, leading to high mortality. Because they can infect internal organs, treatment is very difficult.

A ciliate that typically causes significant skin and gill damage is the marine parasite *Cryptocaryon irritans*, which has been reported in cultured Japanese flounder (Kaige and Miyazaki 1985; Jee *et al.* 2000) and turbot (Devesa *et al.* 1989). It has the same life cycle and pathology as the freshwater ciliate *Ichthyophthirius multifiliis* (ich), an important pathogen of many freshwater fish. *Cryptocaryon* burrows into the epithelium, where it is protected from the external environment, making this parasite difficult to treat with water-borne compounds.

Several species of the ciliate *Trichodina*, including *T. hippoglossi*, *T. jadranica*, *T. borealis*, and *T. murmanica*, have been reported from numerous species of cultured winter flounder, yellowtail flounder, Atlantic halibut, and dab (MacKenzie *et al.* 1976; Nilsen 1995; Barker *et al.* 2002; Arthur *et al.* 2004). Trichodinids remain on the surface of the skin and gills, and thus are much less pathogenic than scuticociliates or *Cryptocaryon*, but in high numbers, may cause severe fin and tail erosion, localized skin ulcers, as well as thickening of the gill tissue. Hughes and Smith (2003) found that trichodinids were more common on the skin than the gills of cultured summer flounder and were generally found in higher numbers on the eyed-side of the fish. Trichodinids have a direct life cycle and can multiply rapidly in aquaculture situations where fish are crowded. The

Table 15.3 Parasites reported in flatfish species.

Parasite(s)	Flatfish species											
	Turbot	Atlantic halibut	Plaice	Yellowtail flounder	Japanese flounder	Fine flounder	Winter flounder	European flounder	Dab	Summer flounder	English sole	Dover sole
Protozoa and Protozoan-like												
Scuticociliatiasis[a]	D(C)				D(C)							
Cryptocaryon irritans					D(C)							
Trichodina sp.	D(C)	A(W)	A(W)	A(W)	D(C)		D(C)	A(W)				
Scyphidia sp.			A(W)									
Amyloodinium ocellatum												
Ichthyobodo sp.					D(C)				A(W)			
Hemoflagellates[b]						A(W)				D(W)		D(C)
Microsporidians[c]		A(C)	D(W)				A(W)	A(W)	D(W)		D(W)	
Myxozoans[d]	D(C)	A(W)	A(W)			D(W)		A(W)	D(W)			
Ichthyophonus sp.			D(W)	D(W)								
Monogeneans[e]		D(C)			D(C)	A(W)		A(W)			A(W)	
Digeneans		A(W)	D(W)			A(W)		A(W)				
Cestodes		A(W)	A(W)			A(W)		A(W)				
Nematodes		A(W)	D(W)			A(W)		A(W)				
Acanthocephalans		A(W)	A(W)			A(W)		A(W)				
Copepods		D(W)				A(W)		A(W)				D(C)
Leeches						A(W)		A(W)				D(C)

A, asymptomatic or no pathology reported; D, causes disease; W, wild fish; C, cultured fish.
[a] Includes numerous genera such as *Uronema*, *Philasterides*, *Miamiensis*, and *Pseudocohnilembus*.
[b] Includes genera such as *Cryptobia and Trypanosoma*.
[c] Includes genera such as *Glugea*, *Microsporea*.
[d] Includes numerous genera such as *Sphaerospora*, *Myxidium*. *Rhabdospora*, and *Enteromyxum*.
[e] Includes numerous genera such as *Gyrodactylus*, *Neoheterobothrium*, and *Entobdella*.

ciliate *Scyphidia adunconucleata* has been reported from skin and gill of plaice (MacKenzie *et al.* 1976).

Amyloodinium ocellatum is a dinoflagellate that also parasitizes the skin and gill epithelium of many marine fish. This parasite has a direct life cycle that includes a feeding (trophont) stage on the host and an infectious free-swimming (dinospore) stage in the water column; the dinospore is the only stage susceptible to drugs. The trophont attaches to the fish by rhizoids, causing major cell damage. *Amyloodinium* has caused significant mortalities in captive summer flounder broodstock (Schwarz and Smith 1998; Hughes and Smith 2003). Another ectoparasitic flagellate of the skin and the gill is *Ichthyobodo*, a pathogen of cultured Japanese flounder and wild dab (Diamant 1987; Urawa *et al.* 1991; Kusakari and Urawa 1990).

Cryptobia (*Trypanoplasma*) *bullocki* is a flagellate blood parasite that causes anemia in captive broodstock summer flounder (Newman 1978; Burreson and Zwerner 1984). It is transmitted via a marine leech (*Calliobdella vivida*). *Cryptobia bullocki* may ultimately infect the gastrointestinal tract where it can cause rectal prolapse (Burreson and Zwerner 1984). Large numbers of an unidentified *Cryptobia* species have also been reported in the intestinal tract and liver of morbid cultured summer flounder (Newman 1978), while *Cryptobia neghmei* has been reported in the blood of fine flounder and lenguado de ojo chico (Khan *et al.* 2001).

15.4.2 Protozoan-like parasites

While microsporidians have typically been considered to be protozoa, recent studies suggest that they are fungi. The microsporidian *Glugea stephani* encysts in the intestinal wall and rectum of wild plaice (Burn 1980) and cultured plaice (MacKenzie *et al.* 1976). Tumor-like masses, described with "x-cells," occur on the body surface or in the gill cavity of wild-caught dab, flathead flounder, and English sole (McVicar *et al.* 1987; Milwa *et al.* 2004). Historically thought to be environmentally- or virally-induced, they have recently been attributed to an undescribed species of protist (Khattra *et al.* 2000; Miwa *et al.* 2004). Their significance for cultured flatfish is unknown (McVicar *et al.* 1987; Khattra *et al.* 2000; Miwa *et al.* 2004).

Myxozoans include many species, and as a group infect a wide range of organs and thus have a diverse array of clinical signs. While the Myxozoa were initially considered to be protozoa, recent evidence shows them to be metazoans (Zrzavy and Hypsa 2003). They are common in wild fish but few have been reported as problems in cultured flatfish. *Sphaerospora irregularis* commonly infects the urinary bladder of plaice (MacKenzie *et al.* 1976) but is apparently not pathogenic. However, *Myxidium incurvatum* and *Rhabdospora thelohani* in the liver and kidney (Anderson *et al.* 1976), and *Enteromyxum scophthalmi* in the gut (Redondo *et al.* 2003) cause disease in cultured turbot.

Ichthyophonus hoferi, recently reclassified as a lower protist, infects captive yellowtail flounder (Rand 1994), inducing the formation of white nodules in the muscle and internal organs including the intestine, liver, and heart (McVicar 1999). While there is no evidence that this organism causes significant mortality in cultured flatfish, the nodules reduce carcass quality in the fillet of wild-caught fish.

15.4.3 Metazoan parasites

The most serious metazoan parasites of cultured flatfish are those that infest the skin and gills, because almost all these parasites have a direct life cycle and can rapidly increase in numbers. Several monogenean worms have caused problems in cultured flatfish, including *Gyrodactylus unicopula* in plaice, *G. pleuronecti* in winter flounder, *Neoheterobothrium hirame* in Japanese flounder, and *Entobdella soleae* in Dover sole (Mackenzie *et al.* 1976; Kirmse 1987; Hayward *et al.* 2001; Barker *et al.* 2002). Monogeneans are irritating and can cause hemorrhage at the site of attachment, leading to secondary microbial infection.

Numerous digenean flukes (digenetic trematodes), tapeworms (cestodes), roundworms (nematodes), and thorny-headed worms (acanthocephalans) have been reported as internal infections of wild flatfish, but the great majority are not a problem in cultured fish because they have an indirect life cycle (requiring an invertebrate intermediate host) and thus usually cannot complete their life cycle on a farm (MacKenzie *et al.* 1976; Burn 1980; Koie 2000). However, there are a few species of digenetic trematodes (i.e., encysted metacercaria of *Cryptocotyle lingua* and *Stephanostomum baccatum*) and adult nematodes (i.e., *Contracaecum aduncun*) that can become established in culture facilities and cause significant pathology and/or mortality (MacKenzie *et al.* 1976).

Various copepods, including gill maggots (ergasilids), sea lice (caligids), and anchor worms (lernaeids) are common skin and gill infestations on wild flatfish. They have not been a serious problem in flatfish culture but sea lice have been a very serious problem in other cultured marine fish and might pose a threat to cultured flatfish in the future (Bergh *et al.* 2001). Parasitic copepods may be transmitted horizontally between fish or via water containing the infectious larval stages (Anstensrud 1992). The sea louse, *Lepeophtheirus hippoglossi*, is common on wild caught halibut used for broodstock. Sea lice have also been observed in halibut cultured in sea cages. Anemia in the Dover sole was partly attributed to the blood-feeding activity of the copepod, *Lernaeocera* sp. (Kirmse 1987). The fish louse (*Argulus*), which is closely related to copepods, has caused severe anemia and localized hemorrhagic skin lesions in captive summer flounder (Hughes and Smith 2003). Their feeding activity may also result in secondary infections, and some species of fish lice can transmit blood-borne parasites, bacteria, and viruses.

Leeches can occur in flounder populations obtained from the wild or exposed to a natural, untreated seawater source. Leeches, such as *Calliobdella vivida* and *Hemibdella* sp., may also transmit blood-borne parasites, bacteria, and viruses (Burreson and Zwerner 1984; Liewes 1984).

15.4.4 Fungus-like infections

The most important pathogens in this group are the water molds, fungus-like organisms that commonly infect many freshwater fish. *Saprolegnia* is the most common genus but a number of other genera are also pathogenic. Water molds often colonize pre-existing lesions on the skin, and therefore are generally considered secondary invaders of primary bacterial or parasite infections. They also often develop due to stress (poor husbandry, trauma, etc.) (Bruno 1995). They are spread by a motile zoospore stage and manifest as a cotton-like white to gray growth on the external surface of the fish. *Aphanomyces invadans* (= *A. piscicida*) rarely affects flatfish at low salinities in western Atlantic estuaries. It typically produces very deep, penetrating ulcers. Water molds do not infect fish in full-strength seawater (Noga 1993; Strongman *et al.* 1997; Kurata *et al.* 2008).

15.5 Noninfectious diseases

15.5.1 Introduction

A number of noninfectious diseases can adversely affect flatfish culture and may have a significant impact on development, growth, and survival (Table 15.4). Nutrition often plays a role in such problems, either directly due to a nutrient deficiency, or indirectly, such as due to inadequate feeding leading to aggression. Nutrient requirements vary considerably among flatfish species, as well as among age classes. A very common clinical sign in many noninfectious diseases affecting flatfish is abnormal pigmentation. This may present as either: a) albinism (or pseudoalbinism), where the fish is totally unpigmented or, has only small pigmented areas on the ocular (up) side; b) ambicoloration, where the fish has a partly or fully pigmented abocular (blind) side; c) hypopigmentation, where the pigmentation on the ocular side is reduced in intensity; or d) spots, where pigmentation on the abocular side is less prominent than with ambicoloration. These color changes do not cause distress to the fish, but may be an indication of insufficient nutrition or suboptimal rearing conditions. Moreover, it significantly reduces carcass value.

Another common clinical sign often having a noninfectious cause is fin erosion, a loss of fin tissue resulting in a ragged or torn appearance to the fins. It has been linked with a number of adverse conditions, including overcrowding, inadequate nutrition, aggression, and other stressors. However, it may also be a clinical sign of certain infections, especially bacteria or parasites. Fin erosion can affect the general welfare of the fish and reduce carcass value; it may also reduce life expectancy of fish used for restocking wild populations due to compromise of swimming ability. Latremouille (2003) suggested the following measures for managing fin erosion in hatcheries: feeding fish to satiation, maintaining adequate water flow, using a rough substrate on the tank bottom, and culturing two species in one system (to reduce intraspecific aggression).

Many noninfectious diseases are linked to metamorphosis. Abnormal larval development may affect general welfare, and ultimately lead to a lower production yield in hatcheries. Skeletal anomalies occur frequently in hatchery-produced flatfish larvae. Severe jaw deformities are normally easy to observe, and have been associated with mortality in brown sole. However, some malformations are difficult to detect by external examination and mild malformations

Table 15.4 Noninfectious diseases in flatfish species.

	Flatfish species																				
	Brown sole	Turbot	Patagonian flounder	Greenback flounder	Starry flounder	Yellow-tail flounder	Marbled sole	Black plaice	Atlantic halibut	Plaice	Senegalese sole	Japanese flounder	Southern flounder	Winter flounder	European flounder	Dab	Summer flounder	English sole	Dover sole	Shotted halibut	"Brazilian" flounder
Hypervitaminosis A									D(C)[a]			D(C)[a]					A(C)[a]				
Hypervitaminosis D												D(C)[a]									
Hypovitaminosis C		D(C)[a]										D(C)[a]									
Unbalanced fatty acids									D(C)[a]			D(C)					D(C)[a]				
Phosphorus deficiency												D(C)[a]									
Cannibalism and/or aggression									D(C)			D(C)[a]					D(C)[a]				
Eye migration abnormalities	D(C)[a]	D(C)[a]				D(C)[a]			D(C)			D(C)	D(C)[a]				DC[a]			D(C)[a]	
Ocular abnormalties		D (C)[a]							D(C)			D(C)[a]									
Skeletal (jaw, etc.) deformities	D(C)[a]	D(C)[a]	A(W)						D(C)		D(C)[a]	D(C)			A(W)		D(C)[a]			D(C)[a]	
Pigmentation abnormalities	A(C)[a]	A(C)[a]	A(W)		A(W)	A(C)[a]			A(C)	A(W)	A(C)[a]	A(C)	A(C)[a]				A(C)[a]			A(C)[a]	A(C)[a]
Liver abnormalities		A(W)		A(C)[a]	A(W)	A(W)		A(W)				A(C)[a]		A(W)	A(W)	A(W)		A(W)	A(W)		
Heart abnormalities									A(C)												
Gonadal abnormalities							A(W)					D(C)[a]			A(W)			A(W)	D(W)		
Digestive organ abnormalities												D(C)[a]									
Kidney abnormalities		A(W)																A(W)			
Epidermal papilloma								A(W)	D(C)							A(W)					
Sunburn		D(C)[a]							D(C)										D(C)[a]		
Fin erosion		A(C)[a]							D(C)			D(C)[a]		A(W)	A(W)				A(W)		
Skin ulcers				A(C)[a]		A(W)		D(C)[a]	D(C)	D(W)				A(W)	A(W)	A(W)			A(W)		

D, causes disease or economic loss; A, asymptomatic; W, wild fish; C, cultured fish.

[a] Experimental results.

Keys to fish species, Dab: *Limanda limanda*; Dab, Long Rough: *Hippoglossoides platessoides*; Flounder, Barfin: *Verasper moseri*; Flounder, Brazilian: *Paralichthys orbignyanus*; Flounder, European: *Platichthys flesus*; Flounder, Fine: *Paralichthys adspersus*; Flounder, Flathead: *Hippoglossoides dubius*; Flounder, Greenback: *Rhombosolea tapirina*; Flounder, Japanese: *Paralichthys olivaceus*; Hirame, olive flounder; Flounder, Patagonian: *Paralichthys patagonicus*; Flounder, Southern: *Paralichthys lethostigma*; Flounder, Starry: *Platichthys stellatus*; Flounder, Summer: *Paralichthys dentatus*; Flounder, Winter: *Pseudopleuronectes americanus*; Flounder, Yellowtail: *Limanda ferruginea*; Halibut, Atlantic: *Hippoglossus hippoglossus*; Halibut, Greenland: *Reinhardtius hippoglossoides*; Halibut, Shotted: *Eopsetta grigorjewi*; Halibut, Spotted: *Verasper variegatus*; Hogchoker: *Trinectes maculatus*; Lenguado de ojo chico: *Paralichthys microps*; Sole, Dover: *Solea solea*; Sole, English: *Pleuronectes* (= *Parophrys*) *vetulus*; Sole, Lemon: *Microstomus kitt*; Sole, Marbled: *Pleuronectes yokohamae*; Sole, Senegalese: *Solea senegalensis*; Plaice: *Pleuronectes platessa*; Plaice, Black: *Pleuronectes obscurus*; Turbot: *Psetta maximus* (*Scophthalmus maximus*).

often have no discernible effects on performance and growth, so long as the fish are held under optimal rearing conditions with frequent and adequate feeding. Differences in the prevalence of jaw deformities among families and half siblings of Atlantic halibut larvae cultured under identical conditions indicate that certain unknown factors may be important in its development (Ottesen and Babiak 2007). Hatchery-produced juveniles for stock enhancement should be of high quality, as deformities may render the fish susceptible to predation and be a disadvantage in competition for food and reproduction.

15.5.2 Nutrient deficiencies

Critical window of feeding

Inadequate nutrition during larval stages and metamorphosis may have a serious impact on development, including eye migration, pigmentation, and survival. The inadequate nutritional quality of rotifers, *Artemia*, or artificial feed may be overcome by supplementing with live marine zooplankton, in particular small calanoid (*Calanus*), copepod nauplii, and copepodids. Apparently there is a critical window of time, that varies among flatfish species, when feeding copepods instead of *Artemia* will facilitate metamorphosis and further development. Thus, in many cases, a high percentage of normally developed juveniles may be obtained by replacing *Artemia* or rotifers with copepods for a short period just before metamorphosis. However, this may not always be feasible in a commercial hatchery as wild copepods are usually only available at certain times of the year.

Fatty acids

The optimal levels and the interactive effects of different fatty acids on development in flatfish are not fully understood. Obviously, there are species differences in susceptibility to nutritional diseases related to fatty acids. Normal body pigmentation in many flatfish is highly dependent upon proper amounts and ratios of essential fatty acids, especially the level of docosahexanoic acid (DHA, 22:6n-3) and the ratio between DHA and eicosapentaenoic acid (EPA, 20:5n-3). Enriched *Artemia* usually has inadequate DHA and EPA and excess arachidonic acid (AA, 20:4n-6) (Hamre *et al.* 2007, and references therein). The DHA:EPA ratio should be at least two (Bell *et al.* 2003), and an elevated ratio improved pigmentation in turbot. DHA may also have a role in vision since Atlantic halibut larvae fed copepods (*Eurytemora velox*) rich in essential lipids, had more rod cells in the retina than those fed with *Artemia* (Shields *et al.* 1999).

Most yellowtail flounder fed rotifers enriched with DHA and AA had abnormal pigmentation compared to those enriched only with DHA (most of which had normal pigmentation), probably because AA is not essential and may have an adverse effect at certain levels. Juvenile yellowtail flounder that were abnormally pigmented had a lower DHA content than normally pigmented fish, confirming the importance of this fatty acid in the pigmentation process

(Copeman & Parrish 2002). The success of eye migration during metamorphosis also increased when yellowtail flounder larvae were fed a diet high in DHA and EPA, as compared to DHA alone, or DHA plus AA (Copeman *et al.* 2002).

However, dietary n-3 highly unsaturated fatty acid (HUFA) enrichment is probably not needed for normal pigmentation of all flatfish species. For example, over 75% of Japanese flounder fed unenriched, DHA-free, *Artemia* nauplii had normal pigmentation. Copeman *et al.* (2002) fed rotifers with only 1.7% DHA to yellowtail flounder larvae and obtained one of the highest normal pigmentation rates (46%) recorded in their study. Further, Senegalese sole larvae fed *Artemia* that were not enriched in n-3 HUFA still had normal development and skin pigmentation as well as high survival, indicating a low requirement for n-3 HUFA during the live feed period (Villalta *et al.* 2008). Conversely, high levels of n-3 HUFA in a Japanese flounder broodstock diet adversely affected egg quality (Furuita *et al.* 2002). Thus, dietary supplemetation of n-3 HUFA should be carefully considered.

Vitamins

Development of normal pigmentation may require neuronal signaling from the eyes to the brain, increasing melanocyte stimulating hormone production and subsequently melanin synthesis. A deficiency in Vitamin A (VA) (a precursor of rhodopsin) will disrupt this signalling, resulting in abnormally pigmented fish. Thus, enrichment of larval diets with VA is needed for normal pigmentation in flatfish larvae (Bolker and Hill 2000). Moren *et al.* (2005) found that VA was absent in *Artemia*, as well as in copepods (*Eurytemora affinis*, *Acartia grani*, and *Centropages hamatus*) collected from a pond, and proposed that the VA requirement in Atlantic halibut larvae was met by converting VA precursors (canthaxanthin and astaxanthin) to VA. Calanoid copepods have high levels of astaxanthin compared to *Artemia* nauplii, which could partly explain why Atlantic halibut larvae that are fed these copepods have a high frequency of normally pigmented juveniles. While VA is an essential vitamin, excessive VA can cause skeletal (jaw, cranium, vertebral column) deformities during early development in many species. Diet supplementation with retinoic acid, the oxidized form of VA, caused ambicoloration, hypopigmentation, and spots in summer flounder, and skeletal deformities in Japanese flounder and summer flounder (Martinez *et al.* 2007).

Live food enriched with Vitamin C (VC) improved pigmentation of turbot larvae (Merchie *et al.* 1996). Juvenile Japanese flounder fed on a diet without VC exhibited typical VC deficiency signs including anorexia, scoliosis, cataract, exophthalmos, and fin hemorrhage. Vertebral deformities and spots developed in juvenile Japanese flounder fed a diet supplemented with excess Vitamin D. Spinal curvature (scoliosis and lordosis) developed in Atlantic halibut juveniles fed with oxidized dietary lipid, which occurs in rancid feed. Degradation of Vitamin E (VE) was suspected (Lall and Lewis-McCrea 2007), but feeding Atlantic halibut and turbot a diet deficient in VE had no effect on growth or survival (Tocher *et al.* 2002) and VE supplementation did not reduce the frequency of abnormalities observed in Atlantic halibut (Lewis-McCrea and Lall 2007).

Other nutritional problems

Thyroid hormone is important for metamorphosis and pigmentation in flatfish. It is possible to manipulate these processes by adding thyroid hormone to the rearing water. However, adding excessively high levels of thyroid hormone causes a high incidence of albinism in Japanese flounder larvae. Taurine is the most abundant amino acid amongst the free amino acids in marine animals and plants. Taurine supplementation improved the feeding behavior as well as growth of juvenile Japanese flounder. Fat cell necrosis syndrome (FCNS), which develops in the dorsal subdermal fat deposits, has been observed in farmed Atlantic halibut; it may be due to an insufficient level of antioxidants in the feed combined with an exposure to sunlight (Bricknell *et al.* 1996).

15.5.3 Behavior

Aggression, stress, and inadequate nutrition may interact during the larval period, leading to high mortality. Inadequate nutrition at first feeding or metamorphosis may cause a large size variation, leading to cannibalism. In summer flounder, a gradual weaning from live feed to formulated feed gave better growth, and weaning when larvae were older improved overall cohort survival (Bengtson, 1999). Cannibalism in summer flounder may also be reduced by shortening the time needed for settling/metamorphosis by treating larvae with thyroid hormone to synchronize settling behavior. Further, cannibalism is reduced by maintaining a uniform size through frequent grading and feeding to satiation; low light levels might also be useful. In juvenile Japanese flounder, intraspecific aggression is an inherently cannibalistic behavior that is highly related to size dominance.

Aggression in larger fish is often manifested as fin nipping. In flatfish, the caudal, pectoral, and dorsal fins are the most frequent targets for fin nipping. Damage to the eyes and fins due to fin nipping are frequently observed in Atlantic halibut in hatcheries. The pectoral fin and the right eye appear to be the most frequent target for aggression amongst Atlantic halibut cultured in onshore tanks. In sea cages, this behavior is seldom encountered. Frequent and adequate feeding often reduces attacks. Flatfish may also display other abnormal behaviors due to aggression. During culture in cages and to some extent in onshore tanks, some Atlantic halibut are surface swimmers, sometimes with part of their head above the water. This may be due to suboptimal rearing conditions or intraspecific aggression.

15.5.4 Physical and chemical stresses

Being benthic species, substrate has a very important influence on flatfish health. Improper substrate has been associated with skin thickening and spots. In some cases, the skin thickening can be so severe that it appears as large, tumor-like masses (Ottesen *et al.* 2007). These skin lesions greatly decrease carcass quality and might also have adverse effects on growth. Substrate-related spots occur in many flatfish species (e.g., Atlantic halibut and Japanese flounder, barfin flounder and summer flounder). A sand substrate, instead of a smooth substrate, reduced spots in Atlantic halibut (Ottesen and Strand 1996). Skin lesions also began to heal when the fish were transferred from a smooth fibreglass substrate to a sand substrate (Ottesen *et al.* 2007). Using microceramic particles instead of a smooth substrate prevented spots in Japanese flounder (Takeuchi, 1999). Spots can also be avoided by holding barfin flounder in white tanks, or on sand, as shown for Atlantic halibut, summer flounder, and Japanese flounder.

Inappropriate lighting in indoor tanks may be responsible for the prevalence of pigment abnormalities in many hatchery-reared populations. Southern flounder that are raised in outdoor ponds rarely exhibit albinism. When southern flounder raised under low light were exposed to increased light intensity one week posthatching, partially albino fish had much more normal pigmentation (Denson and Smith 1997). Excess light can be deleterious: Sunburn can occur in Atlantic halibut cultured in shallow raceways, and in shallow (3-m depth) sea cages with insufficient UV protection. Cataract is where the normally transparent lens of the eye becomes opaque; when severe, it causes blindness. It is common in Atlantic halibut and might be caused by excess UV light, but other possible causes (e.g., other environmental factors, antioxidant deficiency of the feed) have not yet been ruled out (Treasurer *et al.* 2007).

Improper temperature can have many adverse effects. Elevated temperature during gametogenesis of broodstock causes reduced egg viability in both turbot (Devauchelle *et al.* 1988) and Atlantic halibut (Brown *et al.* 2006). High temperature is also detrimental to normal development. In Atlantic halibut, high ($>9^\circ$C) temperature increased the prevalence of jaw deformities in yolk sac fry; low (<29 ppt) salinity also induced this abnormality. Suboptimal low temperature causes albinism and ambicoloration in brown sole (Aritaki & Seikai 2004). Atlantic halibut transferred from 8 to 1°C tolerated this sudden decrease much better than turbot, indicating that, as expected, environmental tolerances must be determined for each species.

Liver lesions, including cancers, have been observed at a high prevalence in some wild flatfish taken from environments contaminated with high levels of environmental carcinogens and other toxins. These lesions are most common in older fish, because they take time to develop and thus are not often observed in younger fish. Hepatic lesions in marine flatfish have been used as biomarkers of contaminant-induced indicators of environmental pollution.

Broodstock stress may have serious effects on reproductive success, including gamete and progeny quality. Netting exposure to air (Mugnier *et al.* 1998), blood sampling, and hand-stripping of turbot increased blood cortisol levels and caused an imbalance in osmoregulation and increased mortality (Mugnier *et al.* 1998). Handling and transportation of fish within and between production sites should be done after fasting and without changing the culture parameters too much. The length of the time that turbot tolerate transport can be increased by reducing the pH, either by raising carbon dioxide levels or by adding acid. This is because ammonia toxicity is the most important factor affecting turbot survival during transport and reducing pH decreases ammonia toxicity (Grøttum *et al.* 1997). However, low pH and elevated CO_2 are also stressful, with toxicity depending on species, ontogenetic stage, and their interaction with other water chemistry factors.

15.6 Health management in flatfish aquaculture

There are several keys to successfully managing disease. First, it is important to work closely with a disease expert. Over time, producers can become familiar with the most common diseases that affect their fish, which will greatly help to head off problems before they cause serious harm. However, when first starting out, it is important to identify an expert that can be called on before disease occurs. This should also include creating a health management plan. This plan need not be complicated and actually should be as simple as possible so that it will be more likely to be followed rather than just put on a shelf. The plan should include common sense procedures for avoiding, and when needed, controlling disease. With the health management plan in place, specific problems can then be targeted as they develop.

Another important aspect of managing disease is to practice biosecurity. Biosecurity is simply preventing the introduction or spread of infectious disease by placing barriers to its transmission. Biosecurity protocols should always be a major part of a health management plan. One of the key means of preventing disease transmission is the effective use of disinfection and quarantine. Biosecurity should also involve segregating stocks according to age (e.g., hatchery, growout, broodstock), time of entry onto the farm (new arrivals should always be quarantined before introducing into the general population), and location (fish should not be moved from one part of the farm to another unless absolutely necessary). It is also important to keep accurate records. More details on biosecurity are provided in Scarfe *et al.* (2005) and Arthur *et al.* (2008).

Another essential goal of health management is to minimize drug use. This is because drugs are expensive and continued use of many drugs can lead to resistance (making the drug useless). There are also serious legal and environmental restrictions on using most drugs. Because of these serious limitations with use of drugs, it is important to try to minimize their need by reducing stress and thus keeping a population as healthy as possible. This includes identifying optimal conditions for culture (nutrition, environment, etc.) and recognizing that there is an intimate relationship between stress and disease. One must be sure that any drug that is used fully complies with all legal, humane, safety (human and fish), and environmental regulations in that specific jurisdiction.

If fish are to be reared for eventual release into the wild, they should be screened for all possible diseases that are known to affect that species. Procedures to screen for specific pathogens are provided in a number of reference sources, such as OIE (2006) and AFS-FHS (2007). It is important to seek expert assistance in performing this task. It is also important to realize that even if a pathogen is not detected on a health screen, this does not provide 100% assurance that the fish population is free of that disease, but rather only establishes a certain percentage likelihood that the population is free of that pathogen. The confidence level that the population is free of that pathogen is dependent upon the number of fish that are sampled in the screen. It is also important to realize that absence of any of the screened pathogens does not mean that the population is "pathogen-free," because other pathogens may be present that are as yet unknown or undetected.

Literature cited

AFS-FHS (American Fisheries Society-Fish Health Section). 2007. *FHS Blue Book: Suggested Procedures for the Detection and Identification of Certain Finfish and Shellfish Pathogens*, 6th edn. Fish Health Section, American Fisheries Society, Bethesda, MD.

Anderson, C.D., Roberts, R.J., MacKenzie, K., and McVicar, A.H. 1976. The hepatorenal syndrome in cultured turbot (*Scophthalmus maximus* L.). *Journal of Fish Biology* 8:331–341.

Anonymous. 2007. *Viral Hemorrhagic Septicemia*. Center for Food Security and Public Health, Iowa State University, Ames, IA, 4 p.

Anstensrud, M. 1992. Mate guarding and mate choice in two copepods, *Lernaeocera branchialis* (L.) (Pennellidae) and *Lepeophtheirus pectoralis* (Muller) (Caligidae), parasitic on flounder. *Journal of Crustacean Biology* 12:31–40.

Aritaki, M., and Seikai, T. 2004. Temperature effects on early development and occurrence of metamorphosis-related morphological abnormalities in hatchery-reared brown sole, *Pseudopleuronectes herzensteini*. *Aquaculture* 240:517–530.

Arthur, J.R., Bondad-Reantaso, M.G., and Subasinghe, R.P. 2008. *Procedures for the Quarantine of Live Aquatic Animals: A Manual*. FAO Fisheries Technical Paper No. 502. Food and Agriculture Organization of the United Nations, Rome, 74 pp.

Arthur, J.R., Cone, D.K., Cusack, R.R., Barker, D.B., and Burt, M.D.B. 2004. Two species of *Trichodina* (Ciliophora:Peritrichida) from cultured flatfishes (Pleuronectiformes) in Atlantic Canada. *Comparative Parasitology* 7:247–250.

Austin, B., and Austin, D.A. 2007. *Bacterial Fish Pathogens: Diseases of Farmed and Wild Fish*, 4th edn. Praxis Publishing, Chichester, UK, 552 pp.

Barker, D.E., Cone, D.K., and Burt, M.D.B. 2002. *Trichodina murmanica* (Ciliophora) and *Gyrodactylus pleuronecti* (Monogenea) parasitizing hatchery-reared winter flounder, *Pleuronectes americanus*: effects on host growth and assessment of parasite interaction. *Journal of Fish Diseases* 25:81–89.

Bell, J.G., McEvoy, L.A., Estevez, A., and Shields, R.J. 2003. Optimising lipid nutrition in first-feeding flatfish larvae. *Aquaculture* 227:211–220.

Bengtson, D.A. 1999. Aquaculture of summer flounder, *Paralichthys dentatus*: status of knowledge, current research and future research priorities. *Aquaculture* 176: 39–49.

Bergh, O., Nilsen, F., and Samuelsen, O.B. 2001. Diseases, prophylaxis and treatment of the Atlantic halibut *Hippoglossus hippoglossus*: a review. *Diseases of Aquatic Organisms* 48:57–74.

Bolker, J.A., and Hill, C.R. 2000. Pigmentation development in hatchery-reared flatfishes. *Journal of Fish Biology* 56:1029–1052.

Bowden, T.J., Bricknell, I.R., Menoyo Luque, D., and Wergeland, H. 2002. Efficacy of different administration routes for vaccination against *Vibrio anguillarum* in Atlantic halibut (*Hippoglossus hippoglossus* L.). *Fish and Shellfish Immunology* 12: 283–285.

Bricknell, I.R., Bruno, D.W., Bowden, T.J., and Smith, P. 1996. Fat cell necrosis syndrome in Atlantic halibut, *Hippoglossus hippoglossus* L. *Aquaculture* 144:65–69.

Brown, N.P., Shields, R.J., and Bromage, N.R. 2006. Influence of water temperature on spawning patterns and egg quality in the Atlantic halibut, *Hippoglossus hippoglossus* L. *Aquaculture* 261:993–1002.

Burn, P.R. 1980. The parasites of smooth flounder, *Liopsetta putnami* (Gill), from the Great Bay Estuary, New Hampshire. *Journal of Parasitology* 66:532–541.

Burreson, E.M., and Zwerner, D.E. 1984. Juvenile summer flounder, *Paralichthys dentatus*, mortalities in the western Atlantic ocean caused by the hemoflagellate *Trypanoplasma bullocki*: evidence from field and experimental studies. *Helgoländer Meeresuntersuchungen* 37:343–352.

Copeman, L.A., Parrish, C.C., Brown, J.A., and Harel, M. 2002. Effects of docosahexanoic, eicosapentaenoic and arachidonic acids on the early growth, survival, lipid composition and pigmentation of yellowtail flounder, *Limanda ferruginea*: a live food enrichment experiment. *Aquaculture* 210:285–304.

Denson, M.R., and Smith, T.I.J. 1997. Diet and light intensity effects on survival, growth and pigmentation of southern flounder, *Paralichthys lethostigma*. *Journal of the World Aquaculture Society* 28:366–373.

Devauchelle, N., Alexandre, J.C., Le Corre, N., and Letty, Y. 1988. Spawning of turbot (*Scophthalmus maximus*) in captivity. *Aquaculture* 69:159–184.

Devesa, S., Barj, J.L., and Toranzo, A.E. 1989. Ulcerative skin and fin lesions in reared turbot, *Scophthalmus maximus*. *Journal of Fish Diseases* 12:323–333.

Diamant, A. 1987. Ultrastructure and pathogenesis of *Ichthyobodo sp.* from wild common dab, *Limanda limanda* in the North Sea. *Journal of Fish Diseases* 10:241–247.

Do, J.W., Cha, S.J., Kim, J.S., Lee, E.J., Nam, S., Choi, H.J., Lee, C.H., Park, M.S., Kim, J.W., Kim, Y.C., and Park, J.W. 2005. Phylogenetic analysis of the major capsid protein gene of iridovirus isolates from cultured flounders *Paralichthys olivaceus* in Korea. *Diseases of Aquatic Organisms* 64:193–200.

Ferguson, H.W., Millar, S.D., and Kibenge, F.S.B. 2003. Reovirus-like hepatitis in farmed halibut (*Hippoglossus hippoglossus*). *Veterinary Record* 153:86–87.

Furuita, H., Tanaka, H., Yamamoto, T., Suzuki, N., and Takeuchi, T. 2002. Effects of high levels of n-3 HUFA in broodstock diet on egg quality and egg fatty acid composition of Japanese flounder, *Paralichthys olivaceus*. *Aquaculture* 210(31):323–333.

Gagne, N., Johnson, S.C., Cook-Versloot, M., MacKinnon, A.M., and Olivier, G. 2004. Molecular detection and characterization of nodavirus in several marine fish species from the northeastern Atlantic. *Diseases of Aquatic Organisms* 62:181–189.

Gauger, E., Smolowitz, R., Uhlinger, K., Casey, J., and Gomez Chiarri, M. 2006. *Vibrio harveyi* and other bacterial pathogens in cultured summer flounder, *Paralichthys dentatus*. *Aquaculture* 260:10–20.

Grotmol, S., Totland, G.K., Thorud, K., and Hjeltnes, B.K. 1997. Vacuolating encephalopathy and retinopathy associated with a nodavirus-like agent: A probable cause of mass mortality of cultured larval and juvenile Atlantic halibut *Hippoglossus hippoglossus*. *Diseases of Aquatic Organisms* 29:85–97.

Grøttum, A., Staurnes, T., and Sigholt, T. 1997. Effect of oxygenation, aeration and pH control on water quality and survival of turbot, *Scophthalmus maximus* (L.), kept at high densities during transport. *Aquaculture Research* 28:159–164.

Gudmundsdóttir, B.K., Lange, S., Magnadóttir, B. 2003. Protection against atypical furunculosis in Atlantic halibut, *Hippoglossus hippoglossus* (L.); comparison of a commercial furunculosis vaccine and an autogenous vaccine. *Journal of Fish Diseases* 26:331–338.

Hamre, K., Holen, E., and Moren, M. 2007. Pigmentation and eye-migration in Atlantic halibut, *Hippoglossus hippoglossus* L., larvae: new findings and hypotheses. *Aquacultural Nutrition* 13:65–80.

Hansen, G.H., Bergh, Ø., Michaelsen, J., and Knappskog, D. 1992. *Flexibacter ovolyticus* sp. nov., a pathogen of eggs and larvae of Atlantic halibut, *Hippoglossus hippoglossus* L. *International Journal of Systematic Bacteriology* 42:451–458.

Hayward, C.J., Kim, J.H., and Heo, G.J. 2001. Spread of *Neoheterobothrium hirame* (Monogenea), a serious pest of the olive flounder *Paralichthys olivaceus*, to Korea. *Diseases of Aquatic Organisms* 45:209–213.

Hoare, R., Hovland, H., Langston, A.L., Imsland, A., Stefansson, S.O., Mulcahy, M., and Wergeland, H.I. 2002. Susceptibility of three different strains of juvenile Atlantic halibut (*Hippoglossus hippoglossus* L.) cultured at two different temperatures to *Vibrio anguillarum* and temperature effect on antibody response. *Fish and Shellfish Immunology* 13:111–123.

Hughes, K., and Smith, S. 2002. Clinical presentations of *Mycobacterium* sp. in summer flounder (*Paralichthys dentatus*) held in re-circulating aquaculture systems. *Virginia Journal of Science* 53:58 (Abstract).

Hughes, K.P., and Smith, S.A. 2003. Common and emerging diseases in commercially-cultured summer flounder, *Paralichthys dentatus*. *Journal of Applied Aquaculture* 14:163–178.

Inglesias, R., Parama, A., Alvarez, M.F., Leiro, J., Fernandez, J., and Sanmartin, M.L. 2001. *Philasterides dicentrarchi* (Chiliophora, Scuticociliatida) as a causative agent of scuticociliatosis in farmed turbot *Scophthalmus maximus* in Galicia (NW Spain). *Diseases of Aquatic Organisms* 46:47–55.

Isshiki, T., Nagano, T., Kanehira, K., and Suzuki, S. 2004. Distribution of marine birnavirus in cultured marine fish species from Kagawa Prefecture, Japan. *Journal of Fish Diseases* 27:89–98.

Isshiki, T., Nishizawa, T., Kobayashi, T., Nagano, T., and Miyazaki, T. 2001. An outbreak of VHSV (viral hemorrhagic septicemia virus) infection in farmed Japanese flounder *Paralichthys olivaceus* in Japan. *Diseases of Aquatic Organisms* 47: 87–99.

Iwamoto, R., Hasegawa, O., Lapatra, S., and Yoshimizu, M. 2002. Isolation and characterization of the Japanese flounder (*Paralichthys olivaceus*) lymphocystis disease virus. *Journal of Aquatic Animal Health* 14:114–123.

Jee, B.Y., Kim, K.H., Park, S.I., and Kim, Y.C. 2000. A new strain of *Cryptocaryon irritans* from the cultured olive flounder (*Paralichthys olivacsus*). *Diseases of Aquatic Organisms* 43:211–215.

Kaige, N., and Miyazaki, T. 1985. A histopathological study of white spot disease in Japanese flounder (*Paralichthys olivaceus*). *Fish Pathology* 20:61–64.

Khan, R.A., Lobos, V., Garcias, F., Munoz, G., Valdebenito, V., and George-Nascimento, M. 2001. *Cryptobia neghmei* sp. n. (Protozoa: Kinetoplastida) in two species of flounder, *Paralichthys* spp. (Pisces: Paralichthydae) off Chile. *Revista Chilena de Historia Natural*. 74:763–767.

Khattra, J.S., Gresoviac, J., Kent, M.L., Myers, M.S., Hedrick, R.P., and Devlin, R.H. 2000. Molecular detection and phylogenetic placement of a microspordian from English sole (*Pleuronectes vetulus*) affected by x-cell pseudotumors. *Journal of Parasitology* 86:867–871.

Kim, J.H., Gomez, D.K., Baeck, G.W., Shin, G.W., Heo, G.J., Jung, T.S., and Park, S.C. 2006. Pathogenicity of *Streptococcus uberis* to olive flounder *Paralichthys olivaceus*. *Fish Pathology* 41:171–173.

Kim, S.M., Cho, J.B., Lee, E.H., Kwon, S.R., Kim, S.K., Nam, Y.K., and Kim, K.H. 2004. Occurrence of scuticociliatosis in olive flounder *Paralichthys olivaceus* by *Phiasterides dicentrarchi* (Ciliophora: Scuticociliatida). *Diseases of Aquatic Organisms* 62:233–238.

King, J.A., Snow, M., Skall, H.F., and Raynard, R.S. 2001. Experimental susceptibility of Atlantic salmon *Salmo salar* and turbot *Scophthalmus maximus* to European freshwater and marine isolates of viral hemorrhagic septicemia. *Diseases of Aquatic Organisms* 47:25–31.

Kirmse, P. 1987. Important parasites of dover sole (*Solea solea* L.) kept under mariculture conditions. *Parasitology Research* 73:466–471.

Koie, M. 2000. The life-cycle of the flatfish nematode *Cucullanus heterochrous*. *Journal of the Helminthological Society of Washington* 74:323–328.

Korsnes, K., Devold, M., Nerland, A.H., and Nylund, A. 2005. Viral encephalopathy and retinopathy (VER) in Atlantic salmon *Salmo salar* after intraperitoneal challenge with a nodavirus from Atlantic halibut *Hippoglossus hippoglossus*. *Diseases of Aquatic Organisms* 68:7–15.

Kurata, O., Munchan, C., Wada, S., Hatai, K., Miyoshi, Y., and Fukuda, Y. 2008. Novel *Exophiala* infection involving ulcerative skin lesions in Japanese flounder *Paralichthys olivaceus*. *Fish Pathology* 43:35–44.

Kusakari, M., and Urawa, S. 1990. Histopathology of the skin of yearling Japanese flounder *Paralichthys olivaceus* infected with the flagellate *Ichthyobodo* sp. *Fish Pathology* 25:59–68.

Lall, S.P., and Lewis-McCrea, L.M. 2007. Role of nutrients in skeletal metabolism and pathology in fish – An overview. *Aquaculture* 267:3–19.

Latremouille, D.N. 2003. Fin erosion in aquaculture and natural environments. *Reviews in Fisheries Science* 11:315–335.

Lewis-McCrea, L.M., and Lall, S.P. 2007. Effects of moderately oxidized dietary lipid and the role of vitamin E on the development of skeletal abnormalities in juvenile Atlantic halibut, *Hippoglossus hippoglossus*. *Aquaculture* 262:142–155.

Li, J., Gao, D., Wang, Q., and Wang, J. 2005. Efficacy of *Vibrio anguillarum* antigen administered by intraperitoneal injection route in Japanese flounder *Paralichthys olivaceus* (Temminck et Schlegel). *Aquaculture Research* 36:1104–1111.

Liewes, E.W. 1984. *Culture, Feeding and Diseases of Commercial Flatfish Species*. A.A. Balkema, Boston, MA, 104 pp.

MacKenzie, K., McVicar, A.H., and Waddell, I.F. 1976. Some parasites of plaice *Pleuronectes platessa* L. in three different farm environments. *Scottish Fisheries Research Report* 4:1–14.

Mamnur, R.M., Nakai, T., and Muroga, K. 1994. An ecological study on *Edwardsiella tarda* from flounder farms. *Fish Pathology* 29:221–227.

Martinez, G.M., Baron, M.P., and Bolker, J.A. 2007. Skeletal and pigmentation defects following retinoic acid exposure in larval summer flounder, *Paralichthys dentatus*. *Journal of the World Aquaculture Society* 38:353–366.

McVicar, A.H. 1999. *Ichthyophonus* and Related Organisms. In: Woo, P.T.K., and Bruno, D.W. (eds) *Fish Diseases and Disorders. Volume 3. Viral, Bacterial and Fungal Infections*. CABI Publishing, New York, pp. 874–892.

McVicar, A.H., Burke, D., Watermann, B., and Dethlefsen, V. 1987. Gill x-cell lesions of dab *Limanda limanda* in the North Sea. *Diseases of Aquatic Organisms* 2:197–204.

Merchie, G., Lavens, P., Dhert, P., Gomez, M.G.U., Nelis, H., De Leenheer, A., and Sorgeloos, P. 1996. Dietary ascorbic acid requirements during the hatchery production of turbot larvae. *Journal of Fish Biology* 49:573–583.

Miwa, S., Nakayasu, C., Kamaishi, T., and Yoshura, Y. 2004. X-cells in fish pseudotumors are parasitic protozoans. *Diseases of Aquatic Organisms* 58:165–170.

Miyazaki, T. 2005. Ultrastructural features of herpesvirus-infected cells in epidermal lesions in larvae of the Japanese flounder *Paralichthys olivaceus*. *Diseases of Aquatic Organisms* 66:159–62.

Moren, M., Gundersen, T.E., and Hamre, K. 2005. Quantitative and qualitative analysis of retinoids in Artemia and copepods by HPLC and diode array detection. *Aquaculture* 246:359–365.

Mugnier, C., Fostier, A., Guezou, S., Gaignon, J.L., and Quemener, L. 1998. Effect of some repetitive factors on turbot stress response. *Journal Aquaculture International* 6:33–45.

Munday, B.L., Kwang, J., and Moody, N. 2002. Betanodavirus infections of teleost fish: A review. *Journal of Fish Diseases* 25:127–142.

Muroga K. 2001. Viral and bacterial diseases of marine fish and shellfish in Japanese hatcheries. *Aquaculture* 202:23–44.

Mushiake K., Nishizawa, T., Nakai, T., Furusawa, I., and Muroga, K. 1994. Control of VNN in striped jack—selection of spawners based on the detection of SJNNV gene by Polymerase Chain-Reaction (PCR). *Fish Pathology* 29:177–182.

Nerland, A.H., Skaar, C., Eriksen, T.B., and Bleie, H. 2007. Detection of nodavirus in seawater from rearing facilities for Atlantic halibut *Hippoglossus hippoglossus* larvae. *Diseases of Aquatic Organisms* 73:201–205.

Newman, M.W. 1978. Pathology associated with *Cryptobia* infection in a summer flounder (*Paralichthys dentatus*). *Journal of Wildlife Diseases* 14:299–303.

Nilsen, F. 1995. Description of *Trichodina hippoglossi* n. sp. from farmed Atlantic halibut larvae *Hippoglossus hippoglossus*. *Diseases of Aquatic Organisms* 21:209–214.

Noga, E.J. 1993. Fungal diseases in marine and estuarine fishes. In: Couch, J.A., and Fournie, J.W. (eds) *Pathology of Marine and Estuarine Organisms*. CRC Press, Boca Raton, FL, pp. 85–109.

Nguyen, H.T., Kanai, K., and Yoshikoshi, K. 2002. Ecological investigation of *Streptococcus iniae* in cultured Japanese flounder (*Paralichthys olivaceus*) using selective isolation procedures. *Aquaculture* 205:7–17.

Oh, M.J., Kim, W.S., Kitamura, S.I., Lee, H.K., Son, B.W., Jung, T.S., and Jung, S.J. 2006b. Change of pathogenicity in Olive flounder *Paralichthys olivaceus* by co-infection of *Vibrio harveyi*, *Edwardsiella tarda* and marine birnavirus. *Aquaculture* 257:156–160.

Oh, M.J., Kitamura, S.I., Kim, W.S., Park, M.K., Jung, S.J., Miyadai, T., and Ohtani, M. 2006a. Susceptibility of marine fish species to a megalocytivirus, turbot iridovirus, isolated from turbot, *Psetta maximus* (L.). *Journal of Fish Diseases* 29:415–421.

OIE (Office International Des Épizooties). 2006. *Manual of Diagnostic Tests for Aquatic Animals*, 5th edn. Office International Des Épizooties, Paris, 358 pp.

Ototake, M., and Matsusato, T. 1986. Notes on Scuticociliata infection of cultured juvenile flounder *Paralichthys olivaceous*. *Bulletin of National Research Institute of Aquaculture* 9:65–68.

Ottesen, O.H., and Babiak, I. 2007. Parental effects on fertilization and hatching rates and development of Atlantic halibut, *Hippoglossus hippoglossus* L., embryos and larvae. *Theriogenology* 68:1219–1227.

Ottesen, O.H., Noga, E.J., and Sandaa, W. 2007. Effect of substrate on progression and healing of skin erosions and epidermal papillomas of Atlantic halibut, *Hippoglossus hippoglossus* (L.) *Journal of Fish Diseases* 30:43–53.

Ottesen, O.H., and Strand, H.K. 1996. Growth, development, and skin abnormalities of halibut, *Hippoglossus hippoglossus* (L.) juveniles kept on different bottom substrates. *Aquaculture* 146:7–25.

Pakingking, R., Jr., Takano, R., Nishizawa, T., Mori, K., Iida, Y., Arimoto, M., and Muroga, K. 2003. Experimental coinfection with aquabirnavirus and viral hemorrhagic septicemia virus (VHSV), *Edwardsiella tarda* or *Streptococcus iniae* in Japanese flounder. *Fish Pathology* 38:15–21.

Rand, T.G. 1994. An unusual form of *Ichthyophonus hoferi* (Ichthyophonales: Ichthyophonaceae) from yellowtail flounder *Limanda ferruginea* from Nova Scotia shelf. *Diseases of Aquatic Organisms* 18:21–28.

Redondo, M.J., Palenzuela, O., and Alvarez-Pellitero, P. 2003. In vitro studies on viability and proliferation of *Enteromyxum scophthalmi* (Myxozoa), an enteric parasite of cultured turbot *Scophthalmus maximus*. *Diseases of Aquatic Organisms* 55:133–44.

Scarfe, D., Lee, C.-S., and OBryen, P. (eds). 2005. *Aquaculture Biosecurity: Prevention, Control and Eradication of Aquatic Animal Disease*. Blackwell Publishing Professional, Ames, IA, 182 pp.

Schram, T.A., and Haug, T. 1988. Ectoparasites on the Atlantic halibut, *Hippoglossus hippoglossus* (L.), from northern Norway – Potential pests in halibut aquaculture. *Sarsia* 73:213–227.

Schwarz, M.H., and Smith, S.A. 1998. *Getting Acquainted with Amyloodinium ocellatum*. CFAST Virginia Cooperative Extension Publication, pp. 600–200.

Shields, R., Bell, G., Luizi, F.S., Gara, B., Bromage, N.R., and Sargent, J.R. 1999. Natural copepods are superior to enriched *Artemia* as feed for halibut larvae, *Hippoglossus hippoglossus*, in terms of survival, pigmentation and retinal morphology: Relation to dietary essential fatty acids. *Journal of Nutrition* 129:1186–1194.

Skall, H.F., Mellergaard, S., and Olesen, N.J. 2000. Isolation of birnavirus serogroup B in wild and aquacultured fish species. *Bulletin of the European Association of Fish Pathologists* 20:229–236.

Skall, H.F., Olesen, N.J., and Mellergaard, S. 2005. Viral haemorrhagic septicaemia virus in marine fish and its implications for fish farming – a review. *Journal of Fish Diseases* 28:509–529.

Soffientino, B., Gwaltney, T., Nelson, D.R., Specker, J.L., Mauel, M., and Gomez-Chiarri, M. 1999. Infectious necrotizing enteritis and mortality caused by *Vibrio carchariae* in summer flounder *Paralichthys dentatus* during intensive culture. *Diseases of Aquatic Organisms* 38:201–210.

Strongman, D.B., Morrison, C.M., and McClelland, G. 1997. Lesions in the musculature of captive American plaice *Hippoglossoides platessoides* caused by the fungus *Hormoconis resinae* (Deuteromycetes). *Diseases of Aquatic Organisms* 28:107–113.

Svendsen, Y.S., and Haug, T. 1991. Effectiveness of formalin, benzocaine, and hyposaline and hypersaline exposures against adults and eggs of *Entobdella hippoglossi* (Muller), an ectoparasite on Atlantic halibut (*Hippoglossus hippoglossus* L.) – Laboratory studies. *Aquaculture* 94:279–289.

Takeuchi, T. 1999. Mechanism and prevention of color abnormality in flounder. *AquaNet* 2:48–52 (in Japanese).

Tanaka, S., Kuriyama, I., Nakai, T., and Miyazaki, T. 2003. Susceptibility of cultured juveniles of several marine fish to the sevenband grouper nervous necrosis virus. *Journal of Fish Diseases* 26:109–115.

Tocher, D.R., Mourente, G., Van Der Eecken, A., Evjemo, J.O., Diaz, E., Bell, J.G., Geurden, I., Lavens, P., and Olsen, Y. 2002. Effects of dietary vitamin E on antioxidant defence mechanisms of juvenile turbot. *Aquaculture Nutrition* 8:195–207.

Treasurer, J.W., Cox, D.I., and Wall, T. 2007. Epidemiology of blindness and cataracts in cage reared ongrown Atlantic halibut *Hippoglossus hippoglossus*. *Aquaculture* 271:77–84.

Urawa, S., Ueki, N., Nakai, T., and Yamaski, H. 1991. High mortality of cultured juvenile Japanese flounder, *Paralichthys olivaceus* (Temminek & Sehlegel), caused by the parasitic flagellate *Ichthyobodo sp*. *Journal of Fish Diseases* 14:489–494.

Villalta, M., Estévez, A., Bransden, M.P., and Bell, J.G. 2008. Effects of dietary eicosapentaenoic acid on growth, survival, pigmentation and fatty acid composition in

Senegal sole (*Solea senegalensis*) larvae during the Artemia feeding period. *Aquaculture Nutrition* 14:232–241.

Wada S., Kurata, O., Hatai, K., Noga, E.J., Dykstra, M.J., and Burke, J.S. 2009. Reovirus-like infection of cultured summer flounder *Paralichthys dentatus*. *Fish Pathology* 44:151–153.

Watanabe, K.I., Nishizawa, T., and Yoshimizu, M. 2000. Selection of brood stock candidates of barfin flounder using an ELISA system with recombinant protein of barfin flounder nervous necrosis virus. *Diseases of Aquatic Organisms* 41:219–223.

Zheng, X.J., Zhan, W.B., Chen, C.Z., and Fang, H. 2005. Studies on the infection and pathogen of *Edwardsiella tarda* in flounder *Paralichthys olivaceus*. *Shuisheng Shengwu Xuebao* 29:31–37.

Zrzavy, J., and Hypsa, V. 2003. Myxozoa, *Polypodium* and the origin of the Bilateria: The phylogenetic position of "Endocinidozoa" in light of the rediscovery of *Buddenbrockia*. *Cladistics* 19:164–169.

Chapter 16

Flatfish as model research animals: metamorphosis and sex determination

Russell J. Borski, John Adam Luckenbach, and John Godwin

Flatfishes are highly diverse and ecologically, economically, and culturally valuable finfish. They are highly priced species, supporting important fisheries throughout the world. However, a number of flatfishes are showing significant declines in the wild, mainly because of overfishing. The desirable characteristics, high value, and wide distribution of flatfishes along with the potential to enhance depleted wild stocks have made their propagation an important priority. While fish culture has been practiced well over the past century, certain aspects of the biology of flatfishes make their culture particularly challenging. Metamorphosis is a complex stage of development in which the pelagic bilaterally symmetrical larva is transformed into an asymmetrical benthic flatfish. Properly controlling metamorphosis, particularly its synchronization and successful completion to produce uniformly sized juveniles, is a crucial step in growing flatfishes in captivity.

Most species of flatfishes also exhibit sexually dimorphic growth patterns by which females grow to substantially larger sizes than males. Males also mature at a smaller body size, often below marketable size, making their production undesirable. Therefore, from an economical standpoint, it is highly advantageous to grow only females. There has been considerable research over the past decade to produce monosex populations of only female flounder through ploidy manipulations, similar to that established for salmon. However, the unusual sex-determining mechanism observed in flatfishes, particularly of the flounders, where gonadal sex is determined by a combination of genetic and environmental influences, has made controlling sex difficult. An understanding of the mechanisms of sex determination in flatfishes is critical if production of all-females stocks is required for improving efficiency of commercial aquaculture operations or if hatcheries are to produce the appropriate sex ratios for environmentally sound stock enhancement. This chapter provides a brief overview of the biology of metamorphosis and sex determination and its regulation in flatfishes. Since flatfishes are among the only species that exhibit "true" metamorphosis and both environmental and genetic sex-determining mechanisms, they offer valuable research models to better understand how these developmental stages are controlled in vertebrates.

16.1 Metamorphosis

The term metamorphosis literally means to change form. Most organisms undergo some degree of transformation during their life history, but some species show a distinct "first" or "true" metamorphosis. Classic examples of organisms that undergo such a metamorphosis are the tadpole that transforms into an adult frog or the free-swimming, symmetrical larval fish that becomes a bottom-dwelling, asymmetrical flatfish. Metamorphosis is best understood in anuran amphibians (i.e., frogs and toads). However, over the past 20 years, metamorphosis in flatfishes has received attention and many of the mechanisms regulating this phenomenon have been elucidated. Of the flatfishes, Japanese flounder or hirame (*Paralichthys olivaceus*) have been studied most extensively, but more recently considerable research has been conducted in species such as Atlantic halibut (*Hippoglossus hippoglossus*) and summer flounder (*Paralichthys dentatus*). Flatfishes exhibit dramatic changes in external and internal morphology, physiology, behavior, and habitat during metamorphosis and are thus strong research models for studying early development and metamorphic transitions in vertebrates. Furthermore, because undesirable metamorphic abnormalities, such as irregular pigmentation (Figure 16.1) or arrested eye migration, are frequently observed in aquaculture operations for flatfishes, it is important to gain a better understanding of metamorphosis and its regulation to prevent such occurrences (Power *et al.* 2008).

Figure 16.1 Hatchery-produced, juvenile southern flounder (*Paralichthys lethostigma*) with varying degrees of irregular pigmentation, commonly observed in flatfish aquaculture.

16.1.1 Morphological and behavioral changes during metamorphosis

Flatfish larvae are pelagic planktivores that inhabit ocean or estuarine waters. Their bodies are bilaterally symmetrical with an eye located on each side of the head. In fact, during this stage, they more closely resemble larvae of other fish species than they do adult flatfishes. Metamorphosis marks the transition from larva to juvenile, which is often subdivided into several stages: premetamorphosis, prometamorphosis, metamorphic climax, and postmetamorphosis (Figure 16.2). Studies suggest that the precise timing of metamorphosis may be better correlated with body size rather than age of the animal and is influenced by water temperature (Chambers and Leggett 1987; Power *et al.* 2008). However, many studies have reported metamorphic events based on days posthatch. Typically, within several weeks of hatching, the dorsal fin rays of premetamorphic larvae elongate, an early indicator of ensuing metamorphosis, and subsequently shorten through metamorphosis (Inui *et al.* 1994; Kawamura and Hosoya 1997). One eye then gradually migrates to the opposite side of the body, making the body asymmetrical in appearance with an ocular and blind side (Figure 16.2). Whether the right or left eye migrates is species-specific. For example, in Japanese flounder, a left-sided species (also referred to as handedness), the right eye migrates to the left side of the body, whereas in Atlantic halibut, a right-sided species, the left eye migrates to the right side.

Concomitant with eye migration, other developmental and behavioral changes associated with the needs of a benthic, predatory lifestyle occur. The dorsal surface of the body becomes opaque with the spread of scales and melanophores, an adaptation that camouflages the ocular side of the flatfish when lying on substrate. Metamorphosing flatfishes also begin to settle, spending

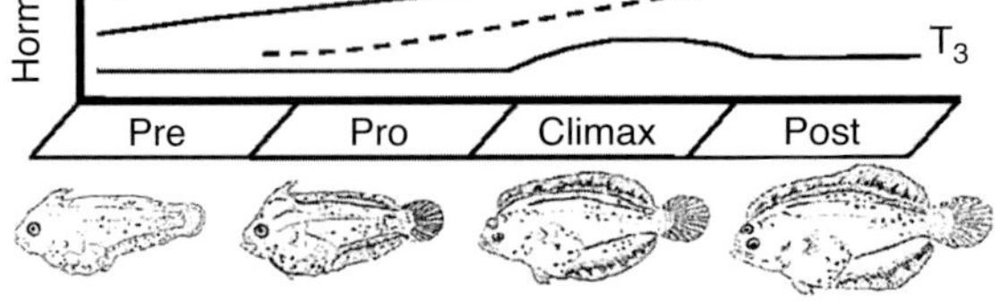

Figure 16.2 Diagram showing the endocrine profile during natural metamorphosis in flounder. Data represent only general patterns in hormone levels and were derived from studies primarily conducted with Japanese flounder, *Paralichthys olivaceus*. Data are from publications by Miwa and Inui (1987a); Miwa *et al.* (1988); Tagawa *et al.* (1990); and de Jesus *et al.* (1991, 1993). The period shown represents approximately 30 days. Thyroid-stimulating hormone (TSH, fine dotted line); cortisol (CORT, long dashed line); thyroxine (T_4, upper solid line); triiodothyronine (T_3, lower solid line); Prolactin (PRL, medium dashed line).

more time on the substrate or tank bottom as metamorphosis progresses, and showing negative phototaxis (Kawamura and Hosoya 1997). Dietary changes also accompany the change in lifestyle from larva to juvenile in wild animals, as the type of food consumed shifts from mainly plankton to small fish and crustaceans.

In addition to these outwardly evident transformations, dynamic changes occur at the tissue, cell, and molecular levels, including remodeling of the cranium and face (Okada *et al.* 2001; Hildahl *et al.* 2008), nervous system (Sæle *et al.* 2006b), musculoskeletal system (Yamano *et al.* 1991; Inui *et al.* 1995; Campinho *et al.* 2007), gastrointestinal tract (Miwa *et al.* 1992; Huang *et al.* 1998), red blood cells (Miwa and Inui 1991), and osmoregulatory tissues (Schreiber 2001). Although it will not be possible to cover all of these research areas, we will touch on a couple of them and discuss exciting recent functional genomic studies aimed at identifying novel factors that may convey sidedness during metamorphosis.

To permit eye migration and asymmetrical feeding, extensive remodeling of the cranium and face occurs (Gibb 1995; Okada *et al.* 2001). This requires that bones soften and degenerate, while others tissues expand asymmetrically to force migration of the eye socket dorsally within the cranium. This migration is thought to be aided by fibroblast proliferation and osteoclast driven mineral deposition ventral to the moving eye (Sæle *et al.* 2006a, 2006b). Recent evidence using subtractive hybridization suggests that the splicing factor arginine/serine rich-3 (SFRS3) gene is upregulated during eye migration in Japanese flounder (Bao *et al.* 2005). This along with locally produced growth factors including those involved in bone morphogenesis may mediate eye migration during flatfish metamorphosis. Douglas and colleagues (2008) have recently developed an Atlantic halibut microarray and identified numerous genes whose products may be involved in regulating metamorphosis including aspects of muscle morphogenesis, digestion, lipid transport, protein transport, cytoskeletal function, and cell motility. Employing functional genomic and proteomic techniques should aid in further elucidating those elements and pathways most critical to tissue transformational processes inherent to metamorphosis, for which muscle development may be best characterized.

Histological analysis of muscle tissue of Japanese flounder showed that marked morphological changes occur during metamorphosis (Inui *et al.* 1995). Premetamorphic larvae possess muscle characterized by thin fibers and basophilic sarcoplasm, which during metamorphosis transforms into adult-type muscle, characterized by thicker fibers and abundant myofibrils. Analysis of muscle proteins demonstrated the presence of three flounder troponin-T isoforms, larval-type, adult-type, and stable isoforms, which have subsequently been studied at the mRNA level in Atlantic halibut (Campinho *et al.* 2007). In Japanese flounder, larval-type troponin-T isoforms were present prior to metamorphosis, but disappeared prior to completion of metamorphosis. Adult-type isoforms appear as the metamorphic climax approaches and remain thereafter (Yamano *et al.* 1991). Similar transitions in myosin light-chain protein isoforms have also been documented (Yamano *et al.* 1994). These metamorphic changes in the musculature of the flounder coincide with the obligatory changes in mobility that accompany transformation from a pelagic larva to an ambush predator.

16.1.2 Endocrine regulation of metamorphosis

Several hormones are thought to play important roles in controlling flatfish metamorphosis. We briefly discuss those most central to regulating metamorphosis in flatfishes and also refer the reader to comprehensive reviews of the subject (de Jesus *et al.* 1993; Inui *et al.* 1994, 1995; Power *et al.* 2008).

Thyroid hormones

Thyroid hormones, thyroxine (T_4) and triiodothyronine (T_3), regulate a range of physiological processes including homeostasis, metabolism, and cellular differentiation and development. Research in anuran amphibians showed that thyroid hormones play a key role in regulating metamorphosis (Dent 1988). The brain–pituitary–thyroid axis (or thyrotropic axis) was thus a logical system to investigate for a role in regulating flatfish metamorphosis. In fishes, thyroid hormones are produced by diffuse thyroid follicles located along the afferent artery in response to thyroid-stimulating hormone (TSH) secreted by the pituitary gland (Norris 2007). Early research with Japanese flounder focused on identifying a possible role of thyroid hormone and TSH in metamorphosis. Researchers exposed premetamorphic flounder to seawater treated with T_4 and found that the fish metamorphosed prematurely, producing miniature juveniles (Inui and Miwa 1985). Other fish were exposed to seawater containing the antithyroidal agent thiourea (TU) that blocks production of thyroid hormone. This treatment produced abnormally large symmetrical larvae. Premetamorphic flounder were then exposed to various doses of T_4 and T_3 in combination with TU. T_3 was found to be several times more potent than T_4 at inducing metamorphic changes in dorsal fin ray length, eye migration, and settling (Miwa and Inui 1987b). T_4 was also shown to markedly stimulate the growth of the gastric glands, while TU treatment suppressed their development (Miwa *et al.* 1992), suggesting that thyroid hormone controls the proliferation of epithelial cells during development in flounder. The morphological and biochemical changes in muscle tissue during transition from the larval to adult form also can be experimentally induced by thyroid hormone or inhibited by TU treatment (Inui *et al.* 1994).

To complement functional studies assessing the effects of exogenous thyroid hormone on metamorphosis, whole-body profiles of thyroid hormone were also shown to increase during natural metamorphosis. Concentrations of T_4 were low during premetamorphosis, rose gradually during prometamorphosis, surged at the metamorphic climax, and then gradually declined postmetamorphosis (Figure 16.2; Miwa *et al.* 1988; Tagawa *et al.* 1990; de Jesus *et al.* 1991). Concentration of T_3 on the other hand remained low throughout pre- and prometamorphosis, only showing a slight rise during metamorphic climax (Tagawa *et al.* 1990). Collectively, studies on the regulation and actions of thyroid hormones suggest that they are important in inducing the tissue transformations characteristic of metamorphosis in flatfishes, a function similar to that in amphibian metamorphosis. The greater potency of T_3 in promoting metamorphosis and the lower whole-body titers of T_3 compared to T_4 during metamorphosis may be a reflection of the higher activity of monodeiodinase enzymes in peripheral tissues where T_4 is converted into the more active thyroid hormone form, T_3. This would explain the low concentrations of T_3 in circulation during metamorphosis (Inui *et al.* 1994; Tagawa *et al.* 1995) and its higher potency relative to T_4. Hence, large increases in T_4 and smaller elevations in T_3 are both likely important in eliciting the confounding number of actions inherent to "true" metamorphosis in flatfishes.

TSH released from the pituitary gland is the primary inducer of thyroid hormones. Immunocytochemistry was used to detect TSH in pituitary sections from spontaneously and artificially-induced metamorphosing Japanese flounder larvae at various stages of development (Miwa and Inui 1987a). Pituitary thyrotropes showed elevated TSH levels at premetamorphosis, which increased further during prometamorphosis (Figure 16.2). Degranulation of the TSH cells occurred at metamorphic climax, while in contrast, thyroid hormone production by thyroid follicles was at peak levels at this time. This suggested that pituitary TSH production subsequently activates the thyroid follicles (Miwa and Inui 1987a). The next step was to examine whether in vivo thyroid hormone release could be stimulated with TSH treatment. Bovine TSH was injected into prometamorphic flounder and resulted in a whole-body spike in T_4 within several hours, and responded in a dose-dependent manner (Inui *et al.* 1989). Two distinguishing metamorphic characteristics in flounder, reduction in length of dorsal fin rays and eye migration, were also significantly accelerated by TSH treatment.

Thyroid hormones exert their actions through binding and activation of thyroid hormone receptors (TRs) for which several isoforms (TRαA, TRαB, TRβ1, and TRβ2) have been cloned in Japanese flounder (Yamano and Miwa 1998). Thyroid receptors are expressed in a wide range of tissues in metamorphosing larvae, with distinct tissue specificity and timing for a given receptor subtype. TR mRNA variants were expressed in pre- and prometamorphic larvae, except for TRαB mRNA, which was low throughout larval development. TRαA mRNA expression increases during premetamorphosis and surges at the metamorphic climax, while TRβ1 and β2 increase similarly, but surge after metamorphic climax. The differential gene expression of receptor variants may enable sequential timing of physiological development elicited through thyroid hormones. For instance, TRβ mRNA expression is prolonged after completion of metamorphosis suggesting that active development of the musculoskeletal system may continue in juvenile flounder. Thus, it would appear that the differential expression of multiple TRs is likely mediating tissue changes elicited by thyroid hormones in a spatially and temporally specific manner during metamorphosis.

Cortisol

Although thyroid hormones primarily regulate metamorphosis in flounder species and presumably other flatfishes, various other hormones may act either synergistically or antagonistically with thyroid hormone. Whole body cortisol concentrations preceded, yet closely paralleled the change in T_4 concentration seen during metamorphosis (Figure 16.2; de Jesus *et al.* 1991, 1993). To test whether cortisol is involved in metamorphosis, prometamorphic flounder were treated with cortisol alone or in combination with thyroid hormones. Cortisol alone had no effect on the shortening of dorsal fin rays. However, when cortisol was added in combination with T_4 or T_3, the rate of dorsal fin ray shortening was significantly accelerated compared to fish treated only with thyroid hormones (de Jesus *et al.* 1990). These results suggest that cortisol may play a permissive role in natural flounder metamorphosis and could act synergistically with thyroid hormones to drive metamorphosis in flounder (Wada 2008).

Prolactin and the growth hormone/insulin-like growth factor axis

The antagonistic effect of prolactin on thyroid hormones is well known in amphibians. Tadpoles exposed to high concentrations of exogenous prolactin experience metamorphic stasis, developing as giant tadpoles (Norris 2007). An antagonistic effect of prolactin on metamorphosis in flatfishes was also shown through in vitro studies on dorsal fin rays of Japanese flounder (Inui *et al.* 1994). Fin rays normally regressed when treated with thyroid hormone, but when fins were treated with thyroid hormone and prolactin together, fin ray regression was retarded. Analysis of whole-body concentrations of prolactin during metamorphosis showed that levels increase steadily during early metamorphosis, reaching peak levels after metamorphic climax (Figure 16.2; de Jesus *et al.* 1993). Additionally, levels of growth hormone (GH), a closely related protein to prolactin, appear to mimic increases in prolactin. Based on the overall endocrine profile during flounder metamorphosis, it appears that prolactin antagonizes thyroid hormone actions, eventually halting metamorphosis.

The GH and insulin-like growth factor (IGF) system is a primary regulator of skeletal growth in fish. Accordingly, they are prominent regulators of cartilage growth and bone deposition. Recent results show that the expression patterns of GH and IGF receptors are closely associated with cranial remodeling during halibut metamorphosis (Hildahl *et al.* 2008). Hence, it would appear that GH and IGF may act to induce cranial–facial remodeling. Immunoneutralization studies using specific antibodies to selectively remove GH or IGFs and their receptors are required to establish whether the growth regulatory hormones are causally linked to cranial morphogenesis during flatfish metamorphosis.

Sex steroids

In 1992, de Jesus *et al.* conducted in vitro studies on dorsal fin ray response to gonadal steroids in premetamorphic flounder. They were interested in possible modulation of thyroid hormone action during metamorphosis by the sex steroids, estradiol and testosterone. Addition of either estradiol or testosterone into the culture medium had little effect on flounder fin rays. However, when fins were treated with estradiol or testosterone combined with thyroid hormone, dorsal fin ray shortening was inhibited relative to controls and those treated with thyroid hormone alone. In additional experiments, prometamorphic flounder were reared in seawater containing either estradiol or testosterone, both of which delayed metamorphosis compared to untreated controls (Inui *et al.* 1994). Unlike the above hormones, whole-body concentrations of sex steroids were detected at very low levels throughout metamorphosis (de Jesus *et al.* 1992) and thus the physiological significance of sex steroids during metamorphosis remains unclear, particularly since the gonads of flounder do not differentiate, and hence may not produce gonadal steroids, until after metamorphosis is completed (Luckenbach *et al.* 2003).

16.2 Sex determination

Flatfishes are also useful animals for studying sex determination for several reasons. As detailed in this section, a number of flounders exhibit both genotypic and environmental sex determination *within* the species. Secondly, sex determination and initial differentiation occur in relatively large settled juveniles so sex responses can be tied to potentially important environmental variables. Third, gynogenetic techniques are established for flounder, allowing the rapid establishment of relatively genetically homogeneous stocks for study. Finally, female flounder of several species grow larger than males so there is significant incentive from an aquaculture standpoint to control sex and produce all-female culture stocks.

The best understood flatfish species from a sex determination standpoint are the flounders of the genus *Paralichthys* that include both the Japanese and Southern (*Paralichthys lethostigma*) flounder (Luckenbach *et al.* 2009). These species and other Paralichthids are important as a fisheries and aquaculture resource in Asia and America. Like virtually all flatfishes, flounder females grow faster and reach a much larger size than males. There has therefore been considerable effort aimed at understanding and manipulating sex determination in these species, both for producing faster growing all-female stocks for aquaculture and the appropriate sex ratios for stock enhancement. Early work in Japanese flounder and now in other Paralicthids indicates these species show an unusual form of sex determination (see Luckenbach *et al.* 2009 for review). Genotypic and temperature-dependent sex determination (GSD and TSD, respectively) are well documented in fishes (Baroiller and D'Cotta 2001; Devlin and Nagahama 2002; Ospina-Álvarez and Piferrer 2008). There is also documentation of genetic influences on TSD responses (Lagomarsino and Conover 1993; Rhen and Lang 1998). For instance, the Atlantic silverside (*Menidia menidia*), a fish native to the east coast of North America, shows variation in temperature responses and responsiveness both within and between populations (Conover and Kynard 1981; Lagomarsino and Conover 1993). What makes flounder unusual is that temperature responsiveness is genotype-specific. Specifically, evidence indicates that XY individuals will become male regardless of the environment experienced during early development. However, the phenotypic sex of XX individuals depends strongly on environment, with effects of temperature being most thoroughly explored (Luckenbach *et al.* 2009). Working with Japanese flounder, Yamamoto (1995) originally showed that sex ratios varied with rearing temperature, but

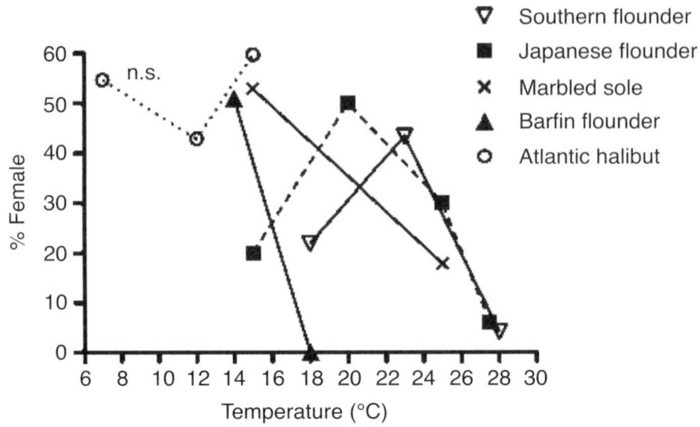

Figure 16.3 Summary of temperature effects on sex determination in flatfishes. Points represent the mean female percentage at each temperature. The figure was reproduced with permission from an article by Luckenbach and colleagues (2009). n.s., no significant effect in Atlantic halibut.

maximal proportions of females were approximately 50% at 18°C with overwhelmingly male-biased sex ratios being observed at warmer and cooler temperatures (Figure 16.3). Hypothesizing that XY animals were developing as males and XX animals were responsive to temperature, gynogenetic methods were employed to generate all XX stocks and then re-examined for sex determination responses to temperature (Yamamoto 1999).

Consistent with the hypothesis that XX animals are responsive to rearing temperature, much higher proportions of females than males were observed in treatments that maximized females in previous work, although these numbers never reach 100% and can be significantly lower. This failure to generate 100% females strongly suggests that either there is significant genetic variation in XX individuals in responsiveness to temperature or that other factors are also capable of affecting sex determination. Both possibilities are consistent with patterns in other species including the southern flounder (for review see Luckenbach *et al.* 2009).

Our team has explored factors affecting sex determination and means of manipulating this process in southern flounder. As with Japanese flounder, rearing temperature during the postsettlement juvenile period has a very strong influence on resultant sex ratios in southern flounder, again with a mid-range temperature (23°C) producing the greatest proportions of females and higher (28°C) and lower (18°C) temperatures producing primarily males (Figure 16.3, Luckenbach *et al.* 2003). We have captured southern flounder on the southeastern coast of the United States where temperatures exceed 28°C during summer and when gonads are undergoing differentiation. This suggests that temperatures causing the development of the male phenotype in genotypic XX females may be ecologically relevant.

Histologically discernible gonadal differentiation in southern flounder occurs beginning at approximately 85 mm total length (TL) and males and females are clearly distinguishable based on gonadal structure by 120 mm TL. This is a larger size at differentiation than has been described for Japanese flounder where ovarian and testicular differentiation begin at approximately 30 and 37 mm TL, respectively (Tanaka 1987; Yamamoto 1995, 1999). This species difference in histologically discernible sex differentiation also is mirrored by a difference in the expression of a key sex-determining enzyme, cytochrome P450 aromatase, which converts testosterone to estrogen. The onset of sexually dimorphic expression of aromatase (higher in developing females than males) occurs much earlier in the Japanese than southern flounder (Kitano *et al.* 1999; Luckenbach *et al.* 2005; discussed further below). This difference in the timing of sex differentiation relative to body size is not understood, but the flounders are potentially a useful system for exploring variations in this key life history variable.

Interestingly, we found that the temperature that produces the maximal number of females (50:50 female to male sex ratios) is also the temperature that induces greater growth rate of southern flounder. Hence, southern flounder cultured at 23°C exhibited double the growth rate relative to those fish grown at the higher temperature (28°C) that induced male development. A significant increase in IGF-I, a potential proxy of growth rate status of fishes (Picha *et al.* 2008), was also found in fish grown at 23°C relative to those maintained at the higher temperature (Luckenbach *et al.* 2007). The differences in growth rate in response to temperature occur after the sex-determining period and were independent of the sex of the fish. Consistent with the Charnov and Bull (1977) hypothesis for environmental sex determination, we postulate that the temperature that yielded poorer growth rates (high temperature) may favor male development as males mature at smaller sizes than females and hence will achieve higher fitness than would be likely for a female when grown at warmer temperatures. Conversely, female reproductive fitness is likely favored by rapid growth and therefore there may be a strong selection for maximizing the number of females at temperatures where high growth is observed (Godwin *et al.* 2003).

Southern flounder have a lower latitude distribution than Japanese flounder and both species inhabit significantly warmer waters than the barfin flounder (*Verasper moseri*) (Godwin *et al.* 2003). Inspection of the temperature response profiles shown in Figure 16.3 suggests a "warm-shifted" response for southern and Japanese flounder relative to barfin flounder and potentially a similar but smaller difference between southern and Japanese flounder. Although the detailed studies necessary to critically test for such a latitudinal difference in TSD have not been performed, this phenomenon has been observed in reptiles with TSD (Rhen and Lang 1998). A shift from TSD to GSD with increasing latitude is observed in the Atlantic silverside (Lagomarsino and Conover 1993). However, differences in temperature responses have not been described across related species to our knowledge and the potential of latitudinal differences in responses within species has not been addressed in flatfishes. These species may be promising model systems for exploring the physiological and molecular bases of such latitudinal differences with the additional benefit that some flatfishes have been tested for TSD and do not apparently exhibit this pattern (e.g., Atlantic halibut, Figure 16.3; Hughes *et al.* 2008).

16.2.1 Other factors influencing sex determination

A variety of factors are known to influence sex determination and differentiation in fishes (Devlin and Nagahama 2002), but it is often not clear whether the environmental influences are ecologically relevant. As previously indicated, genotypic XX populations of Japanese flounder may not all develop as females even at the temperature that maximizes female development, suggesting other environmental variables or individual variation may occur with sex determination. Recent results suggest factors other than temperature may regulate sex determination in other flounder as well, and in ways that are selectively advantageous. For example, tank color can strongly affect sex determination in southern flounder with a light blue background inducing a significantly greater proportion of individuals to develop as males than darker backgrounds (Turner 2008; in preparation). This may be relevant to the natural situation as flounder rely on crypsis to both function as successful ambush predators and also to avoid detection by larger predators.

Whether the responses to temperature or tank color are mediated by a common factor, perhaps associated with a stress response, is unknown. Cortisol is a glucocorticoid steroid secreted from the interrenal gland of fish in response to stress. The steroid increases glucose as part of the "fight-or-flight" response. We are currently assessing whether cortisol might mediate some of the effects of environmental variables influencing sex determination in flounder.

16.2.2 Physiological mediation of sex determination responses

How might the effects of environmental factors or physiological mediators such as cortisol be linked to the sex determination response? As with many other species of fish, sex determination in flounder can be strongly influenced by steroid hormone manipulations and endogenous steroid patterns during sexual development, which is consistent with a key role for these hormones. Much of the attention in this area has focused on the rate-limiting enzyme in estrogen formation from androgens, aromatase. Unlike tetrapods, fishes have both a gonadal and brain form of aromatase (Piferrer and Blázquez 2005). Attention has focused on the gonadal form of this enzyme (also referred to as *cyp19a1a*) in relation to TSD with any potential role of the brain form remaining largely unknown. As noted above, increases in aromatase expression coincide with female differentiation in both Japanese and southern flounder and this increase is associated with an increase in estrogen concentrations in Japanese flounder (Kitano *et al.* 1999; Luckenbach *et al.* 2005). Similar increases in aromatase gene expression during ovarian differentiation have been recently reported in the half-smooth tongue sole (*Cynoglossus semilaevis*) as well (Deng *et al.* 2009). We found that early elevations of aromatase gene expression in southern flounder accurately predicted female sex differentiation in an experiment where fish were raised in captivity with one group sampled for gonadal aromatase expression around the time of sex determination and a

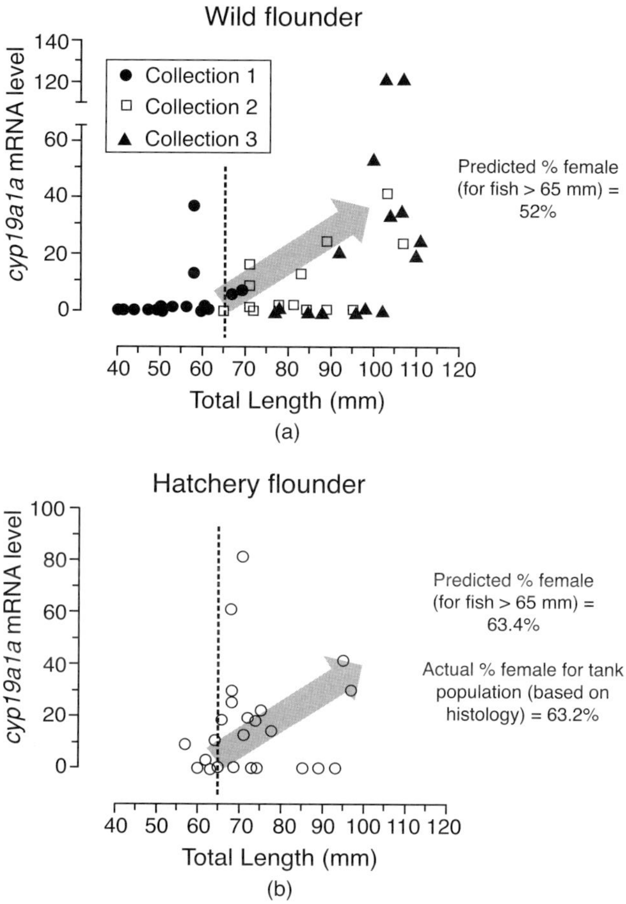

Figure 16.4 P450 aromatase (*cyp19a1a*) mRNA levels in gonads of wild juvenile southern flounder (a) and gonads of hatchery-reared southern flounder (b) as determined by quantitative PCR. The range of bottom temperatures at capture sites for wild flounder collections 1, 2, and 3 were 23.2–24.1°C, 24.6–26.7°C, and 29.1–29.3°C, respectively. The gray arrows denote the increase in aromatase expression in differentiating females beginning at ≈65 mm total length (dotted line). Expression of *cyp19a1a* was normalized to elongation factor-1 alpha (*ef1a*) expression. The figure was reproduced from Luckenbach and colleagues (2009).

second group from the same experimental population sampled later after gonadal sex could be unambiguously assigned based on gonadal histology (Luckenbach *et al.* 2005; Figure 16.4). Agreement between the two sexing approaches was excellent. With regard to TSD, high temperature exposure suppressed aromatase mRNA levels in XX Japanese flounder (Kitano *et al.* 1999) providing a direct link between temperature and aromatase regulation and subsequent sex determination.

What else might regulate aromatase expression? Cortisol is one candidate. The promoter region of the gonadal form of aromatase has binding sites ("response elements") for a variety of steroid hormone receptors and other regulators of transcription. A discussion of factors regulating aromatase expression is beyond the scope of this chapter, but has been recently reviewed (Luckenbach *et al.* 2009). The key point here is that because of the temperature-dependent expression of aromatase and the favorability of the system in terms of generating all-XX populations for study, flounder are an excellent system for studying ecologically relevant environmental modulation of aromatase expression as well as the role of potential physiological mediators of these effects.

16.3 Conclusion and future research directions

After completion of metamorphosis a once pelagic larva has transformed into a benthic, carnivorous, ambush predator, probably the most striking change associated with flatfish metamorphosis is migration of one eye to the opposite side of the body. However, other dynamic changes occur externally and internally, including changes in organs, tissues, and at the cellular and molecular levels. Although the overall picture is not complete for flatfish metamorphosis, it is clear from studies summarized above that TSH and the thyroid hormones play a prominent role in regulating this process. Thyroid hormones appear to be responsible for transformation of various systems of the body, including the gastrointestinal tract and musculature.

The upstream, presumably brain/hypothalamic and environmental and nutritional factors that regulate TSH in fishes, and the subsequent metamorphic rise in thyroid hormone, as well as those elements that regulate eye migration require identification and further study. Continued research in this area will aid in identifying factors that are responsible for abnormal metamorphosis commonly observed in aquaculture operations and synchronizing metamorphosis, settlement, and development of flatfishes in captivity.

Flatfishes show a strong sexual dimorphism in growth rate with females achieving substantially larger sizes than males. From an aquaculture point of view, it is favorable to culture all female stocks if one is to improve production efficiency of flatfish culture. Likewise, stock enhancement of some flatfishes is ongoing or under consideration in order to restore depleted wild stocks. It is important, therefore, that hatcheries produce juveniles of appropriate sex ratios, presumed to be 50:50 female to male, for sound enhancement practices. Flatfishes exhibit either pure GSD or a combination of GSD and environmental sex determination with the XX genotype in Paralichthids being prone to the influence of the environment. An understanding of those environmental cues that regulate sex determination is clearly required if we are to properly control sex ratios to maximize the female phenotype in all XX genotype populations generated through meiogynogenesis or to minimize differentiation of the male phenotype in natural populations containing the XX genotype. The mechanisms through which environmental variables alter aromatase, a key enzyme regulating female sex differentiation, and by default male differentiation, remain unclear. Flatfishes provide an excellent model system for identifying the potentially common element(s) mediating aromatase responsiveness and environmental sex determination in vertebrates.

16.4 Acknowledgments

We would like to thank all of those who contributed to the southern flounder research partly covered in this review including Dr. Harry V. Daniels, Ryan Murashige, Andrew J. Morgan, Poem M. Turner, Lea W. Early, Ashley H. Rowe, and others. We also appreciate those agencies that have supported our work on flatfish biology and sex determination in fishes including North Carolina Sea Grant, Salstonstall-Kennedy Program of the National Marine Fisheries Service, and USDA Cooperative State Research, Education, and Extension Service (CSREES), National Science Foundation, and the National Institutes of Health.

Literature cited

Bao, B., Yang, G., Liu, Z., Li, S., Wang, Z., and Ren, D. 2005. Isolation of SFRS3 gene and its differential expression during metamorphosis involving eye migration of Japanese flounder *Paralichthys olivaceus*. *Biochimica et Biophysica Acta* 1725:64–70.

Baroiller, J.F., and D'Cotta, H. 2001. Environment and sex determination in farmed fish. *Comparative Biochemistry Physiology* 130C:399–409.

Campinho, M.A., Silva, N., Nowell, M.A., Llewellyn, L., Sweeney, G.E., and Power, D.M. 2007. *Troponin T* isoform expression is modulated during Atlantic Halibut metamorphosis. *BMC Developmental Biology* 7:71.

Chambers, R.C., and Leggett, W.C. 1987. Size and age at metamorphosis in marine fishes – an analysis of laboratory-reared winter flounder (Pseudopleuronectes-americanus) with a review of variation in species. *Canadian Journal of Fisheries and Aquatic Sciences* 44(11):1936–1947.

Charnov, E.L., and Bull, J.J. 1977. When sex is environmentally determined. *Nature* 266:828–830.

Conover, D.O., and Kynard, B.E. 1981. Environmental sex determination: interaction of temperature and genotype in a fish. *Science* 213:577–579.

de Jesus, E.G., Hirano, T., and Inui, Y. 1991. Changes in cortisol and thyroid hormone concentrations during early development and metamorphosis in the Japanese flounder, *Paralichtys olivaceus*. *General and Comparative Endocrinology* 82:369–376.

de Jesus, E.G., Hirano, T., and Inui, Y. 1992. Gonadal steroids delay spontaneous flounder metamorphosis and inhibit T_3-induced fin ray shortening in vitro. *Zoological Science* 9:633–638.

de Jesus, E.G., Hirano, T., and Inui, Y. 1993. Flounder metamorphosis: its regulation by various hormones. *Fish Physiology and Biochemistry* 11:323–328.

de Jesus, E.G., Inui, Y., and Hirano, T. 1990. Cortisol enhances the stimulating action of thyroid hormones on dorsal fin-ray resorption of flounder *in vitro*. *General and Comparative Endocrinology* 79:167–173.

Deng, S.-P., Chen, S.-L., Xu, J.-Y., and Liu, B.-W. 2009. Molecular cloning, characterization and expression analysis of gonadal P450 aromatase in the half-smooth tongue sole, *Cynoglossus semilaevis*. *Aquaculture* 287:211–218.

Dent, J.N. 1988. Hormonal interaction in amphibian metamorphosis. *American Zoologist* 28:297–308.

Devlin, R.H., and Nagahama, Y. 2002. Sex determination and sex differentiation in fish: an overview of genetic, physiological, and environmental influences. *Aquaculture* 208:191–364.

Douglas, S.E., Knickle, L.C., Williams, J., Flight, R.M., and Reith, M.E. 2008. A first generation Atlantic halibut *Hippoglossus hippoglossus* (L.) microarray: application to developmental studies. *Journal of Fish Biology* 72:2391–2406.

Gibb, A.C. 1995. Kinematics of prey capture in a flatfish, *Pleuronichthys verticalis. The Journal of Experimental Biology* 198:1173–1183.

Godwin, J., Luckenbach, J.A., and Borski, R.J. 2003. Ecology meets endocrinology: environmental sex determination in fishes. *Evolution and Development* 5:40–49.

Hildahl, J., Power, D.M., Björnsson, B.T., and Einarsdóttir, I.E. 2008. Involvement of growth hormone-insulin-like growth factor I system in cranial remodeling during halibut metamorphosis as indicated by tissue- and stage-specific receptor gene expression and the presence of growth hormone receptor protein. *Cell Tissue Research* 332:211–225.

Huang, L., Schreiber, A.M., Soffientino, B., Bengston, D.A., and Specker, J.L. 1998. Metamorphosis of summer flounder (*Paralichthys dentatus*): thyroid status and the timing of gastric gland formation. *Journal of Experimental Zoology* 280:413–420.

Hughes, V., Benfey, T.J., and Martin-Robichaud, D.J. 2008. Effect of rearing temperature on sex ratio in juvenile Atlantic halibut, *Hippoglossus hippoglossus. Environmental Biology of Fish* 81:415–419.

Inui, Y., and Miwa, S. 1985. Thyroid hormone induces metamorphosis of flounder larvae. *General and Comparative Endocrinology* 60:450–454.

Inui, Y., Miwa, S., and Yamano, K. 1994. Hormonal control of flounder metamorphosis. In: Davey, K.G., Peter, R.E., and Tobe, S.S. (eds) *Perspectives in Comparative Endocrinology*. National Research Council of Canada, Ottawa, Canada, pp. 408–411.

Inui, Y., Tagawa, M., Miwa, S., and Hirano, T. 1989. Effects of bovine TSH on the tissue thyroxine level and metamorphosis in prometamorphic flounder larvae. *General and Comparative Endocrinology* 74:406–410.

Inui, Y., Yamano, K., and Miwa, S. 1995. The role of thyroid hormone in tissue development in metamorphosing flounder. *Aquaculture* 135:87–98.

Kawamura, K., and Hosoya, K. 1997. Larval morphometry of the Japanese flounder, *Paralichtys olivaceus. Ichthyological Research* 44(4):389–398.

Kitano, T., Takamune, K., Kobayashi, T., Nagahama, Y.Y, and Abe, S-I. 1999. Suppression of P450 aromatase gene expression in sex-reversed males produced by rearing genetically female larvae at a high water temperature during a period of sex differentiation in the Japanese flounder (*Paralichthys olivaceus*). *Journal of Molecular Endocrinology* 23:1–10.

Lagomarsino, I.V., and Conover, D.O. 1993. Variation in environmental and genotypic sex-determining mechanisms across a latitudinal gradient in the fish, *Menidia menidia. Evolution* 47:487–494.

Luckenbach, J.A., Borski, R.J., Daniels, H.V., and Godwin, J. 2009. Sex determination in flatfishes: mechanisms and environmental influences. *Seminars in Cell and Developmental Biology* 20:256–263.

Luckenbach, J.A., Early, L.W., Rowe, A.H., Borski, R.J., Daniels, H.V., and Godwin, J. 2005. Aromatase cytochrome P450: cloning, intron variation, and ontogeny of gene expression in southern flounder (*Paralichthys lethostigma*). *Journal of Experimental Zoology* 303A:643–656.

Luckenbach, J.A., Godwin, J., Daniels, H.V., and Borski, R.J. 2003. Gonadal differentiation and effects of temperature on sex determination in southern flounder (*Paralichthys lethostigma*). *Aquaculture* 216:315–327.

Luckenbach, J.A., Murashige, R., Daniels, H.V., Godwin, J., and Borski, R.J. 2007. Temperature affects insulin-like growth factor I and growth of juvenile southern flounder, *Paralichthys lethostigma. Comparative Biochemistry Physiology* 146A:95–104.

Miwa, S., and Inui, Y. 1987a. Histological changes in the pituitary-thyroid axis during spontaneous and artificially-induced metamorphosis of larvae of the flounder *Paralichtys olivaceus. Cell and Tissue Research* 249:117–123.

Miwa, S., and Inui, Y. 1987b. Effects of various doses of thyroxine and triiodothyronine on the metamorphosis of flounder (*Paralichtys olivaceus*). *General and Comparative Endocrinology* 67:356–363.

Miwa, S., and Inui, Y. 1991. Thyroid hormone stimulates the shift of erythrocyte populations during metamorphosis of the flounder *Journal of Experimental Zoology* 259:222–228.

Miwa, S., Tagawa, M., Inui, Y., and Hirano, T. 1988. Thyroxine surge in metamorphosing flounder larvae. *General and Comparative Endocrinology* 70:158–163.

Miwa, S., Yamano, K., and Inui, Y. 1992. Thyroid hormone stimulates gastric development in flounder larvae during metamorphosis. *Journal of Experimental Zoology* 261:424–430.

Norris, D.O. 2007. *Vertebrate Endocrinology*, 4th edn. Academic Press, New York.

Okada, N., Takagi, Y., Seikai, T., Tanaka, M., and Tagawa, M. 2001. Asymmetrical development of bones and soft tissues during eye migration of metamorphosing Japanese flounder, *Paralichthys olivaceus. Cell and Tissue Research* 304(1): 59–66.

Ospina-Álvarez, N., and Piferrer, F. 2008. Temperature-dependent sex determination in fish revisited: prevalence, a single sex ratio response pattern, and possible effects of climate change. *PLoS ONE* 3:e2837.

Picha, M.E., Turano, M.J., Beckman, B.R., and Borski, R.J. 2008. Endocrine biomarkers of growth and applications to aquaculture: a minireview of growth hormone, insulin-like growth factor (IGF)-I, and IGF-binding proteins as potential growth indicators in fish. *North American Journal of Aquaculture* 70:196–211.

Piferrer, F., and Blázquez, M. 2005. Aromatase distribution and regulation in fish. *Fish Physiology and Biochemistry* 31:215–226.

Power, D.M., Einarsdóttir, I.E., Pittman, K., Sweeney, G.E., Hildahl, J., Campinho, M.A., Silva, N., Sæle, O., Galay-Burgos, M., Smáradóttir, H., and Björnsson, B.T. 2008. The molecular and endocrine basis of flatfish metamorphosis. *Reviews in Fisheries Science* 16:95–111.

Rhen, T., and Lang, J.W. 1998. Among-family variation for environmental sex determination in reptiles. *Evolution* 52:1514–1520.

Sæle, Ø., Silva, N., and Pittman, K. 2006a. Post-embryonic remodeling of neurocranial elements: a comparative study of normal versus abnormal eye migration in a flatfish, the Atlantic halibut. *Journal of Anatomy* 209:31–41.

Sæle, Ø., Smáradóttir, H., and Pittman, K. 2006b. Twisted story of eye migration in flatfish. *Journal of Morphology* 267:730–738.

Schreiber, A.M. 2001. Metamorphosis and early larval development of the flatfishes (Pleuronectiformes): an osmoregulatory perspective. *Comparative Biochemistry and Physiology B (Biochemistry and Molecular Biology)* 129:587–595.

Tagawa, M., de Jesus, E.G., and Hirano, T. 1995. The thyroid hormone monodeiodinase system during flounder metamorphosis. *Aquaculture* 135:127–129.

Tagawa, M., Miwa, S., Inui, Y., de Jesus, E.G., and Hirano, T. 1990. Changes in thyroid hormone concentrations during early development and metamorphosis of the flounder, *Paralichthys olivaceus. Zoological Science* 7:93–96.

Tanaka, H. 1987. Gonadal sex differentiation in flounder, *Paralichthys olivaceus. Bulletin of National Research Institute of Aquaculture* 11:7–19 (in Japanese; with English abstract).

Turner, P.M. 2008. Effects of light intensity and tank background color on sex determination in southern flounder (*Paralichthys lethostigma*). MSc thesis, North Carolina State University, Raleigh, NC.

Wada, H. 2008. Glucocorticoids: mediators of vertebrate ontogenetic transitions. *General and Comparative Endocrinology* 156:441–453.

Yamamoto, E. 1995. Studies on sex-manipulation and production of cloned populations in hirame flounder, *Paralichthys olivaceus*. *Bulletin of the Tottori Prefecture Fisheries Experimental Station* 34:1–145 (in Japanese, with English summary).

Yamamoto, E. 1999. Studies on sex-manipulation and production of cloned populations in hirame, *Paralichthys olivaceus* (Temminck et Schlegel). *Aquaculture* 173:235–246.

Yamano, K., and Miwa, S. 1998. Differential gene expression of thyroid hormone receptor α and β in fish development. *General and Comparative Endocrinology* 109:75–85.

Yamano, K., Miwa, S., Obinata, T., and Inui, Y. 1991. Thyroid hormone regulates developmental changes in muscle during flounder metamorphosis. *General and Comparative Endocrinology* 81:464–472.

Yamano, K., Takano-Ohmuro, H., Obinata, T., and Inui, Y. 1994. Effect of thyroid hormone on developmental transition of myosin light chains during flounder metamorphosis. *General and Comparative Endocrinology* 93:321–326.

Chapter 17

Behavioral quality of flatfish for stock enhancement

John Selden Burke and Reji Masuda

Efforts to enhance wild populations through the release of animals reared in captivity have been hampered by low survival rates. Low survival of hatchery reared (HR) marine fishes in the wild, appears to be caused by deficits in behavioral quality that render stocked juveniles particularly vulnerable to predators (Blaxter 1976; Olla *et al.* 1994). Similar problems occur in attempts to enhance terrestrial animal populations. Supplementation of an endangered prairie chicken population (*Tympanuchus cupido attwateri*) met with marginal success due to high predation rates on pen-reared birds, deficient in predator avoidance behavior and flight endurance (Hess *et al.* 2005). Field studies of the fate of HR fishes suggest that much of their mortality occurs shortly after stocking (Brown and Laland 2003). Low survival of stocked fishes may be partially due to stress and the release of inherently poor quality individuals (Wales 1954); however, the ecological viability of all HR fish may be compromised to some extent, due to the behavioral impact of development within the hatchery environment. While ensuring high survival in captivity, the artificial environment of the hatchery appears to induce development of morphological and behavioral traits poorly suited to natural conditions (Olla *et al.* 1994; Masuda and Tsukamoto 1998; Stoner and Glazer 1998).

A variety of commercially important flatfish species are cultured for release to the wild and advances in culture and stocking technique have resulted in successful enhancement programs (Kitada *et al.* 1992; Stoettrup *et al.* 2002). The largest flatfish stock enhancement effort is in Japan where approximately 25 million Japanese flounder, *Paralichthys olivaceus,* are stocked annually by prefectural and national hatcheries (Tomiyama *et al.* 2008). Survival of stocked Japanese flounder is spatially and temporally variable, but even in the programs demonstrated to be economically viable, work on improving the efficiency and flexibility of stock enhancement is needed (Tomiyama *et al.* 2008). Culture research has made significant progress in improving the efficiency and quality of juvenile seed production (Seikai 1998); however, striking an economic balance between production efficiency and juvenile behavioral quality is likely to be particularly challenging. Our goal is to identify likely mechanisms responsible for development of behavioral deficits of cultured flatfish and consider methods for improving behavioral quality. Throughout we emphasize the importance of understanding the ecological characteristics of the system to be stocked and the behavior of wild juveniles in it. We view such knowledge as critical to identification and correction of behavioral deficits of fish cultured for enhancement. Specific questions we address include:

1. Why do deficits in behavioral quality develop in the hatchery and what aspects of the rearing process are likely to cause them in flatfishes?
2. How do flatfish enhancement programs cope with the behavioral quality problem?
3. What approaches appear promising in improving behavioral quality and thus survival?

17.1 Behavioral quality and the hatchery environment

Behavioral deficits of HR flatfish presumably result from the process of domestication, a developmental phenomena which occurs because of the genetic changes over generations as well as the environmental stimulation and experience of the individual in captivity (Price 1984). Stock enhancement programs have sought to minimize genetic difference between animals reared for enhancement and the target wild population in recognition that domestication may reduce capacity for survival in the wild and the potential for surviving domesticated animals to alter the gene pool (Allendorf and Phelps 1980; Price 1999). Evaluation of the impact of practical hatchery practices on the genetic composition of HR Japanese flounder showed that genetic variation generated by tank spawning was significantly lower than expected and selective operations such as grading further reduced genetic variation of juveniles used for enhancement (Sekino *et al.* 2003). Careful breeding strategies have been developed that provide HR juveniles of genetic variation comparable to the wild stock (Asahida *et al.* 2003); however, development within the hatchery environment is expected to result in expression of a domestic phenotype despite underlying "natural" genetic variation.

The impact of captive conditions on phenotypic expression and the magnitude of the impact on behavioral quality will vary relative to the duration and ontogenetic timing of exposure. Domestication probably starts during the larval rearing period and progresses with a corresponding decline in behavioral quality relative to wild fish, with development (Tsukamoto *et al.* 1999). The negative impact of captivity on behavioral quality can be expected to be particularly important during the morphologically and behaviorally flexible juvenile stage (Tsukamoto *et al.* 1999; Kihslinger and Nevitt 2006). Impact may also vary among individuals and species relative to their inherent behavioral and physiological flexibility. For species such as Japanese flounder, whose wide geographic distribution indicates considerable flexibility, captive conditions may readily cause the development of traits appropriate to the hatchery environment. While such traits may be advantageous relative to the hatchery's artificial social, structural, and feeding environment, they are unlikely to be suited to conditions in the wild.

Hatchery conditions may lack or obscure stimuli that act as cues in development of instinctive patterns of behavior important to survival in nature. Crowded conditions, frequent human intervention, and limited dietary, structural, and environmental variability experienced within the hatchery heighten response thresholds and desensitize animals to subtle changes in their environment (Price 1999). Crowded conditions generally associated with flatfish culture result in juveniles experiencing relatively constant social interactions, which may reduce the threshold for response to other individuals. Such a reduction in responsiveness though generally appropriate for hatchery conditions, can be dangerous in both the hatchery and the wild. Crowding can result in development of size hierarchies and high mortality rates due to cannibalism in flatfish culture tanks (Burke *et al.* 1999). Grading of juveniles is effective in reducing mortality within the hatchery (Dou *et al.* 2004), but may be counterproductive when rearing fish for release as encountering individuals that vary in size and intent is likely important for learning to cope with predators. An alternative to restricting size distribution in the hatchery is provision of a refuge from cannibals. Mortality of small HR Japanese flounder was significantly reduced when sand was provided in tanks with mixed size groups of juveniles (Dou *et al.* 2000).

The simple structure of the hatchery environment may fail to provide flatfish the stimulation required to develop sheltering skills essential for survival in the wild. Experimental work has shown that HR summer flounder buried in sediment less, took significantly longer to exhibit cryptic coloration and spent significantly more time in the water column than wild fish (Kellison *et al.* 2000). The observation that HR fish spend extended periods in the water column suggests that they lack the requisite caution of a juvenile whose size ensures the existence of a multitude of potential predators. Lack of exposure to the physical substrata needed for development of cryptic skills may suppress a natural instinct for concealment in HR flatfish. Hard substrata typical of juvenile culture tanks differ fundamentally from the soft bottom areas early juvenile flatfish select in nature (Burke *et al.* 1991). HR juveniles thus fail to experience the ontogenetic transition to a soft-bottom environment, an experience that may be fundamental to development of cryptic behavior and the caution appropriate to their size. As development in the hatchery proceeds, accumulated experience in the absence of shelter may override an inherent tendency toward caution in flatfish. In contrast to HR fish, wild Japanese flounder exhibit cautious behavior likely to minimize exposure to predators (Furuta 1996). When feeding, wild Japanese flounder minimized spatial and temporal exposure by rapid forays into the water column returning to the same location from which their strike originated. These observations suggest that wild juveniles are well orientated relative to their surrounding as they return to their previous (safe) sheltering location. In comparative observations, HR flounder moved slowly in the water column and returned to the bottom at some distance from their original location (Figure 17.1). Such bold behavior, focused exclusively on feeding is likely a poor strategy for flatfish whose form and behavior in the wild is adapted to concealment from and detection of both predators and prey.

Given the limited environmental variation of the hatchery environment it is not surprising that behavior of HR juveniles is focused on food at the expense

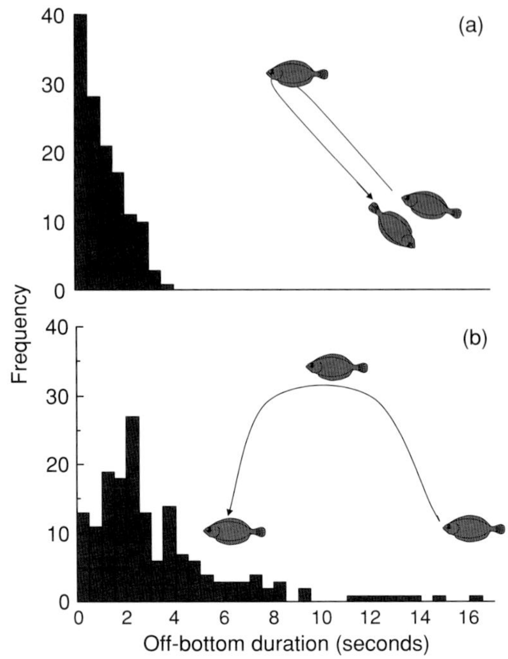

Figure 17.1 Frequency distributions of off-bottom duration in wild (a) and hatchery-reared (b) Japanese flounder juveniles while feedings. (Redrawn from Furuta (1998).) Inserted drawings represent typical swimming patterns of each origin of juveniles.

of behaviors relevant to predator avoidance. Over time, regimented feeding regimes typical of the hatchery environment are likely to condition flatfish to narrowly focus attention relative to the location and timing of feeding and the selection of food (prey). Early in the culture period, flatfish often demonstrate remarkable flexibility relative to prey selection. During larval culture, summer flounder (*Paralichthys dentatus*) larvae transition from rotifers to brine shrimp nauplii before they are trained to feed on an artificial diet (Burke *et al.* 1999). In contrast to larvae, HR juveniles receive a regimented artificial diet whose monotony in terms of quality and temporal and spatial availability is likely to be a poor preparation for feeding in the wild. Japanese flounder larvae appear to have an endogenous feeding rhythm with peak feeding in the morning and a secondary peak in the evening (Dou *et al.* 2000). Such a circadian feeding pattern is likely to be reinforced in HR juveniles as it is consistent with the daily hatchery routine. In addition to prey availability, feeding of wild flatfish juveniles is expected to be structured relative to predator abundance and feeding may be entrained to a tidal or lunar cycle (Olla *et al.* 1972; Lockwood 1980). For some HR stocks, entrainment to an environmentally appropriate cycle that conditions released fish to forage when their risks are low and the probability of encountering appropriate prey is high, may improve survival rate.

The observed lack of balance in feeding and predator avoidance behavior of HR juveniles is consistent with the belief that predation is the principal agent of mortality for released juveniles; however, poor feeding performance probably represents an important underlying reason for the high susceptibility of HR juveniles to predators. Based on landings, stocked flounder appear to have a slower growth rate than wild fish (Tomiyama *et al.* 2008) suggesting that HR flounder feeding skills are inferior to those of wild conspecifics. Poor feeding skills may increase susceptibility to predation by increasing exposure due to increased activity associated with inefficient feeding and extension of the time spent within a given size class. Yamashita *et al.* (1994) demonstrated that small (<9 cm) HR Japanese flounder had a low probability of survival in the wild. Otolith analysis indicated that those small individuals that did survive exhibited rapid growth suggesting that limiting the duration spent within this vulnerable size class was critical. Although physiological variation among recently stocked fish may contribute, variation in the efficiency of transition to wild prey is likely a key factor in surviving the initial critical encounter with wild conditions. Modifications to the hatchery environment and to feeding practices that increase the variability of feeding experience and focus attention on immediate surrounding seem likely to provide benefits in predator avoidance and feeding performance during the critical period following release.

17.2 Tactics for reducing the impact of behavioral deficits

17.2.1 Current approach

Stock enhancement tactics to address low survival rates include stocking large fish relative to the preference of wild predator and optimization of where and when to stock. Yamashita *et al.* (1994) provided evidence that survival of stocked Japanese flounder was dependent on size at release, and 10 cm TL was suggested as a cost-effective stocking size (Tomiyama *et al.* 2008). This tactic is presumed to succeed because stocked fish greater than 9 cm TL have outgrown the period of maximum vulnerability to resident predators (Yamashita *et al.* 1994). In addition to the size of juveniles at stocking, adjustment of stocking time and location relative to local patterns of predator and prey distribution has successfully been used to improve survival (Furuta 1996).

Problems associated with stocking large juveniles suggest enhancement with smaller juveniles could be more efficient. Enhancement efforts based on releasing large juveniles requires a relatively high recapture rate to achieve economic efficiency due to high rearing costs. For example, stocking of 10 cm juveniles in Fukushima Prefecture requires a recapture rate of 15%, which is difficult to obtain (Tomiyama *et al.* 2008). Also, troubling is the possibility that longer periods of captivity are likely to amplify the behavioral difference between wild and stocked fish as these differences increase with age at release (Tsukamoto *et al.* 1999). While behavioral deficits of small juveniles may be less extreme, their vulnerability due to size will require careful selection of stocking times and locations based on detailed knowledge of the target stock's life history. Such knowledge should provide an ecological basis for modifications to rearing practices and conditioning to prepare small juveniles behaviorally for release in the wild.

17.3 Life history considerations

A wild population's early life history, particularly behaviors that exploit opportunities and minimize the impact of dangers posed by the system provide insight to the survival strategy of a particular stock. Inherent to a particular survival strategy is the behavioral condition at transitional stages in the life history. This behavioral condition is determined by experience during preceding life history stages and may precondition wild fish to limit their distribution in the wild to those areas that selection has equipped them to exploit. HR fish, even when stocked at appropriate locations and times, arrive with preconditioning better suited for commercial aquaculture than the complex physical and biological demands of the wild.

The notion of preconditioning suggests that understanding the behavioral development of wild fish during the larval and early juvenile stages may be critical to understanding behavioral skills important for juvenile survival. For example, a variety of commercially important flatfish stocks develop a tidal pattern of activity during the late larval stage that allows utilization of tidal currents for transport to inshore nursery grounds. During this period, feeding rates are low suggesting that the initial function of this behavior is efficient transport to nursery grounds. This pattern of activity does not develop in stocks residing in systems that lack a consistent tidal signal or in HR flatfish indicating that transitioning wild larvae entrain to a tidal rhythm based on environmental cues (Burke *et al.* 1998, Figure 17.2). Though the initial ecological consequence of this pattern of activity appears to be directed transport, tidal patterns of activity may persist during later stages influencing distribution patterns relevant to both feeding and predator avoidance. Recently settled summer flounder (*P. dentatus*) move with the tide allowing the utilization of intertidal areas that serve as a refuge from predation and provide rich feeding grounds for a variety of juvenile flatfishes (Toole 1980). Early juvenile summer flounder invade dense *Spartina alterniflora* marshes during the flood tide where they feed on intertidal prey and emerge during the falling tide before marsh grass stands drains completely (J.S. Burke personal observation). Such observations indicate that transitioning wild summer flounder larvae arrive at estuarine nursery grounds preconditioned to exploit the feeding and refuge opportunities that the intertidal provides. HR flounder, lacking an appropriate behavioral rhythm are unlikely to utilize intertidal habitat efficiently after release. Entrainment of HR flatfish to an appropriate tidal rhythm prior to their release might be an effective means of providing appropriate preconditioning. The wild entrainment of flatfish to a tidal pattern of activity appears to depend on a complex of factors or zeitgeber (Boehlert and Mundy 1988). Tidal rhythms of activity of wild summer and southern flounder persist under constant conditions in the laboratory (Burke *et al.* 1998) indicating that if such a rhythm were conditioned in the hatchery, it would persist in HR fish after release, facilitating entrainment to the natural zeitgeber.

17.4 Environmental enrichment

Recognizing that behavioral development of wild fish relies on complex cues and that the simple psycho-sensory nature of the hatchery environment can

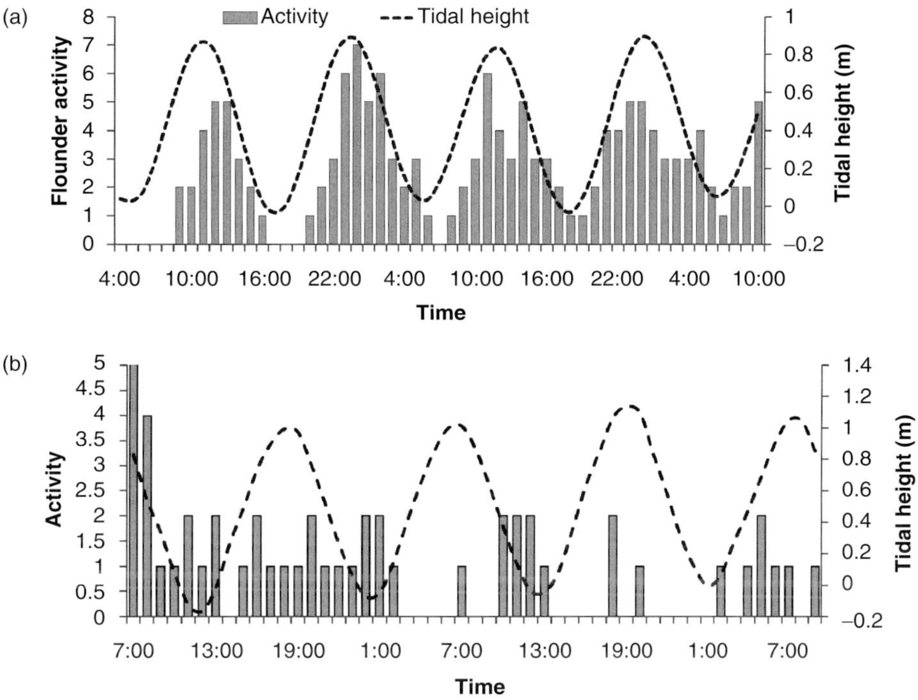

Figure 17.2 Comparison of activity patterns in the laboratory of settlement stage wild (a) and hatchery reared (b) Paralichthid flounder at 0 lx and constant temperature. Tidal heights observed in the adjacent estuary for the days of the experiments are plotted for comparison to activity.

cause behavioral deficits (Olla *et al.* 1998), one approach to improve survival of HR fish after release is enrichment of the rearing environment. While any enrichment of the normally barren rearing environment can be expected to influence behavior of HR fish, enrichment with surrogates for natural structure with which wild juveniles normally are associated may cue development of instinctive behavioral patterns critical to survival. Pioneering work with salmon showed that fish reared in hatchery systems that incorporated aspects typical of their nursery areas exhibited significantly higher survival in the wild than did control fish reared in conventional hatchery systems (Maynard *et al.* 1996). While causal factor(s) could not be determined, reduced predation rates due to improved cryptic coloration and enhanced response to movement were considered most important to higher survival.

Laboratory experiments show that rearing of juveniles in enriched environments can result in development of morphological, physiological, and behavioral traits that differ from juveniles reared in conventional hatchery systems. Salmonid juveniles from enriched rearing environments had longer dorsal fins than did control fish (Berejikian and Tezak 2005). Juveniles from enriched systems dominated control hatchery fish in behavioral experiments and grew faster when these experimental groups were mixed (Berejikian *et al.* 2000). Structurally enriched environments appear to foster exploratory behavior in juveniles (Lee and Berejikian 2008) and greater behavioral flexibility. Juveniles reared in enriched environments were able to generalize from one live prey type to another while fish reared in simple environments were not (Brown *et al.* 2003). In salmonids, there appears to be a physiological explanation for such behavioral differences as variation in rearing environment can result in differences in brain morphology commonly attributed to selection (Marchetti and Nevitt 2003; Kihslinger and Nevitt 2006). Salmon alevins reared in systems that incorporated stones developed brains with significantly larger cerebella and exhibited corresponding differences in locomotory behaviors than did genetically similar fish reared in conventional tanks (Kihslinger and Nevitt 2006).

Work with marine fishes suggests that an enriched rearing approach might be particularly effective with benthically oriented species. Cod (*Gadus morhua*) reared in enriched environments spent more time in shelter, were less active, and showed stronger antipredator responses than did fish reared without access to heterogeneous spatial cues (Salvanes and Braithwaite 2005). Work that compared behaviors of enriched and traditionally reared reef fish juveniles revealed behavioral and presumably physiological differences that can reasonably be expected to improve survival of fish raised in enriched conditions (Burke unpublished data). Enriched conditions in these experiments consisted of a sand substrate and a natural element that juveniles of the species were associated with in the wild: seagrass for a juvenile grunt, the pigfish, *Orthopristis chrysoptera*; oysters for a juvenile Serranid, the black sea bass, *Centropristis striata*; and rubble for a juvenile Sparid, the red porgy, *Pagrus pagrus*. We observed no negative effects of enriched rearing on growth or survival of the three species; however, juvenile pigfish grew faster in enriched, than in conventional tanks. In feeding behavior trials conducted in the dark, the ability of juvenile black sea bass reared in enriched tanks to locate food faster than conventionally reared sea bass appeared due to more acute olfactory skills (Figure 17.3a). Results of sea bass shelter utilization trials, conducted at a variety of light levels suggested that the enriched rearing environment may provide greater experience under low light conditions or stimulate ocular development. Sheltering experiments with a novel structure (plastic pipe fitting) suggested the ability to generalize from one type of shelter to another was enhanced in fish from enriched tanks compared to those reared in simple ones (Figure 17.3b).

During enriched rearing trials with reef fishes, development of both simple and complex behavior patterns that can be expected to be critical to survival in the wild were observed. Nocturnal observations showed that juvenile red porgies in enriched tanks oriented to specific benthic features and consequently exhibited significantly less nocturnal movement than red porgies in simple tanks (Figure 17.3c). The preconditioning of porgies to this simple orientation skill should reduce exposure to nocturnal predators after release. Juvenile sea bass in enriched culture tanks dug pits in the sand, which they defended from intruding conspecifics. Initially the location of pits appeared to be random; however, over time, sea bass "learned" to dig pits next to oyster clumps or rocks, creating burrows that presumably provided greater shelter and could be more easily defended. Extended periods of exposure to natural structural elements have also been shown to be necessary for development of critical behaviors in flatfish. Burying efficiency (the proportion of the body covered with sand after a single burial attempt) and ability to change pigmentation improved with exposure to

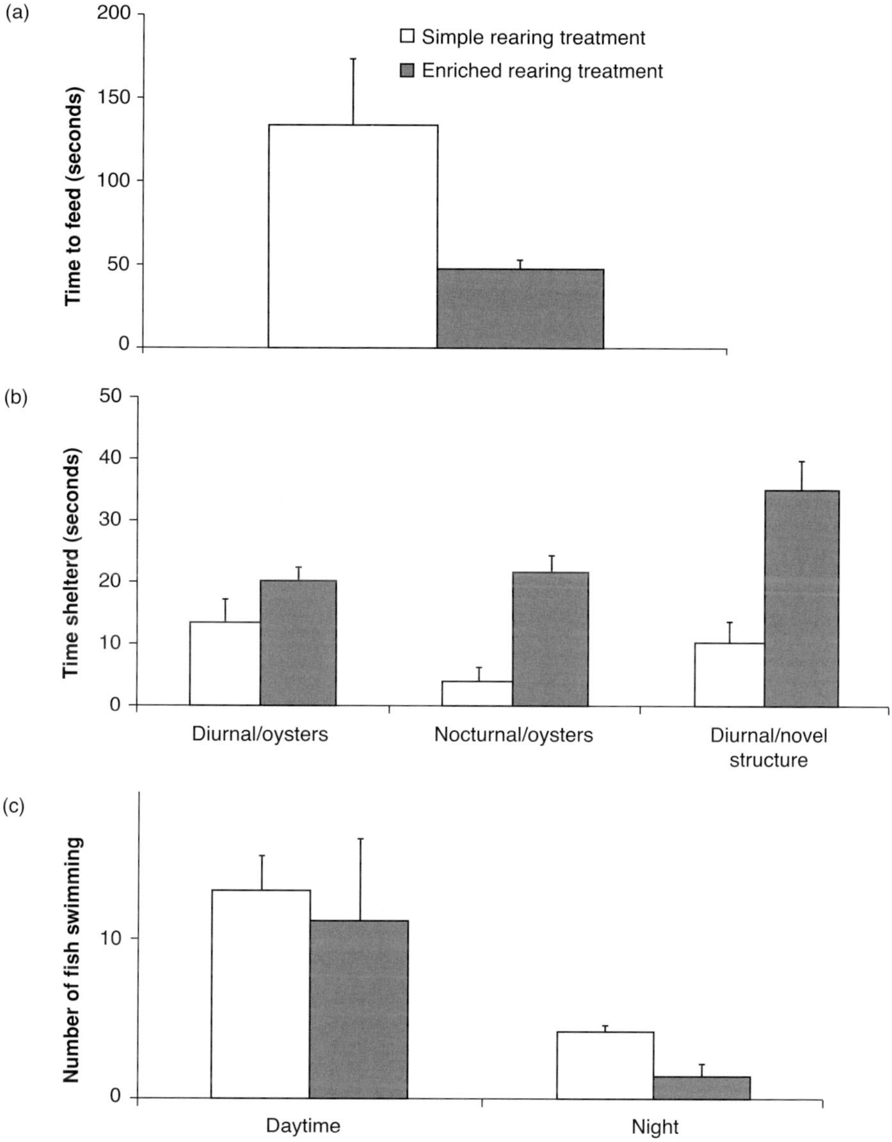

Figure 17.3 Results of laboratory experiments and hatchery observations comparing behaviors of fishes reared in simple and structurally enriched hatchery environments. (a) Black sea bass, mean time to find food at 0 lx. (b) Black sea bass, mean time spent sheltering at oyster clumps at nocturnal and crepuscular light levels and at a novel structure (plastic pipe fitting) under crepuscular light levels. (c) Red porgy, mean number of fish observed swimming in simple and enriched culture tank during day and night.

natural conditions in reared sole (*Solea solea*). Although the motivation of sole to bury was innate, 12 days of exposure to sand were required for HR sole to bury with the same efficiency as wild fish (Ellis *et al.* 1997). Collectively, these observations suggest that given sufficient exposure to the olfactory, optical, and structural variation of enriched tanks juveniles develop more complex, adaptively significant behavior than they would in conventional rearing systems.

17.5 Nutritional factors and foraging

Although during a particular developmental stage, each fish species has a size window of prey items to pursue, prey selection will be fine-tuned through a process of learning. In the hatchery, fish are usually provided with pellets, whereas wild fish (and thus released juveniles) need to feed on live prey. One major problem is that HR fish might be unintentionally trained to feed on prey items specific to hatchery. Consequently, HR fish sometimes prefer nonbiotic materials such as glass beads and stones (Ellis *et al.* 2002).

Another potential problem is that fish may have certain critical period during which their capability of learning new prey items is highest (Browman 1989). Indeed, according to Makino *et al.* (2006), striped knifejaw, *Oplegnathus fasciatus*, has the highest learning capability when they are at 7 cm body length. They suggested that this is the typical size of their recruitment from offshore pelagic to inshore rocky reef habitat so that their requirement of learning prey items is highest. Similar phenomenon has been reported in Pacific threadfin, *Polydactylus sexfilis*, that showed relatively high performance in learning trials at 50 and 90 mm fork length compared to either smaller or larger fish (Masuda and Ziemann 2000). Although there is no relevant study in flatfish, it is likely that flatfish juveniles also have certain period with high learning capability. Releasing fish on either side of such a behaviorally elastic stage may result in poor feeding success after release. For flatfish, the period following settlement when juveniles switch from pelagic to benthic prey requires flexibility and learning capability. Yamashita and Aritaki (2010) reported recapture rates of Japanese flounder from different prefectures and found that juveniles released larger than 10 cm did not necessarily have higher recapture rates than did juveniles released at smaller sizes. If 10 cm juveniles have outgrown the period of maximum behavioral elasticity, their domestication may impair learning after release.

Adaptability of feeding in new environments may suffer due to inadequate development of the central nervous system (CNS). A major biochemical component of the CNS is docosahexaenoic acid (DHA) and other highly unsaturated fatty acids. Masuda *et al.* (1999) demonstrated that dietary DHA is incorporated into the CNS in yellowtail, *Seriola quinqueradiata*, and found that DHA deficiency induces the lack of schooling behavior in this species (Masuda *et al.* 1998). Fatty acids requirement in flatfishes are typically reported based on growth, survival, and pigmentation (Bell *et al.* 2003). These symptoms of deficiency are visible, whereas insufficient development of CNS would be superficially ambiguous but may well result in low adaptability in the wild environment. Some authors reported the required DHA level based on morphological characters, but we suggest that these values could be underestimated when we target high adaptability of released fish. Furuita *et al.* (1998) compared growth and brain development of Japanese flounder larvae provided with oleic acid (OA), eicosapentanoic acid (EPA), or DHA, and found that larvae fed with OA had significantly smaller relative volume of the cerebellum. Behavioral deficiency of flatfishes reared with suboptimal fatty acid composition is yet to be studied.

17.6 Predator avoidance

The major antipredator tactic of flatfishes is burrowing in sand along with cryptic coloration. This capability is generally inferior in HR fish compared to wild counterpart, although it could be improved after some period (Figure 17.4).

Inferior coloration and/or burrowing behavior in HR flatfishes have been reported in sole *S. solea* (Ellis *et al.* 1997); Japanese flounder (Furuta 1998); summer flounder, *P. dentatus* (Kellison *et al.* 2000); and winter flounder

(a)
(b)
(c)

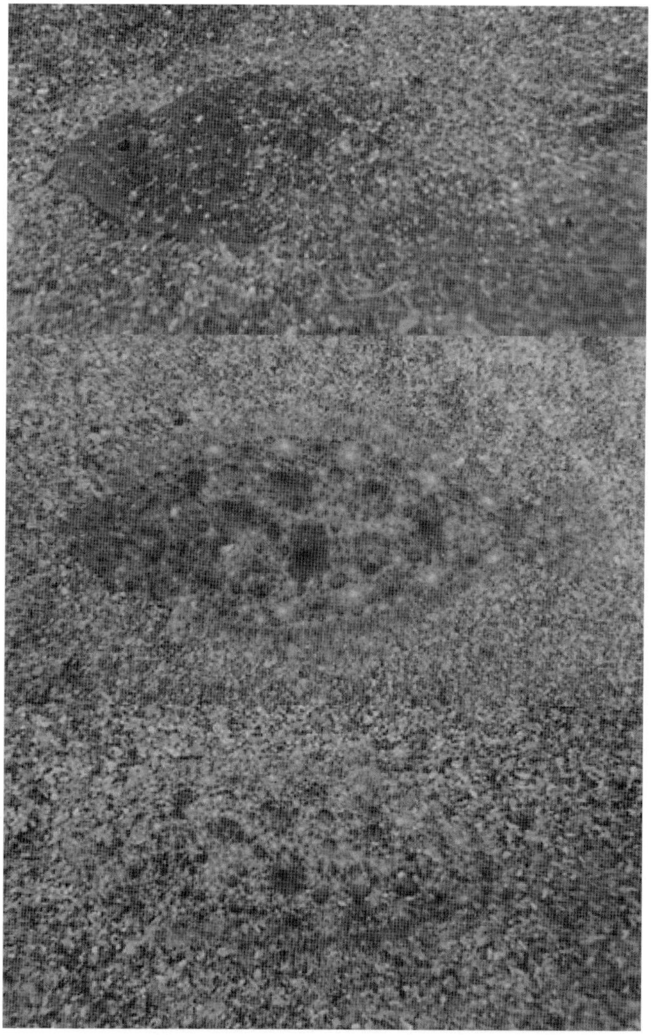

Figure 17.4 Japanese flounder juveniles in natural waters. (a) A hatchery-reared (HR) juvenile at 30 minutes after release in Wada Beach, Fukui, Japan, in June 2003. The fish was burrowing but its body color was not matched with the substrate. (b) A HR juvenile at 4 hours after release. The body color was better matched to the substrate compared to (a). (c) A wild Japanese flounder juvenile found near the individual of (b).

Pseudopleuronectes americanus (Fairchild and Howell 2004). Ellis *et al.* (1997) observed adaptation of cryptic coloration in HR sole and found that adaptation of color value (lightness) took 4–7 days, whereas that of chroma (density of color) and hue took 33 and 69 days.

Predation can occur both during daytime and at night. Miyazaki *et al.* (1997) conducted laboratory observation of HR Japanese flounder juveniles both day and night and found that fish buried themselves into sand in the daytime, but approximately 60% of fish emerged from the sand at night. This may explain high predation by crustacean predators such as swimming crab, *Charybdis japonica*, in released juveniles (Sudo *et al.* 2008) and sandy shore crab, *Mutata lunaris*, in captivity (Hossain *et al.* 2002). Hossain *et al.* (2002) demonstrated the feasibility of training HR flounder to avoid crustacean predators by direct exposure to small and thus ineffective crustacean predators or indirect exposure to large crabs which were separated from flounders by a fence.

On-site acclimatization before release, combining substratum acclimatization to improve capability of burrowing and cryptic coloring and the training against predators, seems promising. The efficiency of such *in situ* acclimatization, however, is controversial. Fairchild *et al.* (2008) reported that acclimatization cage for the release of winter flounder also attracted predators. One possible solution would be a release accompanying SCUBA divers or snorkelers (Figure 17.5), because potential predators of released fish generally avoid divers. Thus, released fish can be protected, while being acclimatized in new environment. This is likely to be plausible at shallow waters and with the help of volunteer local divers.

Figure 17.5 Release of Japanese flounder juveniles with the help of a SCUBA diver. June 2000, Wada Beach, Fukui, Japan.

Arai *et al.* (2007) found that Japanese flounder, in addition to learning from individual experience, can learn predator avoidance through observational learning, i.e., learning through observing the success or failure of other individuals. They conditioned flounder juveniles to predation using either direct exposure to a predator or allowing juveniles to observe predatory attacks on conspecifics. Comparing their behavior to naïve fish, Arai *et al.* (2007) found that both predator-exposed and predator observed fish were better at avoiding a predator than were naïve fish (Figure 17.6a). Fish trained with either method also showed less off-bottom feeding when they were fed on pellets (Figure 17.6b), thus their cautiousness was maintained in the context of feeding without the presence of predation pressure.

Observational (or social) learning is beneficial because naïve fish can quickly and efficiently acquire locally adaptive behavior from more knowledgeable

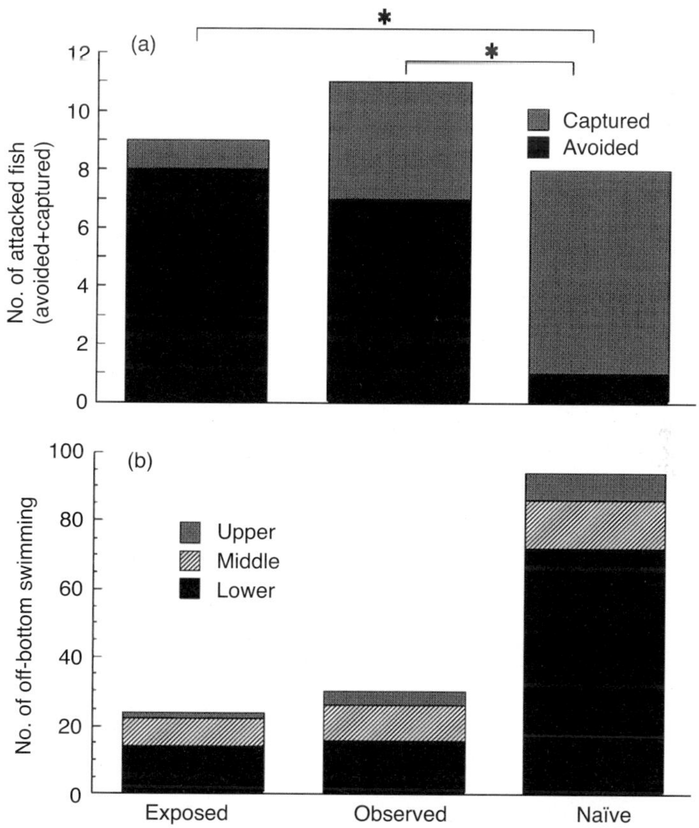

Figure 17.6 Predator avoidance and off-bottom swimming in Japanese flounder juveniles at 6 hours after the first encounter with a predator (sculpin). Flounder juveniles were either exposed to a predator, observed predation on con-specifics in a neighboring tank, or naïve to a predator. (a) Numbers of Japanese flounder juveniles that were captured and showed avoidance when they were attacked by a sculpin. (b) The number of off-bottom swimming individuals classified into three levels of swimming height. (Redrawn from Arai *et al.* 2007.)

individuals. With this method HR fish could be trained *en masse* to recognize predators (Brown and Laland 2003).

Kelly and Magurran (2003) generalized that visual predator recognition skills are largely built on unlearned predispositions, whereas olfactory recognition typically involves experience with conspecific alarm cues. Studies on olfactory conditioning are mostly in freshwater fishes including carp and salmonids (Brown 2003), and equivalent studies in marine fishes including flatfishes are required.

17.7 Behavioral indicators

It is important to determine behaviors that provide useful indicators of the quality of fish in the context of stock enhancement (Tsukamoto *et al.* 1997; Masuda and Tsukamoto 1998; Le Vay *et al.* 2007). Adaptive coloration and borrowing are the prerequisite antipredator behaviors in flatfish juveniles. Ryer *et al.* (2008) conducted predation experiment using wild juveniles of English sole, *Parophrys vetulus*, northern rock sole, *Lepidopsetta polyxysta*, and Pacific halibut, *Hippoglossus stenolepi*, under different color of substrate. They confirmed that juveniles with mismatching substrate color had higher predation by visual predators. Interestingly, fish with mismatched color were more active and less likely to burrow, thus they were more vulnerable to predation. They also suggested that cryptic coloration of flatfish may not be entirely volitional but is controlled both neurologically and physiologically.

Even if HR flatfish develop similar coloration to wild fish, juveniles can be vulnerable to predation because of exposure due to their feeding behavior. Furuta (1996, 1998) revealed that HR Japanese flounder spend longer times in the water column during feeding compared to the wild flounder (Figure 17.1).

He indicated that this is the time they are most vulnerable to predation. Therefore, off-bottom feeding behavior can be a behavioral indicator to evaluate quality of Japanese flounder juveniles. Although feeding from the surface is also typical in the HR individuals of salmon (Huntingford 2004), it is more problematic for flatfish compared to round fish, considering their vulnerability to predation. To improve the feeding behavior of Japanese flounder, juveniles are fed from the bottom rather than from the surface in some nursery facilities.

It would be convenient to find an indicator in larvae to predict future behavioral quality of juveniles. Ohm-posture, a typical shivering behavior of Japanese flounder, is a good candidate for such an indicator (Sakakura 2006). Sakakura (2006) identified Japanese flounder larvae which showed ohm-posture (Figure 17.7), marked otoliths with alizarine complexone, and compared their behavior with those which did not show ohm-posture. He found that fish which showed ohm-posture during the larval stage tended to be more aggressive after the settlement. Because flounder juvenile is basically solitary, aggressive behavior is likely to be a suitable behavioral characteristic in the wild. Development of ohm-posture has also been confirmed in spotted halibut, *Verasper variegates* (Sabate *et al.* 2008). It is however yet to be confirmed whether relatively aggressive individuals that have shown ohm-posture in their larval stage really

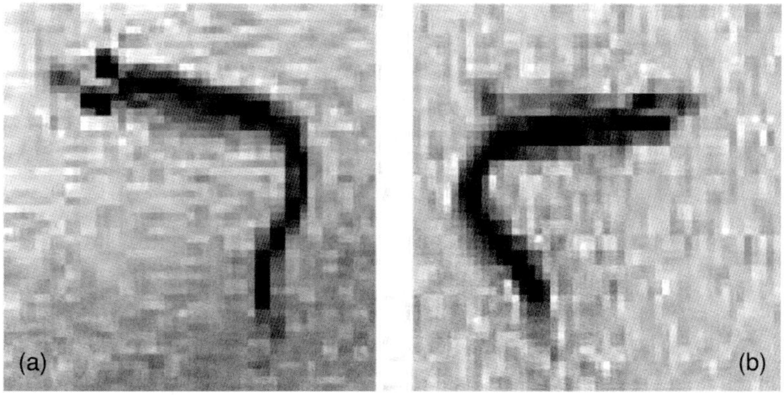

Figure 17.7 Video-captured photographs of typical ohm-postures in Japanese flounder larvae. The posture can be either left-sided (a) or right-sided (b). (The photographs were provided by the courtesy of Y. Sakakura [also see Sakakura and Tsukamoto 2002].)

have higher recruitment to fishery, because they could suffer more predation mortality during their "aggressive" feeding compared to less aggressive siblings.

17.8 Conclusion and recommendations

Laboratory results and field release trials suggest that the behavioral quality of stocked fish can be an important determinant to their survival rate. The low survival of HR juveniles in the wild indicates that efforts to reduce the impact of domestication through conditioning them for wild conditions might have a substantial impact on survival rates. Increased susceptibility to predation has been hypothesized to represent the most serious obstacle to survival of HR fish. Observations of domesticated flatfish indicate they lack the caution required for a group that relies on cryptic behavior for protection from predators. Given their relatively limited maneuverability in the water column, the tendency of HR flatfish to remain in the water column (Furuta 1996, 1998: Miyazaki *et al.* 1997; Kellison *et al.* 2000) must be a serious liability. Combine this apparently aberrant activity pattern with their poor skills relative to burial and cryptic coloration and one might conclude that conditioning that increases caution would be of highest priority. While such conditioning might be generally beneficial, the importance of caution to the survival strategy of flatfishes appears to differ even among morphologically similar species (Lemke and Ryer 2006). Curiosity is also likely to be an important behavioral trait. The observation that behavior might be driven by curiosity in the wild (Figure 17.8), and that reduced levels of exploratory behavior are associated with simple rearing conditions in laboratory experiments (Figure 17.3b), indicates that curiosity could be crucial to developing coping behaviors to variable wild conditions. Some balance between bold and shy phenotypes is likely optimal, though this balance must differ markedly for intrinsically shy (e.g., soles) and bold (e.g., halibut) species. Crucial behavioral patterns can also vary among different stocks of the same species (Burke *et al.* 1998) suggesting

Figure 17.8 Video-captured photograph of Pacific halibut (*Hippoglossus stenolepi*) sub-adults investigating a drop camera in a 25 m water column of a fjord in the Kenai Peninsula Alaska.

that knowledge of a particular flatfish stock's juvenile ecology is indispensable to developing a behavioral conditioning program for juveniles reared to enhance a particular stock. Despite the complexity of these problems, the economic and cultural importance of flatfish stocks suggests that behavioral quality problems like the physiological ones addressed by pioneering culturists (Seikai 1998) will be overcome. As behavioral quality and survival of stocked flatfishes increase, greater care will be required to avoid trophic and genetic impact to wild stocks. More attention to carrying capacity of systems will be required (Yamashita *et al.* 2006) especially in structurally impacted systems where coastal habitat restoration may represent a more efficient means of enhancement than stocking.

Literature cited

Allendorf, F.W., and Phelps, S.R. 1980. Loss of genetic variation in a hatchery stock of cutthroat trout. *Transactions of the American Fisheries Society* 109(5):537–543.

Arai, T., Tominaga, O., Seikai, T., and Masuda, R. 2007. Observational learning improves predator avoidance in hatchery-reared Japanese flounder *Paralichthys olivaceus* juveniles. *Journal of Sea Research* 58:59–64.

Asahida, T., Shinotsuka, Y., Yamashita, Y., Saitoh, K., Hayashizaki, K., and Ida, H. 2003. Influence of hatchery protocols on mitochondrial DNA variation in Japanese flounder juveniles. *Journal of the World Aquaculture Society* 34(2):121–132.

Bell, J.G., McEvoy, L.A., Estevez, E., Shields, R.J., and Sargent, J.R. 2003. Optimizing lipid nutrition in first-feeding flatfish larvae. *Aquaculture* 227(1–4):211–220.

Berejikian, B.A., and Tezak, E.P. 2005. Rearing in enriched hatchery tanks improves dorsal fin quality of juvenile steelhead. *North American Journal of Aquaculture* 67(4):289–293.

Berejikian, B.A., Tezak, E.P., Flagg, T.A., LaRae, A.L., Kummerow, E., and Mahnken, C.V.W. 2000. Social dominance, growth, and habitat use of age-0 steelhead

(*Oncorhynchus mykiss*) grown in enriched and conventional hatchery rearing environments. *Canadian Journal of Fisheries and Aquatic Sciences* 57(3):628–636.

Blaxter, J.H.S. 1976. Reared and wild fish – how do they compare? In: Persoone, G., and Jaspars, E. (eds) *Proceedings of the 10th European Symposium of Marine Biology.* Universa Press, Wetteren, Belgium.

Boehlert, G.W., and Mundy, B.C. 1988. Roles of behavioral and physical factors in larval and juvenile fish recruitment to estuarine nursery. In: Weinstein, M.P. (ed.) *Workshop on Larval Fish and Shellfish Transport.* Ocean Springs, MS (USA). American Fisheries Society Symposium, Bethesda, MD, pp. 261–281.

Browman, H.I. 1989. Embryology, ethology and ecology of ontogenetic critical periods in fish. *Brain, Behavior and Evolution* 34: 5–12.

Brown, C., Davidson, T., and Laland, K. 2003. Environmental enrichment and prior experience of live prey improve foraging behaviour in hatchery-reared Atlantic salmon. *Journal of Fish Biology* 63(Suppl. 1):187–196.

Brown, C., and Laland, K.N. 2003. Social learning in fishes: a review. *Fish and Fisheries* 4(3):280–288.

Brown, G.E. 2003. Learning about danger: chemical alarm cues and local risk assessment in prey fishes. *Fish and Fisheries* 4(3):227–234.

Burke, J.S., Miller, J.M., and Hoss, D.E. 1991. Immigration and habitat selection of *Paralichthys dentatus* and *Paralichthys lethostigma* larvae and juveniles in an estuarine nursery ground, North Carolina, USA. *Netherlands Journal of Sea Research* 27(3/4):393–405.

Burke, J.S., Seikai, T., and Tanaka, Y. 1999. Experimental intensive culture of summer flounder, *Paralichthys dentatus. Aquaculture* 176(1999):135–144.

Burke, J.S., Ueno, M., Tanaka, Y., Walsh, H., Maeda, T., Kinoshita, I., Seikai, T., Hoss, D.E., and Tanaka M. 1998. The influence of environmental factors on the early life history patterns of flounders. *Journal of Sea Research* 40(1998):19–32.

Dou, S., Masuda, R., Tanaka, M., and Tsukamoto, K. 2004. Size hierarchies affecting the social interactions and growth of juvenile Japanese flounder, *Paralichthys olivaceus.* Aquaculture 233(1–4):237–249.

Dou, S., Seikai, T., and Tsukamoto, K. 2000. Feeding behaviour of Japanese flounder larvae under laboratory conditions. *Journal of Fish Biology* 56(3):654–666.

Ellis, T., Howell B.R., and Hughes R.N. 1997. The cryptic responses of hatchery-reared sole to a natural sand substrate. *Journal of Fish Biology* 51(2):389–401.

Ellis, T., Hughes, R.N., and Howell, B.R. 2002. Artificial dietary regime may impair subsequent foraging behaviour of hatchery-reared turbot released into the natural environment. *Journal of Fish Biology* 61(1):252–264.

Fairchild, E.A., and Howell, W.H. 2004. Factors affecting the post-release survival of cultured juvenile *Pseudopleuronectes americanus. Journal of Fish Biology* 65(Supplement A):69–87.

Fairchild, E.A., Rennels, N., and Howell, W.H. 2008. Predators are attracted to acclimation cages used for winter flounder stock enhancement. *Reviews in Fisheries Science* 16(1–3):262–268.

Furuita, H., Takeuchi, T., and Uematsu, K. 1998. Effects of eicosapentaenoic and docosahexaenoic acids on growth, survival and brain development of larval Japanese flounder (*Paralichthys olivaceus*). *Aquaculture* 161(1–4):269–279.

Furuta, S. 1996. Predation on juvenile Japanese flounder (*Paralichthys olivaceus*) by diurnal piscivorous fish: field observations and laboratory experiments. In: Watanabe, Y., Yamashita, Y., and Oozeki, Y. (eds) *Survival Strategies in Early Life Stages of Marine Resources.* A. A. Balkema, Rotterdam, the Netherlands, pp. 285–294.

Furuta, S. 1998. Comparison of feeding behavior of wild and hatchery-reared Japanese flounder, *Paralichthys olivaceus*, juveniles by laboratory experiments. *Nippon Suisan Gakkaishi* 64(3):393–397.

Hess, M.F., Silvy, N.J., Griffin, C.P., Lopez, R.R., and Davis, D.S. 2005. Differences in flight characteristics of pen-reared and wild prairie-chickens. *Journal of Wildlife Management* 69(2):650–654.

Hossain, M.A.R., Tanaka, M., and Masuda, R. 2002. Predator-prey interaction between hatchery-reared Japanese flounder juvenile, *Paralichthys olivaceus*, and sandy shore crab, *Matuta lunaris*: daily rhythms, anti-predator conditioning and starvation. *Journal of Experimental Marine Biology and Ecology* 267(2002):1–14.

Huntingford, F.A. 2004. Implications of domestication and rearing conditions for the behaviour of cultivated fishes. *Journal of Fish Biology* 65(Supplement A):122–142.

Kellison, G.T., Eggleston, D.B., and Burke, J.S. 2000. Comparative behavior and survival of hatchery reared versus wild summer flounder (*Paralichthys dentatus*). *Canadian Journal of Fisheries and Aquatic Science* 57(2000):1870–1877.

Kelly, J.L., and Magurran, A.E. 2003. Learned predator recognition and antipredator responses in fishes. *Fish and Fisheries* 4(3):216–226.

Kihslinger, R.L., and Nevitt, G.A. 2006. Early rearing environment impacts cerebellar growth in juvenile salmon. *Journal of Experimental Biology* 209(2006):504–509.

Kitada, S., Yaga, Y., and Kishini, H. 1992. Effectiveness of a stock enhancement program evaluated by a two-stage sampling survey of commercial landings. *Canadian Journal of Fisheries and Aquatic Science* 49(8):1573–1582.

Lee, J.S.F., and Berejikian, B.A. 2008. Effects of the rearing environment on average behaviour and behavioural variation in steelhead. *Journal of Fish Biology* 72(7):1736–1749.

Lemke, J.L., and Ryer, C.H. 2006. Risk sensitivity in three juvenile (Age-0) flatfish species: does estuarine dependence promote risk-prone behavior? *Journal of Experimental Marine Biology and Ecology* 333(2):172–180.

Lockwood, S.J. 1980. The daily food intake of O-group plaice (*pleuronectes platessa* L.) under natural conditions. *Journal du Conseil Permmanent International pour l'Exploration de la mer* 39(2):154–159.

Le Vay, L., Carvalho, G.R., Quinitio, E.T., Lebata, J.H., Ut, V.N., and Fushimi, H. 2007. Quality of hatchery-reared juveniles for marine fisheries stock enhancement. *Aquaculture* 268(1–4):169–180.

Makino, H., Masuda, R., and Tanaka, M. 2006. Ontogenetic changes of learning capability under reward conditioning in striped knifejaw *Oplegnathus fasciatus* juveniles. *Fisheries Science* 72(6):1177–1182.

Marchetti, M.P., and Nevitt, G.A. 2003. Effects of hatchery rearing on brain structures of rainbow trout, *Oncorhynchus mykiss. Environmental Biology of Fishes* 66(2003):9–14.

Masuda, R., Takeuchi, T., Tsukamoto, K, Ishizaki, Y., Kanematsu, M., and Imaizumi, K. 1998. Critical involvement of dietary docosahexaenoic acid in the ontogeny of schooling behaviour in the yellowtail. *Journal of Fish Biology* 53(1998):471–484.

Masuda, R., Takeuchi, T., Tsukamoto, K., Sato, H., Shimizu, K., and Imaizumi, K. 1999. Incorporation of dietary docosahexaenoic acid into the central nervous system in the yellowtail *Seriola quinqueradiata. Brain, Behavior and Evolution* 53(1999):173–179.

Masuda, R., and Tsukamoto, K. 1998. Stock enhancement in Japan: review and perspective. *Bulletin of Marine Science* 62(2):337–358.

Masuda, R., and Ziemann, D.A. 2000. Ongotenetic changes of learning capability and stress recovery in the Pacific threadfin *Polydactylus sexfilis* juveniles. *Journal of Fish Biology* 56(5):1239–1247.

Maynard, D.J., Flagg, T.A., Mahnken, C.V.W., and Schroder, S.L. 1996. Natural rearing technologies for increasing postrelease survival of hatchery-reared salmon. *Bulletin of the National Research Institute of Aquaculture* 2(1996):71–77.

Miyazaki, T., Masuda, R., Furuta, S., and Tsukamoto, K. 1997. Laboratory observation of the nocturnal activity of hatchery-reared juvenile Japanese flounder *Paralichthys olivaceus*. *Fisheries Science* 63(2):205–210.

Olla, B.L., Davis, M.W., and Ryer, C.H. 1994. Behavioural deficits in hatchery-reared fish: potential effects on survival following release. *Aquaculture and Fisheries. Management* 25(Suppl. 1):19–34.

Olla, B., Davis, M., and Ryer, C. 1998. Understanding how the hatchery environment represses or promotes the development of behavioral survival skills. *Bulletin of Marine Science* 62(2):531–550.

Olla, B.L., Samet, C.E., and Studholme, A.L. 1972. Activity and feeding behavior of the summer flounder (*Paralichthys dentatus*) under controlled laboratory conditions. *Fisheries Bulletin of the National Oceanic and Atmospheric Administration* 70(4):1127–1136.

Price, E.O. 1984. Behavioral aspects of animal domestication. *Quarterly Review of Biology* 59(1984):1–32.

Price, E.O. 1999. Behavioral aspects of animals undergoing domestication. *Applied Animal Behaviour Science* 65(1999):245–271.

Ryer, C.H., Lemke, J.L., Boersma, K., and Levas, S. 2008. Adaptive coloration, behavior and predation vulnerability in three juvenile north Pacific flatfishes. *Journal of Experimental Marine Biology and Ecology* 359(1–3):62–66.

Sabate, F.S., Sakakura, Y., and Hagiwara, A. 2008. Comparison of behavioural development between Japanese flounder (*Paralichthys olivaceus*) and spotted halibut (*Verasper variegatus*) during early life stages. *Journal of Applied Ichthyology* 24(2008):248–255.

Sakakura, Y. 2006. Larval fish behavior can be a predictable indicator for the quality of Japanese flounder seedlings for release. *Aquaculture* 257(1–4):316–320.

Sakakura, Y., and Tsukamoto, K. 2002. Onset and development of aggressive behavior in the early life stage of Japanese flounder. *Fisheries Science* 68(4):854–861.

Salvanes, A.G.V., and Braithwaite, V.A. 2005. Exposure to variable spatial information in the early rearing environment generates asymmetries in social interactions in cod (*Gadus morhua*). *Behavioral Ecology and Sociobiology* 59(2005):250–257.

Seikai, T. 1998. Japanese flounder seed production from quantity to quality. In: Howel, W.H., Keller, B.J., Park, P.K., McVey, J.P., Takayanagi, K., and Uekita, Y. (eds) *Proceedings of the twenty-sixth U.S.-Japan Aquaculture Symposium*. University of New Hampshire Sea Grant Program, Durham, NH, pp. 5–17.

Sekino, M., Saitoh, K., Yamada, T., Kumagai, A., Hara, M., and Yamashita, Y. 2003. Microsatellite-based pedigree tracing in a Japanese flounder *Paralichthys olivaceus* hatchery strain: implications for hatchery management related to stock enhancement program. *Aquaculture* 221(1–4):1–4.

Stoettrup, J.G., Sparrevohn, C.R., Modin, J., and Lehmann, K. 2002. The use of reared fish to enhance natural populations: A case study on turbot *Psetta maxima* (Linne, 1758). *Fisheries Research* 59(1–2):161–180.

Stoner, A.W., and Glazer, R.A. 1998 Variation in natural mortality; implications for queen conch stock enhancement. *Bulletin of Marine Science* 62(2):427–442.

Sudo, H., Kajihara, N., and Fujii, T. 2008. Predation by the swimming crab *Charybdis japonica* and piscivorous fishes: a major mortality factor in hatchery-reared juvenile Japanese flounder *Paralichthys olivaceus* released in Mano Bay, Sado Island, Japan. *Fisheries Research* 89(1):49–56.

Tomiyama, T., Watanabe, M., and Fujia, T. 2008. Community-based stock enhancement and fisheries management of the Japanese flounder in Fukushima, Japan. *Reviews in Fisheries Science* 16(1–3):146–153.

Toole, C.L. 1980. Intertidal recruitment and feeding in relation to optimal utilization of nursery areas by juvenile English sole (*Parophrys vetulus*: Pleuronectidae). *Environmental Biology of Fishes* 5(4):383–390.

Tsukamoto, K., Kuwada, H., Uchida, K., Masuda, R., and Sakakura, Y. 1999. Fish quality and stocking effectiveness: behavioral approach. In: Howell, B.R. Moksness, E., and Svasand, T. (eds) *Stock Enhancement and Sea Ranching*. Fishing News Press, London, pp. 205–218.

Tsukamoto, K., Masuda, R., Kuwada, H., and Uchida, K. 1997. Quality of fish for release: behavioral approach. *Bulletin of National Research Institute of Aquaculture* Supplement 3: 93–99.

Wales, J.H. 1954. Relative survival of hatchery and wild trout. *Progressive Fish Culturist* 16(3):125–127.

Yamashita, Y., Kurita, Y., Yamada, H., and Takahashi, K. 2006. Estimate of optimum stocking density of hatchery-raised Japanese flounder juveniles in Ohno Bay, Northeastern Japan. *Bulletin of Fisheries Research Agency (Japan)* Supplement 5: 169–173.

Yamashita, Y., Nagahora, S., Yamada, H., and Kitagawa, D. 1994. Effects of release size on survival and growth of Japanese flounder *Parlichthys olivaceus* in coastal waters off Iwate Prefecture, northeastern Japan. *Marine Ecology Progress Series* 105(3):269–276.

Yamashita Y., and Aritaki Z., 2010. This book.

Chapter 18
Summary and conclusions

Wade O. Watanabe and Harry Daniels

18.1 Life history and biology

Flatfish are various marine or brackish water fishes in the order Pleuronectiformes. Commonly known as flounder, the Pleuronectiformes consist of many species comprising six families, including species that support important recreational and commercial fisheries and are among the commercially important groundfish in the world that have been harvested since the early nineteenth century. These marine demersal carnivores are found in all of the world's oceans in cold, temperate, and tropical seas, ranging from shallow bays to deep-water habitats, with most species found in habitats ranging from near-shore to depths of about 100 m on the continental shelf. The Atlantic halibut is exceptional, occurring at depths of over 2,000 m (Brown, this volume). Although most of the cultured species migrate from near-shore feeding grounds to spawning offshore spawning areas, at least one species (i.e., winter flounder) has isolated populations that migrate inshore to estuaries or embayments for spawning (Fairchild, this volume). While many species utilize these nursery grounds for the first 2 years of life before migrating offshore, *Paralichthys olivaceus* ("olive" or "Japanese" flounder) spend only about 2 months in their nursery ground before starting their offshore migration.

18.1.1 Metamorphosis

Metamorphosis or the change from symmetric larvae inhabiting the water column to asymmetric juvenile flatfish favoring the benthic habitat is a unique characteristic of the group that presents unique challenges to the flatfish culturist. Among other extreme morphological, physiological, and behavioral changes, one eye translocates to the other side of the head and positions itself next to the eye on the other side (Borski *et al.*, this volume). During metamorphosis, the upper (ocular) side becomes pigmented, and the lower (anocular or blind) side, light colored, and there are changes in dentition and fin placement.

Metamorphosis can be a difficult period in flatfish culture (Borski *et al.*, this volume). Adult flatfish are almost always lacking a swim bladder.

18.1.2 Cultured species

Taxonomically, flatfish comes from six families, five of which contain species that are studied for research, stock enhancement, and aquaculture in many parts of the world and climatic regimes. These include Pleuronectidae (righteye flounders), Bothidae (lefteye flouders), Paralichthyidae (large-tooth flounders), Rhombosoleinae (greenback flounders), and Scopthalmidae (turbots). Among those species covered in the various chapters of this book, members of the righteye flounder family include the winter flounder *Pseudopleuronectes americanus* and Atlantic halibut *Hippoglossus hippoglossus* from the northwestern Atlantic coast. Members of the lefteye flounder family include the summer flounder *Paralichthys dentatus*, the southern flounder *P. lethostigma* from the Eastern United States, and *P. olivaceus* (also known as hirame or Japanese flounder in Japan, and bastard halibut or olive flounder in Korea) from the northwestern Pacific Ocean, and Chilean flounder *P. adspersus* and *P. microps* from North of Peru to the central Chilean coast. The California halibut *P. californicus* from the Pacific U.S. Coast and Baja California in Mexico does not show a preference for the side of eye development with both left or right-eyed individuals equally common. The greenback flounder *Rhombosolea tapirina* is found in southern Australia and New Zealand. The Scopthalmid family contains the turbot (*Scophthalmus maximus* Rafinesque 1810: *Psetta maxima* Linnaeus, 1758) distributed in European waters from the Mediterranean Sea and Black Sea to the Baltic Sea and Norwegian coast.

18.1.3 Morphology

With the exception of the Atlantic and Pacific halibut, which reach exceptionally large sizes (up to 4.7 m length and 318 kg live weight for Atlantic halibut) (Brown, this volume), flatfishes in general are not large fish. In the wild, average length and weight at age 3 years vary considerably, ranging from 30 cm and 400–600 g for the greenback flounder (Hart, this volume) to 30–40 cm and 2–3 kg for turbot (Person-Le Ruyet, this volume). All of these cultured flatfish show clear sexual dimorphism in growth, with females growing faster and reaching larger size than males.

18.1.4 Fishery

Many species are important food fishes and commercial landings for most of these species reached peaks from the 1920s to 1950s and then declined sharply to only a fraction of their maximum tonnages by the 1990s and have since been of lower economic importance when compared to historical highs. In addition to stringent fishing regulations, research on the culture of these species for market and for stock enhancement was implemented in response to declining fishery landings and the expectation that marine finfish aquaculture would help to reduce fishing pressure and help rebuild natural populations and seafood supplies.

18.1.5 Environmental requirements

An advantage for aquaculture is that many of these flatfish species inhabit estuarine conditions for part of their life cycle and are able to tolerate a wide range of temperatures and salinities. Southern flounder adults have been captured in a range of 0–36‰ salinity. The winter flounder is freeze-resistant, able to synthesize antifreeze proteins and secrete them into their blood during the winter (Fairchild, this volume). Optimal growth temperatures range from 7 to 14°C for Atlantic halibut, from 14 to 19°C for turbot, and 21–24°C for the summer and southern flounder, and 20–25°C for *P. olivaceus* (olive or Japanese flounder).

18.1.6 Reproduction

Flatfish of interest for aquaculture are generally serial (i.e., multiple clutch group synchronous) spawners, producing multiple batches of eggs during the spawning season in intervals of 3–4 days. While most of these species produce small pelagic eggs ranging from 0.83 to 1.1 mm in diameter, eggs of the Atlantic halibut are relatively large for a marine fish (3 mm diameter) and they are bathypelagic during development, floating close to the ocean floor and are neutrally buoyant at 36‰. The winter flounder, on the other hand, produces demersal eggs 0.80 mm in diameter and these adhesive eggs are treated with diatomaceous earth to prevent clumping (Fairchild, this volume).

Total fecundity in flatfish is high, ranging from 400,000 to 5 million eggs per female, with an average of 1.0–2.0 million eggs/kg female. Because of their large size, relative fecundity of Atlantic halibut females is lower at around 50–160,000 egg/kg. At hatching, yolk-sac larvae range from 1.7 and 2.0 mm TL in Chilean flounder to 3.0 mm in turbot.

18.2 Broodstock husbandry

Wild-caught flounder may be captured in nets or by hook and line. In many species mortality following capture can be significant as wild-caught fish are slow to take food and do not easily wean to formulated diets. Size regulations and sex-specific differences in habitat can also result in capture of adults favoring only one sex. It may be advantageous to anticipate requirements for mature broodstock and catch large numbers of small juveniles, which can be slowly weaned to compound diets and cultured to maturity. From a biosecurity standpoint, broodstock flounder need to be carefully screened for diseases such as nodavirus (Nervous Necrosis Virus) which can be transmitted vertically to eggs, larvae, and juvenile fish (Noga *et al.*, this volume). Good biosecurity practices, including the quarantine of new fish in a separate facility until health status is ascertained, are imperative to prevent the introduction of pathogens. Broodstock origin affects zygote characteristics and development, since fish from anthropogenically contaminated areas produced eggs with a high incidence of abnormalities and there are latitudinal differences in growth rates as demonstrated for winter flounder (Fairchild, this volume).

Most flatfish hatcheries still use either wild-caught or F1 generation broodstock. Selective breeding is recognized as being of high importance to commercial viability and development for flatfish farming, but for practical reasons, rigorous breeding programs have not been implemented for any farmed species. The space and labor requirements for selecting and holding multiple families apart to prevent inbreeding of broodstock tax a farm and are cost-prohibitive. Informal selection of broodstock from F1 and F2 populations based on growth or appearance is conducted on some farms and research facilities, but the production gains resulting from these efforts are not well documented. In Japanese hatcheries, where flounder fingerlings are used for stock enhancement as well as for commercial growout, wild caught flounder are used for spawning in order to maintain the genetic diversity of juveniles produced for restocking.

Domestication of flatfish for farming has reached its highest level of progression in the turbot farming industry, where research on applied genetics is carried out by commercial farms, and little or no information exists in the public domain. Broodstock management programs for turbot were established since the early to mid-1990s in France and Spain using both wild and farmed fish of genetically distinct lineages (Person-Le Ruyet, this volume). In China, the first few generations of fish originated from United Kingdom, while subsequent batches were selected from their progeny. Most broodstock were adult fish selected from the growout of hatchery-produced juveniles during the early stages of the industry. Hatchery-reared individuals with high growth rate or specific characteristics (e.g., pigmentation and external appearance) were initially selected. Currently, broodstock are imported as hatchery-produced juveniles from France, Denmark, Iceland, and Norway to avoid inbreeding. The selection of brooders is a subjective process depending on the expertise of hatchery technicians who prioritize normal body shape and color, vigorous swimming and feeding behavior, no history of diseases during growout, and representative of as many different populations as possible (Lei and Liu, this volume). In the Atlantic halibut, however, mature hatchery reared (F1) fish have been genotyped using microsatellite markers to establish pedigree for future breeding programs (Brown, this volume). Microsatellite DNA is also being used in the United States to link the fastest-growing summer flounder juveniles to their parents when all the juveniles were reared communally in a tank (Bengtson and Nardi, this volume). Chinese scientists have recently conducted experimental crosses of summer flounder and *P. olivaceus* (olive or Japanese flounder) (Bengtson and Nardi, this volume), and similar crosses between summer flounder and southern flounder are being studied in the United States.

Breeding objectives are different when hatchery-reared juveniles are to be released into natural waters for purposes of stock enhancement. Genetic diversities of juveniles produced from cultured *P. olivaceus* broodstock were significantly lower than those from local wild-caught broodstock (Seikai *et al.*, this volume). Furthermore, because Japanese flounder farms rely on volitional spawning, a reduced number of contributing broodstock is unavoidable (Seikai *et al.*, this volume). To maximize genetic diversity of hatchery juveniles for stock enhancement purposes, it is recommended to use wild fish as broodstock, maximize broodstock number, and use fertilized eggs collected over several days.

18.2.1 System design and requirements

Tanks of a minimum of 2–4 m^3 are used for broodstock of some flatfish species (e.g., greenback flounder) (Hart, this volume), while larger tanks of 15–40 m^3 are used for larger species (e.g., Atlantic halibut) (Brown, this volume). In general, low light levels, temperature regimes that mimic the natural environment, and good water quality are used to help ensure successful acclimation of broodstock. For turbot, natural spawning occurs exclusively in very large and deep tanks (volume = 40 m^3, depth = 1.65 m) with a sandy substrate using broodstock adapted to captivity for at least 2 years (Person-Le Ruyet, this volume). However, when fish must be manipulated for hormone induction and strip spawning, smaller tanks ranging from 20 to 30 m^3 and 1 m deep are used, with stocking densities of 3–6 kg/m^3. In the United States, southern flounder broodstock are held for hormone-induced strip spawning in 4.76 m^3 tanks at a density of 2.7–3.3 individuals/m^2 bottom area (2.7–3.3 kg/m^3) and at a sex ratio of around 1:1 (Daniels *et al.*, this volume). In Japan, flounder *P. olivaceus* broodstock are held for volitional spawning at a low density of 0.5–2 individuals/m^2 (2–5 kg/m^2 bottom area) and at a sex ratio of around 1:1 (Seikai *et al.*, this volume).

Flow-through or recirculating seawater are preferred for holding broodstock for most species, but recirculating tanks are increasingly being used as they permit better control of temperature than flow-through tanks. Depending on source, seawater is usually sand-filtered before being pumped into the tank; deep well seawater is sometimes clean enough to bypass sand filtration. Boilers and chillers are used to maintain temperature within the optimum range for the broodstock. To reduce energy costs, deep well water is mixed with natural seawater and partial water reuse systems are used.

Since all flatfish species require near full-strength seawater for spawning and larval rearing, hatchery facilities are typically situated in coastal regions where a source of high quality seawater is readily available. However, the first commercial hatchery for the southern flounder in North Carolina is located in an inland area without access to seawater. This hatchery, which has already successfully produced hundreds of thousands of flounder fingerlings and subadults, uses low-salinity (~0.5 g/L) groundwater amended to full-strength seawater salinity (34 g/L) with commercial sea salts. Since artificial seawater is costly to prepare, this inland hatchery is forced to be conservative in the use of make-up water, typically only adding freshwater at 1–2% of the volume daily to compensate for losses from evaporation and the backwashing of filters. In operation for over a year, this facility is demonstrating that marine finfish fingerlings can be successfully produced using recirculation aquaculture technology in inland locations where a continuous source of seawater is not available.

18.2.2 Diet and nutrition

The natural diet of flatfish includes a wide variety of fish, mollusks, and crustaceans. Feed given to broodstock should ideally be in a dry form that can be formulated to contain all of the essential nutrients and that can be easily stored, but for most species of flatfish, there is inadequate knowledge of the nutritional requirements of broodstock to develop formulated feeds, and replacing raw or frozen ingredients is therefore a continuing problem. This is especially true of wild-caught flatfish, which are difficult to wean onto prepared diets. Fish that have been raised in captivity on formulated diets are generally fed a commercial pelleted feed containing a minimum of 55% protein and 12–15% fat, which are usually fortified with vitamins. Complications arise when tank space is limited and unweaned wild-caught fish and pellet trained broodstock must be held in the same tank. Broodstock that have been acclimated to spawn in captivity and weaned to formulated feeds are therefore very valuable and are difficult to replace (Brown, this volume.).

In Asia, moist pellets are manufactured on-site by mixing a powdered commercial premix with the trash fish prior to extrusion through a feed processor. In China, turbot broodstock are usually fed chopped trash fish and moist pellets boosted with commercial or farm-specific mixtures (e.g., vitamins and HUFA). Fresh or frozen fish in broodstock diets is a suspected vector for pathogens such as viral diseases as nodavirus (VNN) and viral hemorrhagic septicemia (VHS) which may be transmitted to eggs, larvae, and juveniles, and is increasing the impetus for research on bloodstock nutritional requirements as a basis for formulated dry pelleted diets (Lei and Liu, this volume; Noga *et al.*, this volume).

A fundamental problem in flatfish broodstock nutrition is that very little nutritional data are available. It is clear from recent literature, however, that essential fatty acids such as n-3 HUFAs (EPA, 20:5n-3 and DHA, 22:6n-3) and the n-6 HUFA (ARA, 20:4n-6) are needed for viable eggs and larvae, but oversupplementation can adversely affect egg quality (Seikai *et al.*, this volume), and the ratios of docosahexaenoic acid (DHA) to eicosapentaenoic acid (EPA) to arachidonic acid (ARA) ratio should be 8:4:1. Unfortunately, the establishment of nutritional requirements for flatfish is made difficult by the reproductive characteristics of these fish (i.e., multiple spawning of batches of eggs), and by many factors that affect spawning success and egg quality (e.g., participation by only a few individuals during volitional spawning, latency to ovulation and timing of strip spawning, hormone dose, temperature, and male gamete quality) which confound interpretation of results (Brown, this volume). The size of the reproductive units make replicated studies with strong experimental designs cost-prohibitive, especially for large species such as the Atlantic halibut.

18.2.3 Photothermal conditioning

To achieve out of season spawning, accelerated photothermal regimes, in which the annual photothermal cycle is compressed from 12 months to only 4–10 months are effective in advancing maturation and timing of spawning of flatfish, so that rematuration and spawning may be achieved in less than 12 months to produce viable embryos over a majority of the year. In southern flounder, accelerated photothermal conditions may be used to advance spawning, but a minimum of 5 months was required for rematuration and spawning, probably because of the time required for postspawning fish to regain the requisite levels of energy and lipids for deposition of yolk into the growing ovary (Daniels *et al.*, this volume). On the other hand, delaying spawning by using extended photothermal cycles has met limited success. In contrast, spawning in Atlantic halibut may be delayed up to 6 months, but advancing spawning more than 3 months/year is difficult (Brown, this volume). In the winter flounder, manipulating photoperiod and temperature did not strongly affect gametogenesis (Fairchild, this volume). By using multiple broodstock, photoperiod manipulation is routinely used to achieve out of season spawning to enable year round production of eggs. In many commercial hatcheries for turbot in China, 5–6 groups of broodstock are maintained under different photothermal protocols to obtain fertilized eggs year round (Lei and Liu, this volume). Elevated temperature during gametogenesis of broodstock reduces egg viability in both turbot and Atlantic halibut.

18.3 Monitoring gonad development

When hormone-induced spawning is used to initiate final maturation and ovulation, knowledge of stage of ovarian development is required to gauge receptivity to hormone treatment. Furthermore, when strip-spawning is used, a knowledge of stage of ovarian maturity is critical to timing of stripping, fertilization success, and egg quality. Ovarian biopsy using a polyethylene cannula requires skill and can injure the reproductive tract and is therefore, mainly used by researchers and not for practical hatchery purposes (Daniels *et al.*, this volume). Alternatively, oocyte sampling using a biopsy pincer can also be used (Person-Le Ruyet, this volume). Backlighting is an inexpensive but very effective method that takes advantage of the flatfishes' body conformation to visualize gonadal development non-invasively. Ultrasound, although expensive, is also being used to sex the fish as well as estimate the stage of development of the gonad. In research and commercial hatcheries (e.g., turbot), external observation of the gonad and the behavior of the broodstock can be used by skilled technicians to gauge gonadal stage and the timing of ovulation of the females (Hart, this volume; Lei and Liu, this volume).

18.3.1 Controlled spawning

Natural spawning

Natural (spontaneous) spawning without hormone induction can also be achieved in many species of flatfish, but usually only after 1–2 years of captivity. Some species (e.g., *P. olivaceus*, California halibut, Chilean flounder) spawn large numbers of fertilized eggs in outdoor tanks without hormonal intervention.

Natural spawning of wild-caught adult broodstock improves after acclimation to captivity for at least two seasons and generally continues for several years (Hart, this volume). For the Atlantic halibut, female broodstock are generally stripped by hand even though natural spawning can occur (Brown, this volume).

Artificial insemination (strip-spawning)

Although natural spawning can yield large numbers of viable eggs, hormone-induced spawning provides better control over the timing and availability of high-quality embryos and is a more reliable method for stocking hatcheries. For the Atlantic halibut and for the turbot (in China), females are induced to mature and ovulate through photothermal manipulation, but strip-spawning and artificial fertilization are used to obtain viable eggs (Brown, this volume; Lei and Liu, this volume). More recently, however, in order to supply fertilized eggs to meet customer demand, turbot hatcheries in China are using hormones to induce synchronous ovulation in multiple females, rather than rely on natural ovulation (Lei and Liu, this volume).

For most species of flatfish, artificial insemination of eggs is performed following strip-spawning of females induced to ovulate using hormones, typically gonadotropic hormone releasing hormone analog (GnRHa) administered in saline injections or through a sustained-release cholesterol pellet, but carp pituitary extract (CPE) and human chorionic gonadotropin (hCG) have also been used successfully (Bengtson and Nardi, this volume). Ovulation of *R. tapirina* has been induced using Ovaprim [D-Arg^6, Pro^9 NEt]-sGnRH and a dopamine antagonist; domperidone) (0.5 mL/kg), LHRHa (50 and 100 μg/kg) in a single intraperitoneal injection (Hart, this volume). Hormones may be used to synchronize the spawning of the broodstock within a specific week by implanting a desired number of fish that exhibit oocytes meeting a critical minimum average diameter. For females that do not meet this minimum, a series of CPE injections has been used successfully to induce spawning (Bengtson and Nandi, this volume).

Flatfish males (especially lefteye flounder) often produce very small quantities of milt and can be difficult to strip (Hart, this volume, Daniels *et al.*, this volume). Although Atlantic halibut milt can be cryopreserved, the application of GnRHa implants has proved to be useful in synchronizing spermiation. In southern flounder, GnRHa implants or injections or hCG have been ineffective in increasing sperm volume or initiating spermiation; so, photothermal conditioning is still the most effective method for obtaining spermiating males. In greenback flounder, implanted GnRHa pellets increased milt volume of spermiating males, but spermatocrit was reduced (Hart, this volume). More research is needed to stimulate spermiation in males.

The use of cryopreserved sperm is becoming more and more common in the turbot farming industry in Europe (Fairchild, this volume), allowing for greater flexibility with regard to the need to hold male broodstock, but further improvements are needed for these techniques to gain widespread application.

18.3.2 Egg collection and incubation

In general, the eggs of the Atlantic halibut have a relatively high specific gravity owing to their high inorganic content and they will sink in full strength seawater. To counteract sinking, eggs are incubated in upwelling tanks (Brown, this volume). In some species (e.g., *P. olivaceus*), hatching rate is higher in a darkened room (Bai and Lee, this volume). Live eggs are typically disinfected before hatching (e.g., using iodized-polyvinylpyrrolidone at 4‰, for 5 minutes). Most flatfish hatcheries use an incubation temperature near spawning conditions to optimize yolk utilization efficiency, but then increase temperature by several degree Celsius within a few days after hatching, presumably to accelerate metabolism, feeding, growth, and survival.

18.4 Larval culture

18.4.1 System design and requirements

Semi-intensive larval culture was originally used for commercial production of turbot in Europe and halibut in Norway through the mid-1990s, where newly-hatched larvae from indoor incubators are transferred at low densities (2–5 larvae/L) to large tanks (50 m^3) previously conditioned to provide prey organisms to sustain the fish until harvest or to outdoor bag enclosures and fed wild plankton and *Artemia* (Person-Le Ruyet, this volume, Brown, this volume). While simple in operation, zooplankton availability is unreliable, risk of exposure to pathogens (e.g., nodavirus [VNN] or infectious pancreatic necrosis [IPN]) is greater, and growth variation and production are unpredictable (Brown, this volume).

Intensive larval culture is the preferred method used by researchers as well as commercial hatcheries for flatfish where larvae are produced in indoor tanks under controlled environmental conditions and live prey and formulated feeds are added daily to the tanks. At a research scale, flatfish larvae have been successfully raised in small tanks (3–15 L) for experimental purposes, but tanks of 1,000–4,000 L are typical for pilot production, stocked at initial densities of 20–30 larvae/L. In commercial hatcheries in Europe and China for turbot and in Japan for *P. olivaceus*, larger indoor tanks (circular, square, or octagonal in shape) ranging from 7 to 80 m^3 are used for intensive rearing, stocked at densities of 10–30 larvae/L.

Flow-through or recirculating

In European turbot hatcheries, flow-through seawater systems are generally used but water can be partly reused (Person-Le Ruyet, this volume). Turbot hatcheries in China use flowthrough systems, using natural seawater or saline well water pretreated by mechanical filtration, aeration, UV disinfection, and temperature adjustment, before use in the rearing tank, with no attempt to recirculate water. In these flow-through systems, sand-filtered seawater is adjusted to optimum temperature (18–19°C) by boiler and then pumped into a head tank before distribution to the larviculture system (Lei and Liu, this volume). Saline well water from deep aquifers is preferred because it is devoid of pathogens.

Researchers as well as commercial aquaculturists are turning increasingly to recirculating aquaculture systems (RAS) for hatcheries, which provide more control of water quality (especially, temperature, gas saturation, and salinity) and more consistent production. In RAS, inflow water is filtered and sterilized with UV or ozone, but some RAS hatcheries are supplied with bacteria free ground water. In Korea, warm water from a power station or heated water is commonly used and recirculation systems have become the method of choice for hatcheries (Bai and Lee, this volume).

18.4.2 Hatchery protocols

In Atlantic halibut, the period from hatching to first feeding, can last up to 50 days depending on temperature (Brown, this volume). During this period, yolk-sac larvae are held in upwelling cylindro-conical incubators at 1–20 larvae/L. Typical survival rates in these incubators range from 50 to 80%. Strict temperature control is used to avoid developmental abnormalities (e.g., jaw deformities) and mortality. Salinity must also be within a narrow range and maintenance of good water quality is required. The larvae are generally kept in near or complete darkness because they are strongly attracted to light at the later stages of this phase. Salinity affects the buoyancy of fertilized eggs and yolk-sac larvae as well as the growth and survival of early larvae before metamorphosis, and salinities above 28‰ are preferred for the larval rearing of most flatfish species.

18.5 Water quality

Most cultured flatfishes exhibit abnormal pigmentation (pseudoalbinism on the ocular side and hypermelanosis on the blind side), and incomplete eye migration in a proportion of hatchery-reared fish and are attributed to both genetic and nutritional factors. While these pigment abnormalities do not affect flesh quality, these fish are commercially inferior (Seikai *et al.*, this volume). For stock enhancement application, fish with hypomelanosis (lack of pigmentation on the ocular side) are more susceptible to predation. The causes of these abnormalities remain unclear but are hypothesized to be related to essential fatty acid nutrition, especially DHA, EPA, and ARA, overall energy intake, iodine and thyroid levels, photoperiod, and temperature (Conklin and Piedrahita, this volume). Suboptimal low temperature caused albinism and ambicoloration in brown sole, and the use of ozone-treated seawater induced a high incidence of pseudoalbinism in *P. olivaceus* (Seikai *et al.*, this volume).

Flatfish hatcheries universally use microalgae for intensive larval rearing ("green water culture"), with algae pastes and algae substitutes preferred over live algae, which is labor and space-intensive and can potentially harbor pathogens. Green water is beneficial during the pelagic larval stages to enhance first feeding success, equalize distribution of larvae, and improve larval growth, and promote normal pigmentation patterns. There is no standard protocol for greenwater culture, and a variety of microalgal species (e.g., *Chlorella*, *Isochrysis*, *Tetraselmis*, and *Nannochloropsis)* and species combinations in either live or preserved (but, intact cells) are used, generally in concentrations of 0.1–1 million cells/mL. California halibut larvae reared in greenwater were larger, easier to wean, and survived (>50% versus 10%) better than larvae reared in clearwater, and even juveniles that were poorly pigmented following metamorphosis became fully pigmented on the eyed side with time (Conklin and Piedrahita, this volume). Research in Atlantic halibut has indicated that powdered clay is a cost effective alternative to microalgae, suggesting that the physical attributes (e.g., light-shading) may be more important than the biological (immunostimulation or antimicrobial) effects, water conditioning, or micronutritional benefits to larvae and live prey organisms, at least for this species. In greenback flounder, however, green water *Tetraselmis suecica* improved the feeding ability of larvae at all turbidity levels from 3 to 5 Nephelometric Turbidity Units (NTU) (Hart, this volume).

18.5.1 Food and feeding

In intensive hatcheries, larvae are generally fed live prey, including rotifers, *Artemia* spp. nauplii before weaning to formulated diets. In Atlantic halibut, ongrown *Artemia* are also fed to larvae before weaning. Despite their small initial size, flatfish larvae have a sufficiently large mouth gape to feed readily on rotifers (*Brachionus plicatilis*), and both L-type (160–320 μm in lorica length) or S-type rotifers (90–210 μm) are fed at densities ranging from 3 to 10 individuals/mL twice daily (Seikai *et al.*, this volume; Daniels *et al.*, this volume). *Artemia* nauplii are fed at initial concentrations of 0.1 to 1 individuals/mL and gradually increase to 0.5–5.0 individuals/mL (Lei and Liu, this volume; Daniels *et al.*, this volume), but some hatcheries base feed calculations on individuals per fish (20–700 individuals/fish) (Seikai *et al.*, this volume). Techniques for producing rotifers using commercial dry food and either semicontinuous culture or batch culture are improving the nutritional and hygienic quality of rotifers and lowering their production costs.

Inadequate nutrition during larval stages and metamorphosis in flatfish may affect normal development, including eye migration and pigmentation, and survival. The nauplii and copepodids of calanoid (*Calanus*) copepods are a preferred first food for larvae, but since they are not easily produced at a large scale, their biochemical composition is used as reference for live prey enrichment products and for the formulation of compound diets (Person-Le Ruyet, this volume). Feeding copepods during a critical period ("copepod window") can enhance normal development, pigmentation, and survival, while minimizing demand. For Atlantic halibut, which have a relatively large mouth size, researchers have achieved good survival, pigmentation, and eye migration with an *Artemia* only diet using commercially available enrichment products and ongrown *Artemia* as the larvae grow.

Before use, rotifers and *Artemia* are nutritionally enriched to satisfy the requirements of larvae for n-3 HUFA (highly unsaturated fatty acids with a carbon chain >20), taurine, and vitamins using either homemade enrichment mixtures (lipids, proteins, vitamins, and minerals) or different commercial emulsions. Specific requirements for flatfish larvae have been shown for ARA, EPA, as well as DHA, and *Artemia* nauplii are nutritionally enriched before feeding with commercially available enrichment products to enhance the concentrations of these micronutrients, though there are no standard protocols, and many products are used. Normal eye migration and body pigmentation in many flatfish is dependent upon proper amounts and ratios of these essential fatty acids, especially DHA and the ratio between DHA and EPA. In turbot, the DHA to EPA ratio should be at least 2:1 (Noga *et al.*, this volume), and an elevated ratio improved pigmentation in turbot. Other studies, however, have shown normal pigmentation and high survival in a majority of fish fed unenriched rotifers or *Artemia* or enriched with low levels of DHA (Noga *et al.*, this volume), suggesting the possible involvement of other factors. In China, the use of commercial live feed enrichments such as DHA Protein Selco® (INVE) and AlgaMac-3050® (Aquafauna, Bio-marine) has reduced the incidence of abnormally pigmented fish to less than 20% for most hatcheries, and below 5% for those with good expertise (Lei and Liu, this volume).

A deficiency in vitamin A (a precursor of rhodopsin) disrupts neuroendocrine signaling from the eyes to the brain to produce melanocyte stimulating hormone and subsequently melanin synthesis, resulting in abnormally pigmented fish. Enrichment of larval diets with vitamin A is therefore needed for normal pigmentation in flatfish larvae. In Japan, numerous studies have been conducted since 1980s to reduce pigmentation anomalies and bone deformities, which decrease market value (Seikai *et al.*, this volume). Enrichment of rotifers and *Artemia* with n-3 HUFA or vitamin A has been effective in preventing pigmentation anomalies on the ocular side, and the incidence of pseudoalbinism has declined to less than 5%. Live food enriched with vitamin C improved pigmentation of turbot larvae (Noga *et al.*, this volume). In greenback flounder, vitamin C deficiency during weaning may also be the cause of shortened opercula and lordosis in juveniles (Hart, this volume).

Supplementation of taurine, the most abundant free amino acid in marine animals and plants, improved feeding and growth of juvenile *P. olivaceus* (Noga *et al.*, this volume). Fat cell necrosis syndrome has been observed in farmed Atlantic halibut and may be caused by inadequate antioxidants in the feed combined with an exposure to sunlight (Noga *et al.*, this volume).

In Chilean flounder, studies on immuno-stimulants indicated that the addition of 5 mg/L of β-glucans (βG) and mannan-oligosaccharides (βG MOS) to the culture water increased the growth and survival of larvae. βG MOS promotes macrophage cells precursors in intestinal epithelium, which is associated with the nonspecific immune system of the fish (Noga *et al.*, this volume).

To ensure that rotifers and *Artemia* presented to the larvae are freshly enriched and that nutrients are not catabolized, larvae are fed 3–4 times daily, flushing uneaten rotifers and *Artemia* from the tanks before adding a newly enriched batch. In European turbot hatcheries, enriched preys are slowly metered into larval rearing tanks using peristaltic pumps to ensure satiety, while avoiding overfeeding (Person-Le Ruyet, this volume). Since rotifers and *Artemia* may be vectors for *Vibrio* spp., care in enrichment and rinsing before feeding is essential.

Flatfish hatcheries consistently use long photoperiods of 18–24 hours, feeding throughout the photophase to produce higher larval growth rates and survival to metamorphosis than those attainable under ambient lighting conditions. Recommended light intensities vary considerably with species, as research has shown that light intensity affects growth and survival in some species (southern and summer flounder) more than in others (e.g., winter flounder). Precise comparisons are difficult, because the size, depth, and color of the rearing tank, density of greenwater used, as well as type of light (natural or artificial) affect quality of illumination to the larvae and prey. Commercial hatcheries for olive flounder in Asia recommend light intensities of 400–600 lx (Bai and Lee, this volume), and similar conditions are used for the summer and southern flounder in the United States (Bengtson and Nardi, this volume, Daniels *et al.*, this volume). In Chinese turbot hatcheries, light intensity at the water surface ranges from 500 to 4,000 lx depending on the type of light employed (natural or artificial).

18.5.2 Formulated feeds

Larval rearing through metamorphosis generally requires from 30 to 40 days in cultured flatfish, but is highly temperature dependent and requires as many as 80 days at 5°C in winter flounder (Fairchild, this volume). Recently metamorphosed flounder are weaned onto dry feeds (200–400 μm, 52–55% protein, and 12–15% lipid) by cofeeding a micropelleted diet (150–450 μm, 52–55% protein, and 12–15% lipid) and *Artemia* for a 2–3 week period and gradually reducing the *Artemia* ration during this period (Daniels *et al.*, this volume, Fairchild, this volume). It is generally assumed that earlier weaning to artificial diets can reduce production costs by allow nutrient optimization for reliable production and simplifying rearing protocols. In winter flounder, larvae as small as 5–6.6 mm TL could be weaned onto a commercial microencapsulated diet with no adverse affects on growth rate or time to metamorphosis provided that a long cofeeding period was used (Fairchild, this volume). In some flatfish (e.g., Atlantic halibut, California halibut), however, postmetamorphic fish wean quickly, and extended cofeeding with *Artemia* is unnecessary (Brown, this volume, Conklin and Piedrahita, this volume).

In European turbot hatcheries, larvae are held at high density (2,500 fish/m^2) in shallow circular or square tanks (0.25–0.50 m depth, 5–10 m^2) during weaning to promote feeding activity. During the process of weaning from live feeds, flatfish larvae are generally fed microdiets in excess to increase a larva's opportunity to feed, leaving large amounts of uneaten feed on the tank bottom. This accumulation of uneaten feed can lower water quality and contributes to stress and disease. Manual siphoning of settled organic matter with the aid of a squeegee is a critical, but labor-intensive and aggravating activity for hatchery staff, since larvae are often siphoned along with the debris and must be returned to the tank. Some hatcheries in the United States and in Chile for other species (southern, winter, and Chilean flounder) transfer larvae to small, mesh net cages suspended inside the rearing tank to enhance feeding, while keeping larvae away from the tank bottom where organic debris accumulates and is easily siphoned out with no disturbance to the larvae (Daniels *et al.*, this volume, Fairchild, this volume, Silva, this volume). Hatchery facilities for Atlantic halibut in Norway and *P. olivaceus* in Japan now incorporate self-cleaning equipment (e.g., slowly-rotating squeegee arms with siphon holes) in the larval rearing tanks to reduce labor (Brown, this volume).

18.5.3 Microbial environment

Inconsistent fingerling production in marine finfish hatcheries is often related to high mortality during the early larval period associated with changes in the microbial environment, especially *Vibrio* and *Aeromonas* (Bengtson and Nardi, this volume). Contamination occurs primarily through live food (rotifers and *Artemia*), and can be reduced by hygienic culture and enrichment practices and by adding probiotics to help reduce proliferation of opportunistic pathogenic bacteria (Bengtson and Nardi, this volume). Microbial conditions are more stable in recirculating hatchery systems (Brown, this volume), as bacterial populations in the biofilters, pipes, and tanks are hypothesized to exert probiotic effects to limit the impact of pathogenic bacteria on first feeding stage larvae (Brown, this volume).

18.5.4 Grading and harvest

Growth variation is considerable during the hatchery phase, and as flounder metamorphose and settle to the bottom over an extended period of time, cannibalism of smaller fish by larger, dominant individuals is common. Grading is important to separate size groups to prevent cannibalism, but timing and approaches are different. In Japanese hatcheries, flounder larvae (*P. olivaceus*) are separated by size into several tanks before settlement at a density of 3,000–6,000 individuals/m^2 (bottom area) (Seikai *et al.*, this volume). In European turbot hatcheries, pelagic larvae are moved with nets from larval tanks to weaning tanks between d20 and d30 ph. In Atlantic halibut hatcheries, smaller fish are culled as they reach metamorphosis; otherwise they continue to underperform throughout the growout stage (Brown, this volume). For southern flounder, weaned juveniles (about 60 dph, 5 cm long, $\sim$0.25 g wt.), are graded by size and stocked into nursery tanks at 700 individuals/m^2.

It has been suggested that by synchronizing metamorphosis, growth variation and cannibalism can be minimized in flatfish hatcheries. Thyroid hormones primarily regulate metamorphosis in flounder species (Borski *et al.*, this volume). In summer flounder, cannibalism may be reduced by treating larvae with thyroid hormone to synchronize settlement and shorten the metamorphic period (Bengtson and Nardi, this volume). Other hormones, may act either synergistically (cortisol) or antagonistically (prolactin) with thyroid hormone (Borski *et al.*, this volume) and research is needed to better understand their mechanisms of actions and how they may be used for the benefit of the larval culturist.

In Japan, survival of *P. olivaceus* to 3.0 cm TL is usually higher than 60% and sometimes exceeds 80% (Seikai *et al.*, this volume), and similar survival (50–70%) is obtained in Korean hatcheries for this species (Bai and Lee, this volume). Overall survival from egg through metamorphosis typically averages 40% for southern flounder (Daniels *et al.*, this volume), and from 50–60% for *P. olivaceus* in Korea. In Europe, survival rates in turbot hatcheries are relatively low, ranging from less than 10% to over 30%, and critical periods are at first-feeding and between d12–15 ph, when *Artemia* feeding begins. A mean annual survival rate of 20% to day 90 ph (1–2 g) is considered to be economically acceptable for commercial hatcheries (Person-Le Ruyet, this volume). In China, survival rate from newly hatched larva to juvenile (2 cm TL) varies from 0 to 40% among hatcheries, and averages 10–20% in large hatcheries with good expertise (Lei and Liu, this volume).

18.5.5 Hatchery economics

Fingerling prices for hatchery-raised flatfish are inversely proportional to level of production technology and of production. In France and Spain, production is around 7–10 million/year, and the industry growth is being constrained by high price of juveniles (€ 1.04–1.10 or USD 1.48–1.57 per fish) mainly due to low larval survival rates. In China, where more than 80 turbot hatcheries are currently in operation, large hatcheries produce 4–5 million juveniles/year, while small facilities produce 100,000 to 200,000 (Lei and Liu, this volume). Annual total production of turbot juveniles in Shandong province along the northern coast of China was 120 million in 2005 (Lei and Liu, this volume). The rapid increase in juvenile production in China caused a sharp drop in price, which in turn stimulated expansion of the farming industry and the markets. In China, production costs and farm gate price for hatchery-reared turbot juveniles (5 cm TL) is currently USD 0.07–0.25 and USD 0.22–0.37 per fish, respectively. To reduce costs, many hatcheries purchase embryos from specialized hatcheries that spawn turbot broodstock. In Japan, where survival rates during the larval culture stage are high (60–80%), the cost to produce 700 thousand juveniles (3.0 cm TL) for stock enhancement is 11 million Japanese Yen (16 JPY or approximately USD 0.17 per individual), with labor costs accounting for 18.9% of the total costs (Seikai *et al.*, this volume). Total fingerling production was 38 million individuals in 2006, with 31 million (81.6%) used for stock enhancement and 7 million (18.4%) used for aquaculture (Seikai *et al.* this volume).

The economics of a closed recirculating system for production of southern flounder fingerlings in the United States showed a breakeven cost of approximately USD 0.34 per fish (2.5 cm TL) with a single batch per year and 40% survival from egg (Daniels *et al.*, this volume). Breakeven costs can be reduced to USD 0.25 per fish by increasing the number of production cycles per year. Analyses for summer flounder hatcheries in the United States have indicated that the major operational costs are energy, skilled labor, and feed (including live and formulated feeds). Depending on scale of operation, location and subsidies, breakeven cost ranges from about $0.25–1.00 for 1–2 g juveniles, or from 2.5 to 10% of production costs of market size fish, assuming they are sold at $10/kg (Bengtson and Nardi, this volume). For the Atlantic halibut, a species with a long hatchery production cycle and few fingerling producers, the market price for hatchery-reared juveniles is relatively high ($5 or more per 5 g fish), and represents a significant hurdle to new growers, but this may potentially be offset by a relatively high market price at harvest (Brown, this volume).

18.5.6 Potential for stock enhancement

The advances made in artificial propagation of marine species such as flatfish in the last three decades have made it possible to supplement wild stocks through stock enhancement to complement traditional fishery management by restricting catches and conserving/restoring habitat. The purpose of flatfish stock enhancement is to stabilize the catch and increase productivity (Yamashita and Aritaki, this volume) by utilizing undercolonized nurseries in which natural rate of recruitment is below what the ecosystem can support (Miller *et al.*, this volume, Burke and Masuda, this volume). The disadvantages of stock enhancement include potential adverse impacts on wild populations due to genetic dilution, disease introduction, depression of wild stocks, and alteration of the community structure of the ecosystem. In Japan, where the largest flatfish stock enhancement effort is underway, the only species which have maintained their former abundance despite heavy exploitation are those which are intensively stocked, including the Japanese flounder (Yamashita and Aritaki, this volume). Around 25 million flounder juveniles are released each year in coastal waters of Japan by prefectural and national hatcheries (Burke and Masuda, this volume). In recent years, the total fishery catch of released flounder was estimated to be around 800 MT, assuming an 11.7% contribution rate (the percentage of hatchery fish in the total number of fish of the same species landed at markets). In comparison, aquaculture production of flounder shows similar levels to the total fishery catch, from 6,000 to 8,500 MT in the 1990s, but declining to 4,591 MT in 2005 due to increase of cheaper imported farmed flounder from China and Korea. In Europe, it has been estimated that 8% turbot recaptures would be sufficient to even out the costs of release based on commercial values of fish (Støttrup and Sparrevohn, this volume). In Korea, where the national government and local governments have supported stock enhancement of olive flounder over the last decade, a stable catch during this period (Bai and Lee, this volume) may indicate that the mass release of juveniles has helped to sustain recruitment to the commercial fisheries.

Outside of Japan and Korea, stock enhancement of flatfish is in the very early stages of development. In Europe, stocking of turbot is still at an experimental stage (Støttrup and Sparrevohn, this volume). In North America, only very limited pilot releases of summer flounder have been made in North Carolina

(Bengtson and Nardi, this volume), and southern flounder aquaculture and stocking efforts have just been initiated in Texas. There is interest among U.S. researchers to undertake stock enhancement work; however, political and financial reasons have impeded progress in this area (Miller *et al.*, this volume). For some flatfish species (e.g., Atlantic halibut), major disadvantages are the cost of rearing fingerlings before release, and relatively slow growth to legal landing size. Hence, no stock enhancement of Atlantic halibut has been attempted to date.

Prerequisites for successful stock enhancement

Long-term experience with stock enhancement of flounder in Japan has shown that successful and responsible stock enhancement has a number of prerequisites. The genetic profile of the wild stock must be defined as a basis of determining the effects of hatchery fish on genetic structure (Seikai *et al.*, this volume). Genetic dilution of wild populations can be prevented by replacing at least 25% of broodstock with new, wild broodstock each year, and by limiting use of broodstock to no more than 4 years. The carrying capacity of the ecosystem and the impact of release fish can best be determined by releasing fish at meaningful scale (50,000–100,000 year), and then studying density-dependent effects on growth or survival, and on the abundance of other species. Diseases of hatchery and wild fish and their detection should be understood, including protocols for screening hatchery fish prior to release.

Hatchery and stocking protocols to increase success

A number of criteria are considered to be critical to enhance the success of release technology: (1) size at release, mortality of released juvenile flounder is caused mainly by predation, and a sufficient size at release will minimize predation (Yamashita and Aritaki, this volume). However, the ability of hatchery fish to learn to adapt to the new natural environment varies with age/size, and there is an optimal size for release (9 or 10 cm) that produces the least cost per gram of net production by the released fish (Yamashita and Aritaki, this volume); (2) release habitat, habitats selected for release should provide refuge from predators and maximum retention of released fish (Yamashita and Aritaki, this volume); (3) release timing, hatchery-cultured flounder should be released when prey are most abundant and predators are lowest; (4) release magnitude, the numbers of hatchery fish released should be based on carrying capacity of the habitat and the ability of hatchery fish to use surplus trophic resources, rather than compete for limited resources against wild conspecifics or other commercially important species to cause their replacement; and (5) release method and conditioning, to determine the return rate, released fish need to be marked to distinguish them from the wild stock.

While the evidence is clear that releasing flounder results in increased catches, it is not generally profitable to stock flounder for commercial harvest at present (Yamashita and Aritaki, this volume). Stocking flounder for sportsfishing may be profitable, since sportsfishermen are willing to pay much more (e.g., through license fees) to catch them (Miller *et al.*, this volume). In Japan, some of the costs of producing fingerlings for stock enhancement are recovered by selling juveniles to private growout operations. The initial cost of a hatchery and the preliminary research to determine if stocking would work are prohibitive to fishermen and might be best supported by the state, since stocking is used as a tool for managing public resources (Miller *et al.*, this volume). In the United States, it is unclear whether the American taxpayer is willing to support stocking efforts absent other funding mechanisms (Bengtson and Nardi, this volume). In Denmark, stock enhancement is funded through licenses for anglers and releases are conducted through the National Coastal Fisheries Management Program, financed through fishing licenses for anglers and recreational fisheries.

18.6 Nursery culture

18.6.1 System design and requirements

Depending on species and location, different strategies are used to optimize growth and survival during the nursery period, a period which ranges from the postmetamorphic stages (1–10 g) to the size at which fish are stocked into production tanks (20–150 g) for growout to marketable sizes. Nursery culture of flatfish is typically conducted in land-based tanks situated in a greenhouse or in an industrial building in close proximity to the hatchery. In the United States, attempts were made to raise summer flounder fingerlings up to 150 g in floating sea cages, but the small fish were unable to survive currents and winter temperatures (Bengtson and Nardi, this volume).

In the United States, summer flounder juveniles are typically raised in the hatchery to about 10 g before being shipped to nursery or growout facilities. In Europe, turbot juveniles leave the hatchery at 3–4 months (1–3 g) and are raised in a nursery up to 5–20 g, but sometimes up to 80–100 g, for a period of about 3–6 months. Flowthrough concrete or fiberglass tanks (10–30 m^2 surface area × 0.5–0.7 m deep) were traditionally used, but RAS systems using high stocking densities (500–1,000 fish/m^2) enable better control of environmental factors (e.g., temperature, salinity, gas saturation) and biosecurity and also reduces heating and pumping costs. To maximize use of space in nurseries for turbot, shallow RAS raceways (0.25 m maximum depth) of various sizes are stacked 3–4 high (Person-Le Ruyet, this volume). Closed RAS systems are also commonly used for broodstock and fingerling production systems for commercial hatcheries in Japan.

Juvenile winter flounder can be stocked at densities as high as 300% (ratio fish ventral area to tank bottom area) with no reduction in growth (Fairchild, this volume). However, high stocking density can elevate blood cortisol and render fish more vulnerable to disease (Fairchild, this volume). Under high stocking densities, photoperiod can influence aggressive behavior; a constant 24L:0D photoperiod promotes growth, but also increases aggression, stress, fin damage, and bacterial infections (Fairchild, this volume).

18.6.2 Nursery protocols

Environmental conditions

Temperature control is critical for flatfish growth, but optimum temperature may change with size. In turbot, the optimal growth range decreases with size from 16 to 22°C for 10 g fish to 16–19°C for 40–50 g fish (Person-Le Ruyet, this volume). An ontogenetic decline in temperature optima also occurs in Atlantic halibut (Brown, this volume) and in *P. olivaceus* (Seikai *et al.*, this volume).

In flounder, sex differentiation is believed to be strongly influenced by temperature around the time of metamorphosis with high culture temperatures favoring male development (Borski *et al.*, this volume). In southern flounder, optimum temperature to produce the highest percentage of females is approximately 23°C, so water temperature should be held as close to 23°C (73°F) as possible during the first month, or until the fish reach 75 mm in length (Daniels *et al.*, this volume). Temperatures significantly higher or lower than 23°C will result in a higher percentage of males in the overall population. High stocking densities may also shift the population toward males.

Since flatfish are to varying degrees euryhaline, salinity is an important consideration for management of nursery and growout facilities. In some species, e.g., turbot, juvenile growth may be slightly enhanced at 20‰. Recently metamorphosed southern flounder are extremely tolerant of low salinity and can be raised in freshwater with high hardness and alkalinity (both greater than 200 ppm) (Daniels *et al.*, this volume). This euryhaline ability provides the culturist with great flexibility in management of inland hatcheries where a continuous source of seawater is not available. Juvenile greenback flounder of 80–190 g are also tolerant of low salinities with growth not impaired down to 15 ‰ (Hart, this volume).

18.6.3 Juvenile diet and nutrition

During the nursery period, turbot juveniles are fed dry pellets (51–52% protein and 12–13% fat) delivered automatically and continuously during the photophase. These diets incorporate fish meal and fish oil as the main protein and lipid sources to avoid HUFA deficiency. During the nursery stage, survival averages over 80% and fingerlings reach 20–30 g in 6 months, with feed conversion rate (FCR) as low as 1.0.

There are no commercially available feeds specifically manufactured for most flatfish species, since nutritional requirements are not well studied, and there is insufficient production demand. For these species, researchers either use commercial diets for coldwater marine species (~50% CP, ~10–15% CL) or manufacture diets in-house (Fairchild, this volume). To reduce diet costs and improve sustainability, many researchers are focusing on alternative protein sources to fish meal. In summer flounder, 40% soybean replacement for fish meal reduced cost/kg of fish produced by 14% (Bengtson and Nardi, this volume). Additional studies showed that 40% soybean replacement with added taurine and phytase provided equivalent growth as the fish meal control (Bengtson and Nardi, this volume).

18.6.4 Grading and harvest

Prior to leaving the nursery, turbot are graded by size using automatic machines designed for grading fruits. The most common market size is about 20 g. At this stage, turbot are often vaccinated against vibriosis and furunculosis, but they can also be vaccinated against diseases caused by *Flexibacter* and *Streptococcus*. Fish are transported to grow out farms by road (specialized international transport companies), or by airline transportation. In southern flounder, fingerling stocking densities of about 700 fish/m^2 are recommended to reduce cannibalism and to promote growth (Daniels *et al.*, this volume). Cannibalism can be controlled by grading by size and frequent feeding. Fingerlings may need to be graded 3–4 times during the few months it takes them to grow from 2 to 10 g. Larger fish do not require such frequent grading.

18.6.5 Behavioral conditioning for stock enhancement

Laboratory and field experiments indicate that the behavioral quality of hatchery fish, particularly susceptibility to predation, is an important determinant of the ability of hatchery-reared fish to persist in the wild after release. Hatchery-reared flatfish exhibit behavioral deficits, presumably resulting from genetic changes over generations (domestication) and environmental experiences in captivity that render them poorly-equipped to survive in the wild (Burke and Masuda, this volume). Even when breeding practices produce hatchery-reared juveniles with genetic diversity comparable to the wild stock, development within a hatchery environment result in expression of a domestic phenotype (Burke and Masuda, this volume). Compared to wild individuals, flatfish raised in a hatchery tend to spend more time swimming, lack caution, have poor concealment skills (e.g., burial and cryptic coloration), and show different feeding behavior (Yamashita and Aritaki, this volume, Burke and Masuda, this volume). Cryptic behavior may also enhance the ability to ambush mobile prey (Støttrup and Sparrevohn, this volume). Hence, when stock enhancement is the goal, the hatchery culturist must balance production efficiency against juvenile behavioral quality that optimizes postrelease survival (Burke and Masuda, this volume). To correct behavioral deficits of fish cultured for enhancement, an understanding of the ecological characteristics of the system to be stocked and the behavior of wild juveniles in it are required (Burke and Masuda, this volume). Researchers agree that postrelease survival may be improved by conditioning fish in the weeks preceding release. A main technique in this regard is to provide the fish with substrate characteristic of the release site to allow them to develop cryptic behavioral skills (burial and pigmentation) and reduce vulnerability to predators (Fairchild, this volume, Støttrup and Sparrevohn, this volume). Other techniques, such as rearing at low density with sandy substratum, use of a diet of live mysids, and predator-exposure have been tested successfully in the laboratory, but field trials are needed to confirm their effectiveness postrelease (Yamashita and Aritaki, this volume).

18.7 Growout

18.7.1 System design and requirements

The unique characteristics of flatfish that must be considered by culturists during the growout phase of production are their preference for the tank bottom and their low level of activity, which can affect tank design and hydrodynamics, stocking densities, and related effects on water quality in the microenvironment of the demersal fish. Basic approaches to flatfish growout are using land-based tanks or raceways or at sea in cages. Many flatfish, such as turbot, halibut, and Japanese flounder are cultured in outdoor land-based tanks or in indoor recirculating tank systems. Tank sizes and shape vary considerably but generally are 6.10–9.14 m (20–30 ft) in diameter with black shade cloth for outdoor systems to provide protection from the sun. Raceways used for Atlantic halibut culture were problematic due to a reduction in water quality along the length of the raceway (Brown, this volume). In general, flounder prefer low light intensities and can develop skin ulcerations when left in tanks exposed to direct sunlight. In most flatfish farms, there is a trend away from air blowers to pure oxygen to accommodate higher stocking densities.

Along the northern coast of China, land-based turbot production tanks are held in greenhouses, preferred for their low cost of construction and ease of temperature control in the winter. The growout facilities for subadult and adult fish are similar to those of juvenile nurseries, except larger fish. Concrete tanks (either 5–6 m circular, or 5–6 m square with rounded corners) are used. Water temperature is the most important factor for site selection for turbot farming in China, and farms are located near a source of saline well water (Lei and Liu, this volume), which is clean, has a chemical composition similar to seawater, and has a suitable temperature range. Since annual temperature fluctuations in coastal waters exceed the tolerance limits for turbot, escaped farmed turbot are unlikely to persist in the natural environment.

A few cage culture operations exist on the southern coast of China where the water temperature permits seasonal production. Cage culture also has the advantage of lower cost of pumping water and facility construction, and faster growth rates than tank-culture. Transfer of turbot cultured in tanks in the north to cages in the south has reduced the cost of turbot culture in China. However, further research work is necessary to develop special flatfish culture cages that can resist the strong wind and currents in the southern coastal waters.

In Europe, turbot are grown in land-based tanks and raceways usually situated in industrial buildings. Tank volume ranges from 25 to 100 m^3 with a maximum water depth of 0.70 m. RAS are quickly replacing flowthrough systems in Europe, but are mechanically and biologically complex and require continuous water quality control (Person-Le Ruyet, this volume). To minimize heating costs, makeup water exchange is limited to 5–10% system volume per day with ground water used where possible. Flat-bottomed cages submerged in coastal areas or floating cages are also used for growout or holding of large turbot prior to marketing and sea cages are being tested in North West Spain.

In Japan, land-based tanks with flowthrough seawater are the primary system for growout of Japanese flounder, representing 75% of production area in 2005. Approximately 300–400 farms throughout Japan with an average of 1,300 m^2 of culture area produce about 16 MT of fish/year, with a stable production efficiency of about 13 kg/m^2/year (Seikai *et al.*, this volume). Typical land-based flounder farms are sited seaside, with tanks either installed indoors or covered with shade cloth. Circular tanks (6–10 m diameter × 0.6–0.8 m deep) are common, but square or octagonal tanks are also used. A few farms use tanks with bottoms covered with sand, which produces fish without hypermelanosis on the blind side to improve market value. RAS were tested for the production of Japanese flounder from the late 1980s to mid-1990s, but were never commercialized due to high capital costs and a declining market price of flounder in Japan.

In Korea, land-based coastal facilities are also used to produce olive flounder in flow-through tank systems. Seawater is pumped directly from the open sea into the head tanks and subsequently supplied to the fish tanks after treatment. Each farm produces an average of 110 MT/year (Bai and Lee, this volume).

In North America, the few commercial flatfish production facilities also use RAS, permitting production in inland areas without a continuous source of seawater and where seasonal temperatures exceed those tolerated by the fish (Daniels *et al.*, this volume). In Mexico, where commercial production of flatfish is beginning, ambient seawater water temperatures of 14–25°C along the Baja peninsula are considered favorable to both summer flounder and California halibut culture in both flowthrough and RAS systems, and startup farms plan to target the large southern California market (Conklin and Piedrahita, this volume).

In North America, Atlantic halibut juveniles may spend the entire growout cycle in a land-based tank system, or may be moved to net pens for final growout to market size. Although shallow tanks are considered to be more cost-effective for flatfish production, Atlantic halibut grew faster in deep (4–10 ft deep) tanks, as shallow water impeded access to pelleted feeds and increased interfish-aggression (Brown, this volume). Atlantic halibut have also been produced in surface cages, generally 3–7 m deep of a variety of designs and materials (steel, plastic, wood with mesh netting), with a rigid base to prevent sagging when stocked with fish (Brown, this volume). Submersible cage designs have also been tested successfully in New Hampshire (Brown, this volume). Sheltered conditions are important for rearing flatfish in pens and cages, because currents and waves cause excessive swimming activity, and shade netting is used to prevent excessive exposure to sunlight, which can cause mortality (Brown, this volume, Bengtson and Nardi, this volume).

18.7.2 Stocking and splitting

Stocking density is important to maximize the use of tank space and water. Even when stocked at very low densities, many flatfish species aggregate in layers on the tank bottom rather than spread out across the available space, apparently an innate behavior associated with concealment in their demersal habitat for both predator avoidance and predation. While this behavioral trait suggests that these fish could potentially be raised under very high stocking densities, water circulation and quality become problematic in and around the layers of these sedentary fish, and water quality (e.g., DO and NH_3) measured in the tank effluent underestimate the detrimental conditions faced by fish on the bottom (Conklin and Piedrahita, this volume). Because of this unique behavioral trait, flatfish culturists often measure stocking density in terms of percentage of bottom coverage (PCA = percent ratio of total fish ventral area to total tank bottom area) or kg/m^2 rather than per unit volume (kg/m^3) as for round fish. For California halibut, better growth was achieved at 100% of the coverage area compared to 200 and 300% PCA.

As flatfish grow and increase in body depth, the maximum recommended stocking density also increases (Person-Le Ruyet, this volume). In Europe, intensive systems for turbot increase stocking densities from about 30–35 kg/m^2 for 300 g fish, to 45 kg/m^2 for 750 g fish, and up to 60–80 kg/m^2 for larger fish, with stocking densities as high as 100 kg/m^2 possible. In China, stocking densities are somewhat lower, but are also increased as fish grow: 2 kg/m^2 at 15 g, 7 kg/m^2 at 50–100 g, and 10–20 kg/m^2 at 600–800 g (Lei and Liu, this volume). For southern flounder, Japanese flounder, Chilean flounder and greenback flounder, stocking density is increased as fish grow to a final density of 15–29 kg/m^2 of 0.6–1.0 kg fish. In *P. olivaceus*, stocking density increases with fish size from around 0.66 kg/m^2 for 1.5 g fish to 19.1 kg/m^2 for 764.2 g fish. Such densities are relatively lower compared to stocking densities used for roundfish (i.e., 60–120 kg/m^3), but can be increased by decreasing tank depth (i.e., increasing tank bottom surface to volume ratio). Reducing tank depth, however, lowers the self-cleaning efficiency in round tanks that require a minimum depth to diameter ratio and reduces the use of vertical space for fish production. Since flatfish do not fully use the water column as do round fish, a major challenge of intensive flatfish production is that of maximizing use of vertical space in facilities that are limited in area.

One way of ameliorating water quality conditions to fish at the bottom is to increase water flow rates. In both California halibut and Japanese flounder, maximum juvenile growth was achieved at 1.0 body length/sec (bl/s) (Conklin and Piedrahita, this volume; Seikai *et al.*, this volume). Higher flow velocities up to 1.5 bl/s did not affect survival, but reduced feed efficiency and growth and increased tail beating, a behavior presumably required to maintain position. More research is needed to improve tank design and flow patterns for commercial flatfish tank culture (Conklin and Piedrahita, this volume).

Raceways have also been used for flatfish culture, and these may be stacked to maximize use of vertical space. In Atlantic halibut, however, raceway culture proved problematic because of a reduction in water quality along the length of the raceway (Brown, this volume). Another method that researchers have used to maximize use of vertical space is to use shelving in conventional (relatively deep) tanks. Except for the Atlantic halibut, which readily occupy in-tank shelves, few reports have indicated that flatfish species voluntarily occupy in-tank shelving.

Furthermore, research is needed on shelf design, hydrodynamics, lighting, and stocking density to maximize feeding, growth, and biomass densities, while maintaining self-cleaning efficiency and water quality.

Growth variation is typically observed during nursery culture and growout of flatfish as well as other finfish species, related in part to interfish aggression and disproportionate acquisition of food by more aggressive individuals. In general, it is recommended to minimize interfish aggression during growout by maintaining size homogeneity, though laboratory data has shown that regular grading in turbot does not promote growth (Person-Le Ruyet, this volume). Hence, regular grading is practiced to reduce these effects. Fish are graded by hand with the aid of hand nets or mesh sorters during the nursery stage and with mechanical graders, grading tables, or automatic machines during growout. Fish are separated into several size classes, generally twice during growout.

18.7.3 Growth and survival

Growth rates documented for most species of flatfish are moderate, vary considerably among species, are highly variable from site to site and are dependent on temperature. Turbot grow quickly in comparison with other species of flatfish. In Europe, turbot farms using heated or geothermal water to maintain temperatures between 14 and 19°C can routinely raise turbot to 1 kg at 18 months and 3 kg at 3 years of age. In China, turbot juveniles about 10 g in body weight can grow to market size of 500 g in 7–9 months in greenhouse systems. In Japan, *P. olivaceus* fingerlings (1–3 g) grow to 0.5 kg in 9–10 months and 1 kg in 14–16 months (Seikai *et al.*, this volume). Fingerling Chilean flounder grew to 0.3–0.5 kg in 20–25 months and 1 kg in 35–37 months (Silva, this volume). From the egg stage, mixed-sex populations of southern flounder reached 600 g in 16 months (Daniels *et al.*, this volume). In China, summer flounder raised in coastal ponds grew from 8 to 750 g in 1 year (Bengtson and Nardi, this volume). In the United States, a primarily all-male summer flounder population raised in an intensive RAS grew from 85 to 440 g in 614 days and a slowing of growth was associated with sexual maturation. California halibut were raised from egg to 1 pound in 3 years, but under suboptimal temperature conditions (Conklin and Piedrahita, this volume). Winter flounder also reached market size in 2–3 years; though, production time can be decreased by rearing the fish at warmer temperatures (Fairchild, this volume).

After the nursery stage, mortality of southern flounder is minimal during the rest of the growout cycle until the fish reach market size. Survival of Japanese flounder throughout the growout period varies from farm to farm and ranges from 60 to 80%.

18.7.4 Environmental conditions

Temperature

In flatfish, optimal environmental conditions for subadults and adults are likely to be different from juveniles, but experimental data on such differences are scarce. In China, water temperature in turbot tanks is maintained from 11 to 18°C year round. In Atlantic halibut, optimum growth decreased with size from 11 to 14°C at 0–20 g, to 10–12°C at 150–400 g, temperature for and 9–11°C from 400 to 1,000 g (Brown, this volume). For *P. olivaceus*, the optimum temperature for growth also decreased from 25 to 20°C as fish grew from juvenile to 500 g (Seikai *et al.*, this volume). High mortality during summer affecting mainly larger size fish is well known in *P. olivaceus* farms in Japan. In China, diseases are also precipitated at high temperature on turbot farms, and water temperature must be controlled.

Salinity

Low-salinity tolerance is an advantage to inland-based culture, since it opens up the possibility of using groundwater or geothermally heated water sources for fish production. From the juvenile stages, most cultured flatfish are euryhaline and can tolerate a wide range of salinities. In southern flounder, for example, growth of fingerlings (~130 g) to an average market size (~600 g) in low-salinity (0.5‰) groundwater was not different from growth in full strength seawater (36‰) (Daniels *et al.* 2007). In some species, growth is faster at salinities lower than full strength seawater. Turbot, for example, can survive and grow in a wide range of salinity, from 12 to 40‰, but the optimal salinity range is 20–32‰. In *P. olivaceus*, growth of 0.5 g fingerlings was highest in 50% seawater compared to 75 and 100% seawater (Seikai *et al.*, this volume), and salinities below 3.3‰ impaired survival of 44 g fish (Seikai *et al.*, this volume). Salinity tolerance may vary with age/size. For example, growth of California halibut early juveniles was unaffected at salinities ranging from 5 to 30‰, but older juveniles were not as adaptable (Conklin and Piedrahita, this volume).

Substrate

As benthic species, substrate has a very important influence on flatfish health. Improper substrate has been associated with skin lesions in many flatfish species (e.g., Atlantic halibut and *P. olivaceus*, barfin flounder *Verasper moseri*, and summer flounder), which decrease carcass quality and may also adversely affect growth. Skin lesions began to heal when Atlantic halibut were transferred from a smooth fibreglass substrate to a sand substrate (Noga *et al.*, this volume) and similar results were reported for summer flounder and *P. olivaceus*. Using microceramic particles instead of a smooth substrate prevented spots in Japanese flounder (Seikai *et al.*, this volume). Spots can also be avoided by holding barfin flounder in white tanks (Noga *et al.*, this volume).

Illumination

Inappropriate lighting in indoor tanks may also contribute to abnormalities in hatchery-reared flatfish. Southern flounder that are raised in outdoor ponds rarely exhibit albinism. When southern flounder raised under low light were exposed to increased light intensity 1 week posthatching, partially albino fish had much more normal pigmentation (Noga *et al.*, this volume). The effects of illumination are complex, related to type of light used, intensity, tank color, and water depth. Excess light can be deleterious; for example, Atlantic halibut cultured in shallow raceways and sea cages with insufficient UV protection received sunburn. Cataract is common in Atlantic halibut and might be caused by excess UV light, although other environmental or nutritional factors may be involved (Noga *et al.*, this volume).

18.7.5 Diet and nutrition

Optimal levels of dietary protein for cultured flatfish species are high, ranging from 45 to 63%, but with efficient feed conversion ratios (FCR, wet/dry) below 1.5:1 for formulated pelleted diets, probably related to naturally low metabolism and a sedentary life style. The high price of formulated feeds can therefore be offset by the efficient conversion of feed into biomass (or live weight). In Europe, extruded pellets formulated for turbot were developed in France during the early 1980s and are now commonly used for commercial growout. These diets have a high protein (50–54% dry matter), and low crude lipid content (about 12%). The dietary energy content of flatfish diets is generally lower than that of other farmed fish species. In the United States, similar CP (50%) and CL (10%) levels have been found optimal for southern flounder (Daniels *et al.*, this volume). In many flatfish species, increasing dietary lipid can have a protein sparing effect, but may also elevate body fat deposition (Conklin and Piedrahita, this volume). In Europe, high lipid (20%) finishing diets are used for turbot when specific markets demand a higher flesh fat content (Person-Le Ruyet, this volume).

In China, feed for the turbot farming industry has changed from raw minced fish to moist pellets to commercial feeds, with disease and pollution being major problems causing this transition. Currently, imported commercial dry pellets are formulated specifically for turbot, and research institutes and companies are developing high-quality domestic diets for turbot and other flatfish in China. The FCR (wet/dry) for turbot in China varies with the types of the feed from about 6 for wet fish, 1.83–3.3 for moist pellet, 0.95–1.5 for domestic dry pellet, and 0.81–1.0 for imported dry pellet.

In Japan and Korea, *P. olivaceus* are fed commercial pelleted diets that have high protein and low lipid levels ranging from 48 to 56% CP and 6–14% CL for the first few months, and then are fed moist pellets and raw fish, either whole or as ingredients for the moist pellets. Information on FCR on commercial farms is not available, but in research facilities FCR (wet/dry) on formulated pelleted diets is about 1.0 (Seikai *et al.*, this volume, Bai and Lee, this volume).

In Japan and Korea, considerable work has been undertaken on substitution of alternative protein and lipid sources to fish meal and fish oil in flounder feeds. In general, studies indicated that a significant proportion of fish meal protein can be replaced by several plant and animal protein sources (e.g., soybean meal, feather meal, meat and bone meal, meat meal, corn gluten meal, malt protein flour, fermented fisheries by-products, and soybean curd residue mixture) in the diet of *P. olivaceus* (Seikai *et al.*, this volume, Bai and Lee, this volume). Combinations of multiple ingredients and inclusion of feeding stimulants are most effective in reducing dietary fish meal protein without amino acid supplements. In North America, work with Atlantic halibut and southern and summer flounder have also demonstrated that a significant fraction (approximately 40%) of the fish meal protein can be replaced with soybean meal protein (Bengtson and Nardi, this volume). Studies to date were obtained from short-term feeding trials with fish of less than 10 g initial weight, and long-term culture trials to marketable stages are needed.

Juvenile *P. olivaceus* fed a diet without vitamin C exhibited typical deficiency signs including anorexia, scoliosis, cataract, exophthalmos, and fin haemorrhage. In Korea and China, increasing attention has been paid to the use of growth and immunity stimulants for flatfish culture, including vitamins, lipid, mineral mixtures, and alginate oligosaccharides for turbot, and glucans, chlorella, aloe, Song-Gang stone, and probiotics as feed additives for *P. olivaceus* (Bai and Lee, this volume).

18.7.6 Diseases

In Japan, pathogenic diseases cause serious economic losses in the flounder (*P. olivaceus*) culture industry, valued at 1.3 billion JPY in 2004, or 17% of the total value of flounder aquaculture (Seikai *et al.*, this volume). Based on the available data, the economic impact of diseases in other cultured flatfish in other countries cannot be accurately estimated, and only the most important diseases are mentioned here. In general, cultured flatfish are susceptible to a host of pathogens commonly afflicting other intensively cultured finfish, and severity and range of pathogens increase with level of intensification and production. An increasing number of viral infections have been reported in a variety of flatfish, including turbot, *P. olivaceus*, and Atlantic halibut. Viral infections, usually severe in young fish, are often asymptomatic in older fish, which transmit the virus vertically to offspring and horizontally to cohorts. Since no drugs or commercial vaccines are available to treat viral infections in any fish, control depends on biosecurity, but this is difficult with flatfish since culturists still depend on wild-caught fish with unknown history of virus exposure. Viral hemorrhagic septicemia virus (VHSV) caused by a rhabdovirus of the genus *Novirhabdovirus* have caused significant losses in turbot hatcheries (Noga *et al.*, this volume), and cultured flounder *P. olivaceus* in Japan and Korea also suffer high mortalities from VHSV. *Vibrio harveyi* (and *V. carchariae*) is the cause of flounder infectious necrotizing enteritis (FINE) and mass mortalities in *P. olivaceus* as well as enteritis and stunting in summer flounder. Epidermal hyperplasia (herpes virus) and nervous necrosis (striped jack nervous necrosis virus) also are found in cultured *P. olivaceus*. In North America, nodaviral infection (genus *Betanodavirus*, family Nodaviridae) in larvae and juveniles is a serious hindrance to halibut culture (Noga *et al.*, this volume). Detection of nodavirus using immunoassay may help to select virus-free spawners (Noga *et al.*, this volume).

Bacterial diseases are precipitated by stress (e.g., overcrowding, low dissolved oxygen, high ammonia, transport, high temperature). In Europe, significant mortalities in flatfish have been caused by the bacterial disease *Edwardsiella tarda*, but no vaccines or effective therapeutants are available. Vibriosis (*Vibrio anguillarum*, or *Listonella anguillarum*) has been reported in turbot and in Chilean flounder (*V. splendidus* and *V. anguillarum*) (Silva, this volume). Disease prevention and treatment protocols differ from farm to farm but with a common theme of limiting the use of antibiotics. Immersion vaccine provides good protection against *V. anguillarum* in Atlantic halibut (Noga *et al.*, this volume).

E. tarda is a persistent bacterial pathogen of *P. olivaceus* in Japan that is believed to originate in terrestrial runoff and causing 30–40% of the total mortality every year, especially in summer when water temperatures increase (Seikai *et al.*, this volume). Streptococcosis (*Streptococcus iniae* and *S. parauberis*) and *Lactococcus (Lactococcus garviae)* are also important pathogens in cultured flounder in Japan and Korea during summer (Seikai *et al.*, this volume, Bai and Lee, this volume). Lowering water temperature and stocking density, increasing dissolved oxygen levels, supplementing diets with vitamin C and E, treatment with antibiotics, and fresh water treatment are used to control diseases. Three chemicals, oxytetracycline hydrochloride, sodium nifrustyrenate, and alkyl trimethyl ammonium calcium oxytetracycline, are approved for treatment of bacterial diseases of flatfish in Japan, and none are available for viral and parasitic diseases.

In the United States, external parasites such as *Argulus* spp. (sea lice) are common in wild broodstock and have caused severe anemia and hemorrhagic skin lesions in captive summer and southern flounder. Marine Ich *Cryptocaryon irritans* is a ciliate that can cause skin and gill damage and also kill a large number of fish rapidly, but can be treated in the euryhaline southern flounder by lowering the salinity of the water below 3‰ (Daniels *et al.*, this volume). Marine Ich has also been reported in cultured Japanese flounder and turbot (Noga *et al.*, this volume). Turbot are highly susceptible to parasites *Trichodina* and *Uronema,* and formalin baths once a month are used to control infection.

18.8 Harvesting, processing, and marketing

In Europe, asphyxia in air or on ice are not appropriate for euthanasia of farmed turbot according to animal welfare protocols, so harvested turbot are chilled rapidly and then bled, but electrocution or a percussive blow to the head are also practiced (Person-Le Ruyet, this volume). Fish are transported on ice to processing units and are usually marketed whole and fresh, but a market for fillets is developing in Europe, and a market for live turbot is developing in Asia and some European cities. Private companies precondition and package live turbot for survival up to 2 days without water (Person-Le Ruyet, this volume) to provide maximum product freshness while reduce shipping costs. Demand is higher than supply, so there is minimal competition between farmed and wild turbot, which are larger and command a higher market price. In France, quality labels (e.g., "turbot label rouge") certify high quality and traceability.

In China, turbot are harvested between 500 and 750 g and are packed in 40-L polyethylene bags containing seawater and then filled with oxygen. Water temperature is maintained between 7 and 8°C during live transport to market, either by truck or airline. The majority of the turbot farmed in China were marketed as live fish for domestic consumption in large metropolitan areas near the east coast of China, but as turbot production has increased and prices decreased, it has spread throughout the country and this exotic species is becoming a popular choice among the general public. In 2006, flatfish farming area in China encompassed about 8,060 thousand m^2, producing 83,000 MT, including over 50,000 MT of turbot. Unfortunately, prohibited nitrofuran metabolites were found in fish from Shanghai markets in 2006, causing consumers to distrust farmed turbot and resulting in a quick decrease in the market and production (Lei and Liu, this volume). Cooperation among the industry, researchers, and government is needed to guarantee the safety of farmed turbot.

In Korea, as the flounder aquaculture industry expands, value-added products will be developed to meet consumers' preferences. In the United States, the market demand for flounder (e.g., winter, summer, and southern flounder) varies seasonally with the availability of wild catch and with size. Very little cultured product has been harvested and processed to date, so there are no standard practices. Most wild flounder are processed into fillets, but as much as one-third of the volume are marketed whole (bled), fresh killed for distribution in the Asian, principally Japanese and Korean markets. High-quality fish bled on ice are sold at \$8.00–12.00/kg for 0.5 kg–1.0 kg fish, \$16.00/kg for 1–2 kg fish, and \$20.00–25.00/kg for >2 kg fish. Many potential market niches for cultured flounder are available, each requiring a different size and presentation (i.e., whole on ice, filleted, live, etc.). Since southern and summer flounder in the United States are very similar in appearance to *P. olivaceus*, they have also been exported to Asian markets. Given the considerable size variation shown by most flatfish during growout, commercial farms would ideally avail markets that require different sizes of fish.

Flounder grown in fresh water will need to be purged prior to sale, as freshwater systems have the tendency to impart off-flavors to fish. However, these flavors can be eliminated by purging the fish for 2–3 days in saltwater (15‰ salinity or greater) (Daniels *et al.*, this volume).

In North America, Atlantic halibut are bled immediately postmortem by incision of a major artery during gutting or removal of gill arches, as the presence of blood veins in the fillet detracts from appearance and taste. The fillet yield of halibut is typically around 55%. The traditional market for wild halibut is based on large fish (5–10 kg) sold fresh and in the form of steaks. However, with the availability of farmed product, fish as small as 750 g are being sold to restaurants at higher prices. The United States currently imports <100 MT/year from Norway and Canada, especially during the off-season for the Pacific halibut fishery in order to fetch the highest prices. Start-up farms are targeting high-value niche markets, including sushi chefs, with live transport to optimize quality. In Mexico, startup farms for California halibut plan to access the nearby large southern California market.

In Australia, the wholesale market price for *R. tapirina* is AUD 6–10/kg (USD 5.03–8.39 per kg, 2.28–3.81 per lb), and the most acceptable market size is 500–800 g with a 95% recovery rate gilled and gutted (Hart, this volume).

Increasing export values of wild caught *R. tapirina* from New Zealand suggest there is potential in Australia to substitute farmed fish for imports and potential for both countries to produce farmed product for export to Asia.

18.8.1 Production economics

Limited published information is available of production costs of commercial flatfish farms; however, the available data are remarkably consistent among species. In China, direct (e.g., stock, feed, labor, maintenance, transportation, etc.) and indirect (e.g., insurance, depreciation, etc.) costs for farmed turbot is estimated at USD 7.00/kg (Lei and Liu, this volume). In Japan, depreciation, feed, and fingerlings generally account for 23, 15, and 10% of the total sales price of fish, respectively, with 13% overhead (Seikai *et al.*, this volume). The most important factors are survival rate and market price, and the break-even points for these factors are 65% and 1,700 JPY/kg (18.44 USD/kg, $8.38/lb), respectively.

In the United States, analyses have been conducted on production economics of summer flounder aquaculture in indoor recirculating growout systems using computer spreadsheet simulation models based on biological and engineering information from pilot-scale field trials at NCSU and UNCW. Based on optimal growth rates obtainable with all-female populations, it is assumed that the fingerlings grow to a harvestable size of 1.5 lbs (682 g) in 14 months (420 days), 20% fingerling mortality after stocking, and 25% of the slowest-growing fish are culled by month 9 (day 270). It is assumed that niche market buyers purchase the fish live at the farm gate for $5.00/lb. Assuming a useful life of 10 years for the facility buildings and equipment, and a financial discount rate of approximately 7% per year, the net present value (NPV) of the net returns over 10 years is approximately $3,024,112 in 2006. The internal rate of return (IRR) of the project is 38%. The discounted payback period (DPP) is 38 months. The breakeven price per pound is $3.98 over the 10-year project. Debt service, fingerling cost, and labor account for 50.0%, 21.2%, and 9.1%, respectively, of total annual costs.

Large-scale farms for halibut in Norway producing 200–300 MT/year are profitable with production costs approaching 5–6 € /kg ($3.3–4.0 per lb) (Brown, this volume). Likeswise, industry analyses show that at sufficient scale (>100 MT/annum), land-based recirculating growout of halibut in Canada is profitable (IRR of 15%) with a breakeven cost of between $7.75 per kg ($3.53 per lb) at 100 MT/year and $7.19 per kg ($3.27 per lb) at 300 MT/year. Predicted costs for sea cage growout of halibut in the United States are between $3.19 per kg ($1.45 per lb) and $4.09 per kg ($1.89 per lb). Some analyses of Canadian halibut farming showed that while sea cage farming of halibut would be profitable (IRR = 9%), a land-based operation using flow-through raceways would not (Brown, this volume). IRRs of 9 and 15%, respectively, were modeled for land-based recirculating production and for sea-based growout based on an ex-farm price of $4.50 per lb.

18.9 Industry status

In Europe, turbot was selected for aquaculture in the early 1970s in the United Kingdom and France due to its value and its high potential growth rate (>3 kg within 3 years) under intensive culture conditions. Turbot production increased from 270 MT in 1987 to 7,633 MT in 2006, with Spain (84%) the main producer, followed by France (18%), Portugal (2.4% MT), and the Netherlands (1.3%) (Støttrup and Sparrevohn, this volume). The primary European market is Spain, with much smaller markets in France, Italy, and Germany. There is market demand for whole fish (about € 9,2 or USD 13.09 per kg in 2006), and fillet markets are developing, but a decrease in market price is expected as turbot production increases in the future. Development of the turbot aquaculture industry in Europe is limited by the high price of juveniles (€ 1.04–1.10 or USD 1.48–1.57 per fish), mainly due to relatively low larval survival, and to limited access to seawater and conflicts with tourism. In France and Spain, production is around 7–10 million juveniles/year (Person-Le Ruyet, this volume).

With technical support from Great Britain, turbot farming was introduced to Chile in the late 1980s with 17 MT produced using juveniles supplied from Europe. Since 1998, annual production in Chile has ranged from 268 to 426 MT, using mainly locally-produced juveniles. Most of the turbot produced in Chile are exported to Asia and the United States. Chile has also adapted technology for the cultivation of the *P. olivaceus* (Japanese or olive flounder) and is conducting research on culture of native flounders (*P. microps* and *P. adspersus*) as well as the Atlantic halibut with eggs, broodstock, and juvenile halibut supplied from Canada.

Turbot was also introduced into China from the United Kingdom in 1992 (Lei and Liu, this volume), with commercial-scale juvenile production by 1999 in the Shandong province along the northern coast, where the first growout systems were built in greenhouses using deep saline well water. Fish were initially marketed live in large cities along the southeast coast for USD 80/kg (USD 36/lb), and within 10 years, developed into one of the main mariculture industries in China, with yearly production of over 50,000 MT. China produced about 76,000 MT of flatfish (six species) in 2005 (Bengtson and Nardi, this volume) of which over 50,000 MT were turbot. In 2003, summer flounder juveniles were shipped from the United States to China, which now has a growing summer flounder industry (Bengtson and Nardi, this volume). Chinese scientists expect that summer flounder culture will thrive in colder water conditions in their country, whereas other species (e.g., southern flounder) will do better in warmer water conditions (Bengtson and Nardi, this volume).

Aquaculture of *P. olivaceus* began in the mid-1970s in Japan, and commercial production increased dramatically in the 1980s from 648 MT in 1983 to 6,000 MT in 1990, reaching a peak of 8,583 MT in 1997, a level that exceeded the annual commercial fishery catch in Japan (Seikai *et al.*, this volume). However, production decreased gradually to 4,592 MT by 2005 due in part to competition from flounder imports from Korea. In Japan, *P. olivaceus* maintains a high market price (over JPY 2,000 or USD 21.56/kg, USD 9.80/lb from 2000 to

2007), 2–3 times higher than for yellowtail *Seriola quinqueradiata* and red sea bream *Pagrus major*.

In Korea, aquaculture production of *P. olivaceus* increased from the 6,733 MT in 1995 to 43,852 MT in 2006 valued at USD 458.9 million; this increase attributable in part to government policies favoring production of high-value species (Bai and Lee, this volume). Most of cultured *P. olivaceus* are consumed in Korea, with some exported to Japan (3,729 MT; USD 50,385,478), United States (32 MT; USD 776,280), and Taiwan (17 MT; USD 223,789). Premium market size is 1 kg body weight, and price is around USD 10–15/kg.

A summer flounder industry began in the United States in 1995 with the development of a commercial hatchery in New Hampshire and several growout facilities, but little product was produced. Juveniles were first exported in 2003 to China, which now has a growing summer flounder industry, and subsequently in 2006 to Mexico, which is also developing an industry (Bengtson and Nardi, this volume). A hatchery for California halibut was constructed in Ensenada, Mexico, a joint effort between a government research and education center, and local commercial interests, to produce as many as 500,000 juveniles/year to support farms to supply the large southern California market (Conklin and Piedrahita, this volume).

In the United States, production of southern flounder in inland RAS in North Carolina is a nascent industry, and fish are being produced in inland, low-salinity, or brackishwater RAS facilities in cooperation with university researchers (Daniels *et al.*, this volume). Fish have been grown to marketable sizes and are currently being distributed to local farmers markets and to restaurants.

Pilot halibut farming efforts are underway in North and South America, including a facility in Hawaii with access to deep cold water , which is producing fish for local markets, and hatchery facilities in Chile are being supplied with eggs, broodstock, and juvenile halibut from private and government entities in Canada (Brown, this volume).

18.10 Summary: industry constraints and future expectations

Turbot farming, transplanted from Europe, has developed into a very important mariculture industry in China and in Chile in less than 20 years. Rapid development in China is attributable, in addition to the favorable biological characteristics of these fish, to suitable land and saline water resources and to effective market strategy, starting with high value live markets and then expanding to broader consumer markets as production technology improved and prices decreased. Production efficiency of turbot farming in China may be improved by decreasing the price of juveniles (18% of total production cost), automation to lower labor and feed cost (16 and 17%, respectively, of total production cost), better prevention and control of disease, and by improving genetics and marketing. As the industry has grown in China, availability of saline well water, pollution of the coastal environment, and product safety have emerged as important constraints to industry growth. The development of closed recirculation farming systems is an important technological requirement for industry expansion along the northern Chinese coast.

In Japan, where labor costs account for 18.9% of the total *P. olivaceus* fingerling production costs (Seikai *et al.*, this volume) laborsaving methods are important to maintain production efficiency. The outbreak of diseases is a serious problem in *P. olivaceus* farms, and the selection of disease-free spawners based on PCR-based detection (e.g., nodavirus VNN) may be effective to avoid transmission from broodstock to larvae and juveniles (Seikai *et al.*, this volume). In Korea, where *P. olivaceus* culture and production have intensified, there are growing concerns about pollution of public waters related to the discharge of effluent from flounder farms into the sea. To reduce nutrient discharge, development of highly digestible diets and methods for treatment of effluent are needed.

In North America, commercial culture of summer flounder, southern flounder, winter flounder, California halibut, and Atlantic halibut is in its early stages of development, and limited government resources for research and for industry development have slowed industry growth. Controlled spawning methods are well developed for these species, but more research is needed in the areas of broodstock nutrition and diet development, difficult to accomplish due to the serial spawning character of flatfish and considerable facility requirements needed to conduct replicated experiments. Selective breeding programs are expensive, and although breeding programs for Atlantic halibut are underway in Scotland, Canada, Norway, and the United States, eggs are still mainly sourced from wild caught broodstock, and published information on performance of F1 stocks is scarce.

Although fingerlings production costs must still be reduced, significant progress has been made in hatchery production of all of these species, and large numbers of juveniles can now be produced. Further research is needed to gain a better understanding of the physiological basis for metamorphic abnormalities, such as abnormal pigmentation (pseudoalbinism on the ocular side and hypermelanosis on the blind side), and arrested eye migration, which decrease market value (Borski *et al.*, this volume). The brain/hypothalamic and environmental and nutritional factors that regulate thyroid stimulating hormone and thyroid hormone secretion, as well as those factors that regulate eye migration require elucidation to enable culturists to synchronize metamorphosis, settlement, and development of flatfishes in captivity (Borski *et al.*, this volume).

Moderate growth rates and extended growout times are main drawbacks for most flatfish species. For commercial growout to be realized at a significant scale, production costs need to be lowered and market prices increased. Since in many flounder species, males grow slower after reaching sexual maturity at smaller sizes and younger ages than females, there is an economic incentive to produce all-female populations for fish farming (Borski *et al.*, this volume). While genetically all-female (XX) populations have been produced using diploid gynogenesis in some flounder (e.g., southern and summer flounder), these fish may still develop morphologically as males, because phenotypic sex is influenced by environmental conditions. The effects of temperature are the most studied to date, but there is evidence that tank color and other factors may also be influential (Borski *et al.*, this volume). Hence, the interaction between genetic and environmental influences on sex determination has confounded the application of this technique for practical culture and requires further elucidation (Borski *et al.*, this volume).

In flounder, the effects of high temperature on sex determination were related to suppressed aromatase (a key enzyme regulating female sex differentiation) mRNA levels in XX Japanese flounder (Borski *et al.*, this volume). Flounder are an excellent system for studying environmental (temperature, stocking density, photoperiod, tank color, and other stressors) modulation of aromatase expression and sex differentiation as well as the role of physiological mediators of these effects. A better understanding of sex determination may the culturist to produce faster-growing all-female stocks for growout, but also the appropriate sex ratios for stock enhancement (Borski *et al.*, this volume). Concurrent to sex control, studies are needed to develop improved cultivars for commercial farming through selective breeding of fast-growing, later-maturing fish or, through hybridization.

Production costs may also be lowered by improving RAS design and energy efficiency, and cage culture systems should also be evaluated in coastal waters where environmental conditions are favorable to both the fish and to the dispersal of wastes. Considering their bottom-dwelling behavior, methods to improve stocking and production densities per unit of tank volume will be important toward reducing production costs in intensive growout systems.

As intensification and production increase, improving disease resistance (vaccination) will be important to prevent related losses. Studies are needed to identify pathogens leading to diseases in flounder, establishing protocols for PCR-based methods for early detection and prevention, as well as for their treatment.

Work to date has shown that significant percentages of fish meal and fish oil in flatfish diets can be substituted with cheaper alternative sources. Work is needed to study the efficacy of promising diets through full marketable stages, their effects on product quality, and to develop diets that are highly digestible to minimize wastes in fish farm effluents.

Significant information on fingerlings production, growout, and economics has been developed from research-scale and pilot demonstration studies, but commercial-scale data is lacking. Commercial-scale demonstration projects, including public–private partnerships are needed to transfer technologies to startup farmers to advance technology and lower production costs, allowing for insightful economic analyses, while minimizing financial risk to the startup grower.

As demonstrated in Asia and Europe for turbot, startup flatfish farms will likely need to access local, high-value niche markets until production costs are lowered, when production can be expanded and prices reduced for broader consumer affordability and market expansion. To improve a farmer's bottom line, a diversification of market demand for products of different sizes will be valuable. There are also opportunities for increasing market price and demand for domestically produced flatfish as consumers become more discerning about the source and quality of their seafood purchases. Quality labels that certify product quality and origin may also command a premium value.

Stock enhancement is considered to be a viable tool for enhancing flatfish stocks for recreational or commercial fisheries (Støttrup and Sparrevohn, this volume). A responsible approach is advocated by all, including a breeding program that ensures the use of healthy fry with genetic variability similar to that of the wild local species, proper objectives, a monitoring of releases using appropriate criteria to measure success, and to ensure that no adverse environmental changes occur. Research emphasis should be directed toward improving field techniques for monitoring, for evaluating sites for releases, improving knowledge on the ecology of each species, and developing models to estimate cost benefits of release including commercial and recreational value. The important factors determining the effectiveness of stock enhancement include suitability and carrying capacity of the release site, ecological studies of the release site, and postrelease monitoring to determine the optimum release strategy and the impact of stocking, optimal release magnitude in relation to the carrying capacity of the release site, determination of the economic efficiency of stock enhancement by the contribution to the net increase in harvest or abundance of the target species, with all stakeholders involved in this process.

Index